Architectural Guide
Moon

Architectural Guide
Moon

Paul Meuser

With contributions from Galina Balashova,
Olga Bannova, Alexander Glushko, Brian Harvey,
Hans Hollein †, and Gurbir Singh

DOM
publishers

Soviet Union

NASA (US)

CSNA (China)

ESA (Europa)

JAXA (Japan)

ISRO (India)

Roscosmos (Russia)

SpaceX (US)

Google Lunar X (International)

SpaceIL (Israel)

Moon Express (US)

PTScientists (Germany)

Lander

Orbiter

Crewed mission

Unintentional crash

Intentional crash

Crash after completed mission

Failure

Mission to far side of the moon

Lunar orbit

Lunar rover

Contents

Stanley Kubrick Archives/ Turner Entertainment Co.

2001: A Space Odyssey: the space probe Aries h[illegible]ring over the lunar settlement Clavius Base

Preface

Paul Meuser

This book documents all artefacts that landed on the moon as part of a lunar mission in the last 60 years. It examines these artefacts under the rubric of architectural design. Yet is it even appropriate to systematise the probes and scrap piles as if they were works of architecture? The answer is a resounding yes! As Hans Hollein provocatively remarked over 50 years ago, architecture does not only take the form of buildings. 'Everything is architecture!'. The Austrian design virtuoso describes in his essay (see pp. 10–13) how even media and emotions can form a space and thereby become architecture. If we think along these lines, even the moon becomes a place of architecture. 'Humans broaden the scope of their experience, find new means of defining the world around them (*Um"welt"*), through far more than built structures' (p. 11).

Two basic assumptions allow us to categorise the probes and space capsules as works of architecture. First, they are objects born in engineers' workshops – objects that, as spatial forms, were constructed in accordance with structural laws and designed, more or less, based on aesthetic principles. Second, the landing sites of most of the objects have been mapped precisely. In other words, every single work has its *genius loci*, even though the work was initially designed without precise knowledge of this place.

It is not too early for us to start discussing lunar monument conservation. After all, the relics of the lunar missions represent a history of technology and space. Luna, Ranger, and Apollo – these and all other missions have allowed us to seriously consider the prospect of colonising the moon, that is, of creating architecture on the moon. This is why this architectural guide explores lunar missions of the past as well as ones in the present and future. And it focuses on the technological aspects of rockets and space engineering, for they are always connected to the question of where the boundaries lie between vision and reality.

This book presents a number of valuable contributions from experts who try to answer precisely this question. The architects Galina Balashova and Olga Bannova both reflect on specific issues of design based on their own experiences: the former on the interior architecture of lunar landing modules of her time, and the latter on the possible structures of a lunar base in the present and future. Alexander Glushko offers a historical account of the Soviet lunar programme, which was extensively planned but never realised. And the two space experts Brian Harvey and Gurbir Singh present in detail the current developments of the Chinese and Indian lunar travel programmes.

Of course, very few people will be able personally to visit the artificial relics that have laid scattered across the moon for decades. However, this architectural guide will allow the reader to take off on a journey of the imagination from their cosy armchair at home.

'Architects must stop thinking purely in terms of built structures.'
Hans Hollein

Everything Is Architecture!

Hans Hollein

Today, traditional definitions of architecture and of its means[1] no longer seem to hold. Our efforts must be directed towards the environment as a whole and towards all media[2] that define it. We must consider the television as well as artificial indoor climates, modes of transport as well as clothes, the telephone as well as housing.
Humans broaden the scope of their experience, find new means of defining the world around them (*Um'welt'*), through far more than built structures. Today, to an extent, everything is architecture.
'Architecture'[3] is one of many kinds of media that can define our behaviour and environment, and it is one of many options for solving particular problems.
Humans create artificial conditions. This is the project of architecture. They physically and psychologically reproduce, transform, and expand their physical and psychological world. They define their 'environment' in the broadest sense of the word. They deploy the means available to them, in accordance with their needs, desires, and dreams, in order to fulfil those very needs, desires, and dreams. They extend themselves and their bodies. They express themselves. Architecture is a medium of communication.
Humans are both self-centred individuals and members of society. This duality defines their behaviour. Ever since they were primitive beings, they have used media to continuously evolve into what they are today, and in so doing, they have broadened the scope and reach of those media. Humans have a brain. Their senses form the basis for how they perceive their environment. Media that define and shape the desired environment are an extension of the human senses. These media fall under the concept of architecture in the broadest sense of the word. In more concrete terms, architecture encompasses the following roles and definitions:

- Architecture is ritualistic: it is monument, symbol, figure, and expression.
- Architecture regulates the warmth of the body, serving as protective housing.
- Architecture defines – establishes – space and environment.
- Architecture conditions people's psychological state.

For thousands of years, people have artificially changed and shaped their environment – not least to protect themselves against the weather – primarily by building, and the built structure has been the most prominent manifestation and expression of humankind. To build has meant to create a three-dimensional structure that delimits space and serves as a protective envelope, instrument, psychological vehicle, and symbol. Advances in science and technology, as well as changes in society and in its needs and requirements, have imposed on us completely different conditions. Other, newer media for defining the environment have emerged.
At first, such media will merely represent technological improvements of conventional principles and enhancements of physical 'building materials'; but with time, they will become immaterial means

'Space capsules and space suits are more perfect than any building. They enable comfort even in the most extreme conditions, while providing maximum mobility.'
Hans Hollein

of defining space. Today, a number of tasks and problems are still solved by traditional methods – by building, that is, by means of 'architecture'. But is 'architecture', in its more conventional form, still the only answer to the many questions we face? Are there no other, more appropriate, media available to us?
In this regard, architects could learn something from the history of military strategies. If defensive strategies had been subjected to the same sluggishness as architecture and its consumers have been, we would still be protecting our cities with towers and fortresses. However, military strategies have, to a large extent, loosened their tie to the built structure and devised new possibilities for managing their tasks.
Similarly, nobody sets out to build a wall around sewers or to construct astronomical devices out of stone, as was once done in Jaipur, India. However, the new media of communication, from the telephone to the radio to the television, bring with them far deeper consequences. The school building may very well cease to exist under certain circumstances in the future and be replaced by new means. Architects must stop thinking purely in terms of built structures.
It should also be mentioned that the weight has shifted from meaning to effect. All architecture has an effect. The way a work is occupied, the way an object is used, in the broadest sense, is important. A building can wholly become a cluster of information, and equally, its message can be conveyed entirely through media of communication (press, television, and the like). In fact, it almost seems irrelevant whether the Acropolis and the pyramids physically exist, since the majority of the general public will never be able to experience them in person and only learn about them through other media – since, indeed, the role of those buildings are based on their informational effect. In other words, a building can be simulated.
The phone box is an early example of how communications media expand the boundaries of architecture. It is a building of minimal size, but it directly encloses a global environment. The flight helmet is another example of a medium that creates a new kind of environment. Tailored more specifically to the human body, and designed with an even compacter form, it is connected to an array of communications devices and thereby extends the senses and sensory organs, bringing them in direct contact with even vaster areas of space. Finally, the development of space capsules, and especially of space suits, has led to more extreme formulations of place in today's 'architecture'. Designed for outer space, they are a kind of 'housing' that offers, far more perfectly than any 'building', full control of body heat, provision of food, management of human waste, and comfort, even in the most extreme conditions, while enabling maximum mobility.
These expanded physical possibilities allow us more clearly to envisage the psychological possibilities of an artificial environment. Once we are no longer bound by the need for built environments (such as a building envelope, weather

Archive Hans Hollein

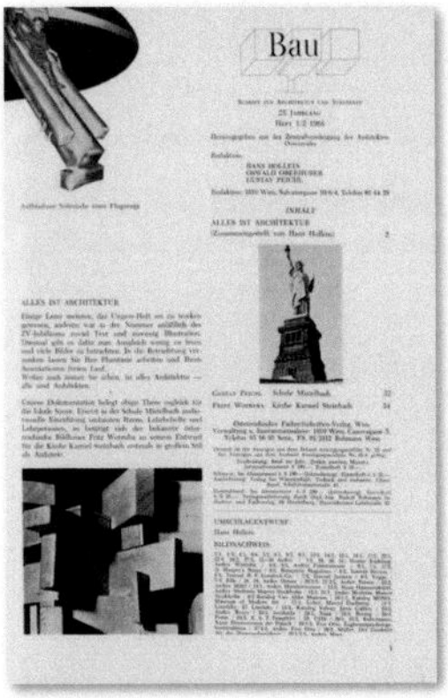
Bau

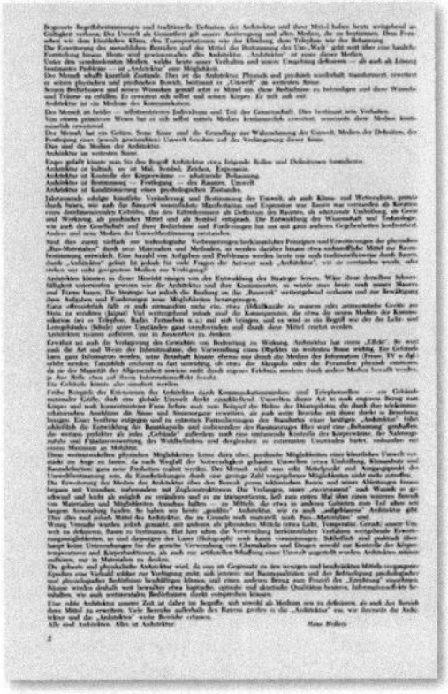

***Bau* magazine, issue 1/2–1968**

protection, and spatial definition), we begin to intimate completely new freedoms. Humans now truly become the focus – and point of departure – for how we define the environment, since we are no longer confined to a small number of prescribed possibilities.

Train designs in particular played a crucial role in expanding architectural media beyond pure tectonic building. The desire to change our environment as quickly and easily as possible and to transport that environment prompted us for the first time to turn our gaze towards a new field of materials and possibilities, to means that had in some fields been in use for a long time. As such, today we have 'stitched' architecture, as we have 'blob' architecture. However, such works are means of architecture that generally still exist in the realm of materiality, of building 'materials'.

Few attempts have been made to define our environment, to define space, with non-physical means – with light, temperature, and smell for example. Of course, there is much we can still do to expand and enhance conventional processes. But laser and holographic technologies, for example, may well lead to unforeseen and unforeseeable possibilities. Finally, there have been virtually no investigations of how we might use chemicals and drugs to control body temperatures and bodily functions and to artificially create an environment. Architects must stop thinking purely in terms of materials.

The means at our disposal today are far greater in scope, and far more diverse, than in previous epochs. As such, built, physical architecture will be able to engage more intensively with spatial qualities, with psychological and physiological requirements, and form a new relationship with the process of building. Spaces will far more purposefully possess tactile, visual, and acoustic qualities. They will convey information and directly address our emotional needs.

We are therefore on the verge of witnessing a rebirth of architecture: the rise of a true architecture of our time that has redefined itself as a medium and expanded the scope of its means. Many different fields, beyond the limits of built structures, take an effect on 'architecture', just as architecture and 'architects' take an effect on other fields.

Everyone is an architect.

Everything is architecture.

Translated and printed here with the kind permission of the private archive of Hans Hollein.

1. Hollein uses the term 'Mittel' throughout this essay. The word encompasses a wide range of meanings, including: methods, measures, resources, tools, instruments, materials, and vehicles. The English term 'means' has been used consistently throughout this translation.
2. The word 'media' was translated from the German word 'Medien'. Hollein uses this term in its most general sense, to describe anything that mediates between people and their environment. 'Media' in this essay therefore includes but is not limited to communication media such as television, radio, telephones, etc.
3. Hollein uses the words architecture and 'architecture' (in inverted commas) as two separate terms. With the former, he refers to architecture in a broader sense, which he sets out to define in this essay. With the latter, he refers to the conventional and traditional definition of architecture, signifying built structures only.

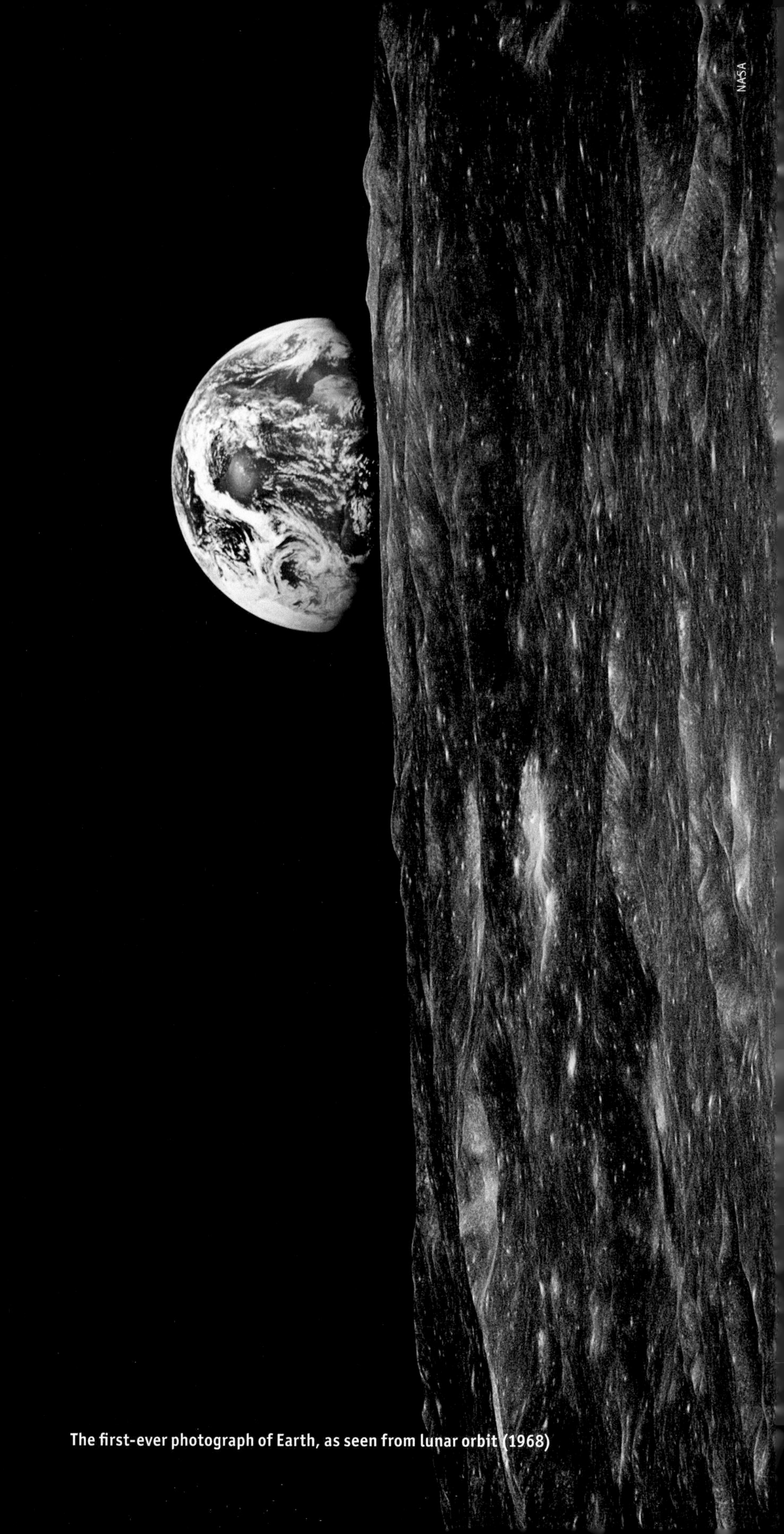
NASA

The first-ever photograph of Earth, as seen from lunar orbit (1968)

Moon, Humanity, and Architecture

Paul Meuser

Some readers may be surprised to be holding in their hands an architectural guide on the moon – even though it has been over 50 years since a person stepped onto the moon for the first time. Of course, there is still no human settlement on the moon or indeed any form of lunar architecture as such. But here we might turn to Hans Hollein's definition of architecture as a 'three-dimensional structure that delimits space and serves as a protective envelope, instrument, psychological vehicle, and symbol' (see p. 11). Thinking along these lines, we can more easily imagine that the several-dozen landing sites on the moon might be entered into a monument conservation list in the future. Humans have left behind artefacts on the moon – traces of designs created by generations of visionary inventors and scientists.

The moon was the first celestial body to be visited by people. But it has played an important role in people's imaginations since long before the Apollo 11 landing. The ancient Romans worshipped Luna, the goddess of the moon, along with her brother, the sun god Sol, and the two entities together represented the cyclical nature of time. Moreover, there is evidence that the sun and moon played an important role in human rituals in cultures dating back even further beyond classical antiquity. The moon has always had a mystical appeal, and it far predates humanity. When humans first developed signs of consciousness some 50,000 years ago, the moon had already been orbiting the Earth for four and a half billion years.

Minneapolis Tribune

Illustration of the Apollo 11 lander from the *Minneapolis Tribune* (1969)

Fictional depictions of lunar architecture: from Lucian of Samosata to Georges Méliès to Stanley Kubrick

The satirical novel *A True Story*, written by Lucian of Samosata in the second century, is the oldest surviving story about a crewed mission to the moon. Since then, people have continued to be fascinated by the moon, and utopian stories about journeys into space have appeared continuously, particularly since the advent of industrialisation. The title character of Rudolf Erich Raspe's *The Surprising Adventures of Baron Munchhausen* (1785) flies to the moon on a cannonball. And the characters in Jules Verne' *From Earth to the Moon* (1865) are launched to the moon on a rocket-shaped projectile. In 1902, Georges Méliès adapted *A Trip to the Moon* (1902), originally an operetta by Jacques Offenbach, into a film, which became the first cinematic work to portray a work of architecture on the moon.

2001: A Space Odyssey: an evening walk with a view of Clavius Base on the moon (1968)

The Palace of the Selenites features oriental decoration in its façade, with the main entrance highlighted by a crescent moon turned sideways. The building looks rather like an excessively decorated pavilion that could easily have been found at the 1889 Paris World's Fair. However, most fictional portrayals of extraterrestrial architecture have featured a more industrial style, stripped of any ornamentation. In Yakov Protazanov's silent film *Aelita* (1924), a classic of early Soviet cinema, the city on Mars is filled with constructivist buildings, arches, and bridges. The futuristic city recalls the etchings of prisons by Piranesi or the imposing structures in Fritz Lang's *Metropolis*. The city's windowless buildings look like the backdrop of a necropolis, and all life has been displaced to the interior spaces. Stanley Kubrick's *2001: A Space Odyssey* reinforces this

Henri de Montaut's illustration in Jules Verne's novel *From the Earth to the Moon* (1868 edition)

Stanley Kubrick Archives/Turner Entertainment Co.

Monumental set design of the Russian avant-garde: *Aelita*, directed by Yakov Protazanov (USSR, 1924)

image of introverted architecture on a celestial body. It features a giant building, the size of a city, engraved into the surface of the moon like a radial tattoo. The Clavius Base, named after a real-life crater on the moon, remains an architectural puzzle. An oversized rosette opens up for spacecraft to land and serves as a gateway between the inside and outside, though it is unclear how people enter the building complex in their space suits.

Fictional representations of lunar architecture seem to have one thing in common: the interior and exterior spaces, the former usually narrow and contorted and the latter usually enormous in scale and proportions, are not designed as a coherent unit. Inside the buildings, there is a claustrophobic labyrinth of corridors and airlocks; outside, there is an oversized city structure that is always shown without any people.

A Trip to the Moon (*Le voyage dans la lune,* 1902), directed by Georges Méliès, France. The film was adapted from Jacques Offenbach's operetta of the same name.

Film poster for Fritz Lang's *Metropolis* (1927)

It is only logical that these depictions consistently divide lunar architecture into deserted outdoor spaces and ergonomically designed interiors. After all, humans can only survive on the moon in artificial, closed systems. As such, we can expect the first generation of lunar architecture in the future to follow the same design principles governing an international space station with cylindrical, inter-connected modules. The only difference will be that the building parts, unlike a space station, will be subject to the moon's gravity and be covered in monotonous lunar rocks.

From the primitive hut to the lunar base

In science fiction, lunar architecture often takes the form of industrial and functional buildings: monotonous, life-sustaining machines that regulate the inhabitants' lives. Yet these housing units do not appear in such a vast scale to be considered instances of Le Corbusier's Unité d'Habitation. Rather, we should think of them as primitive huts on the moon. After all, architecture begins the moments humans create a form of housing for their survival.

NASA/Norman Rockwell

Man's First Step on the Moon, **Norman Rockwell, oil on canvas (1966)**

Primitive hut, frontispiece of *Essai sur l'architecture* (1755), Marc-Antoine Laugier

On the moon, the concept of the 'minimum dwelling'[1] takes on an entirely new meaning. Technical devices are required to optimally use the limited amount of space – indeed, those devices are essential for making survival in the lunar environment possible in the first place. So it is no surprise when moonbases depicted in drawings resemble structures made of tubing and cables or when they evoke the image of postmodern 'high-tech' architecture and the socialist modernism of the twentieth century.

However, even with the successful Apollo missions far behind us, we are a long way away from seeing the first permanently inhabited lunar base. A future generation of moon walkers will be none too pleased about having to squeeze through a minimalist escape hatch just to get to work. Nor will a lunar base designed like a caravan be suitable for stays lasting several months at a time. These constraints make the prospect of a new architectural form on the moon that much more exciting.

The moon landing in 1969 was achieved against the backdrop of an ideological struggle between the East and the West. But today, one could argue that any state that sees itself as 'modern' wishes to take part in an international lunar programme. India, China, Japan, and Israel now each run an ambitious national space programme. And various countries in Europe are collaborating on a

iStockphoto (Gwengoat)

Renzo Piano, Richard Rogers, et al.: Centre Pompidou, Paris (1977)

iStockphoto (petrovalilya)

Abraham Miletsky, N. I. Slogozka, et al.: Hotel Salute, Kiev (1984)

space programme as part of the European Space Agency (ESA). Space technology, whether for crewed or uncrewed missions to the moon or to Mars, has once again become part of a race for political influence and global significance.

The typology of lunar architecture

The many attempts by various rival nations to establish the first settlement on the moon have led to something like a typology of lunar architecture. This can be divided into four types (see p. 289).

At a congress in Beijing in December 2018, Guo Linli, from the Institute of Manned Space System Engineering, presented a number of building systems that could be used to establish a lunar base. The space engineer divided those systems into three categories: (i) rigid, inflatable cabin structures; (ii) a combination of rigid and flexible modules, and (iii) more conventional structures, constructed with a 3D printer, with lunar regolith serving both as building material and cover for the habitation modules beneath.

While some contemporary space architects are busy designing buildings for the lunar surface, others are working on lunar orbital stations, which could be used not only to send supplies to the lunar base but also to launch further crewed missions to Mars. A number of private companies are involved, which means two important elements of life-sustaining systems will not be ignored: architecture and design.

1. German: 'die Wohnung für das Existenzminimum'. Cf. The second congress of the International Congresses of Modern Architecture (CIAM).

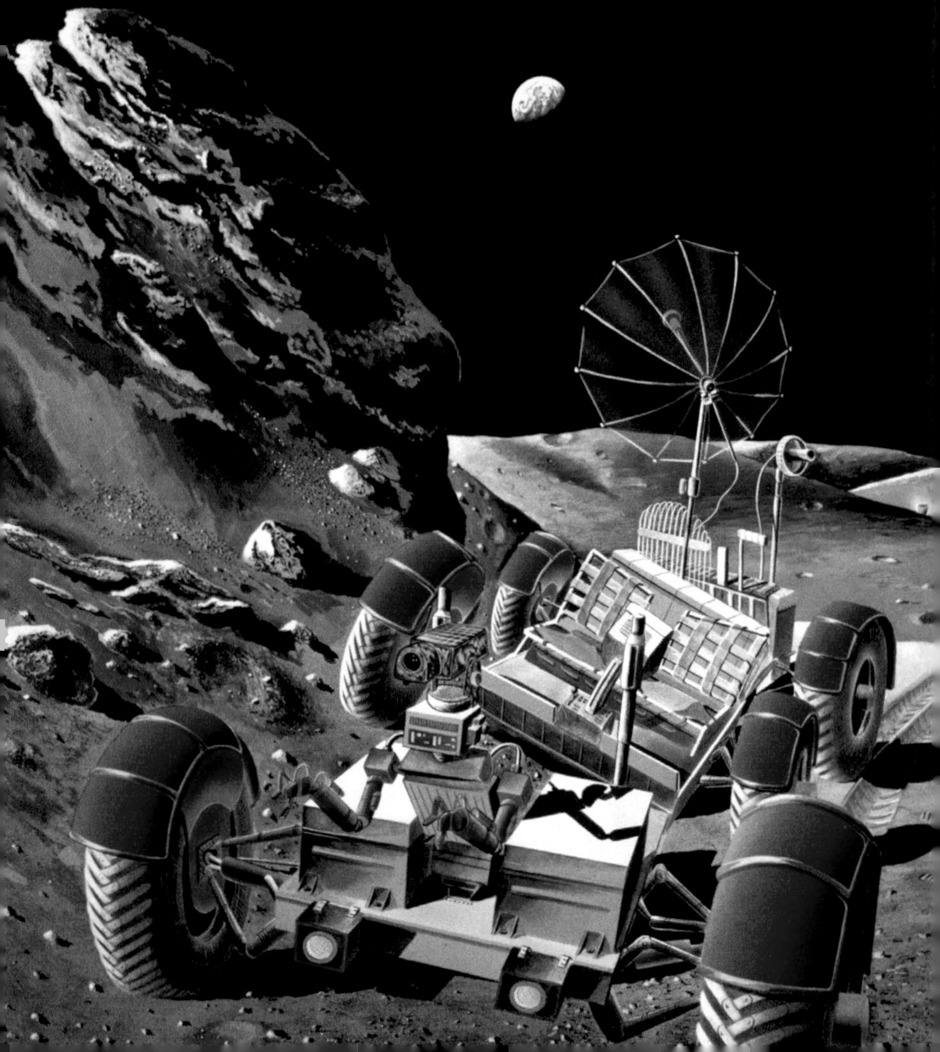

NAS

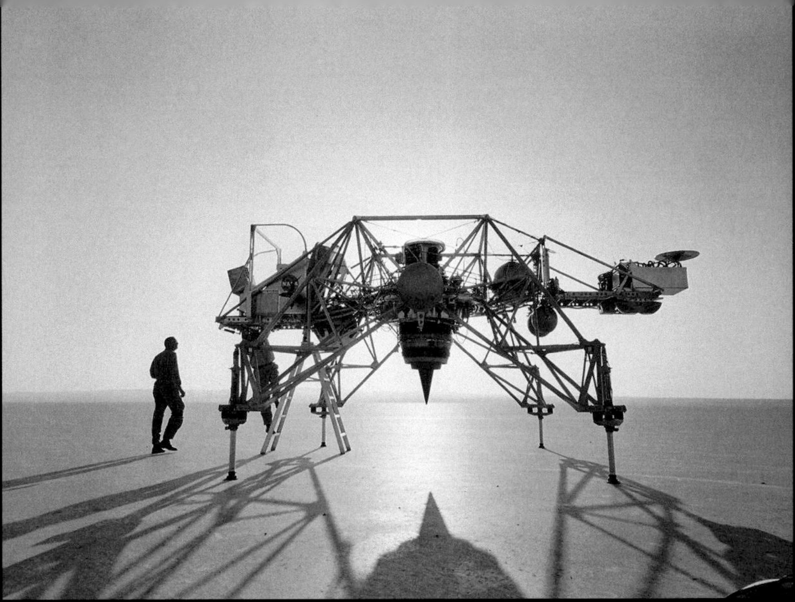

NASA's Lunar Landing Training Vehicle, shortly before a test flight over a salt lake in Nevada. The construction obeys its very own laws of statics.

NASA / Dennis M. Davidson

Study by California Space Institute, on behalf of NASA, for cylindrical habitation and work modules in a moon crater (1986)

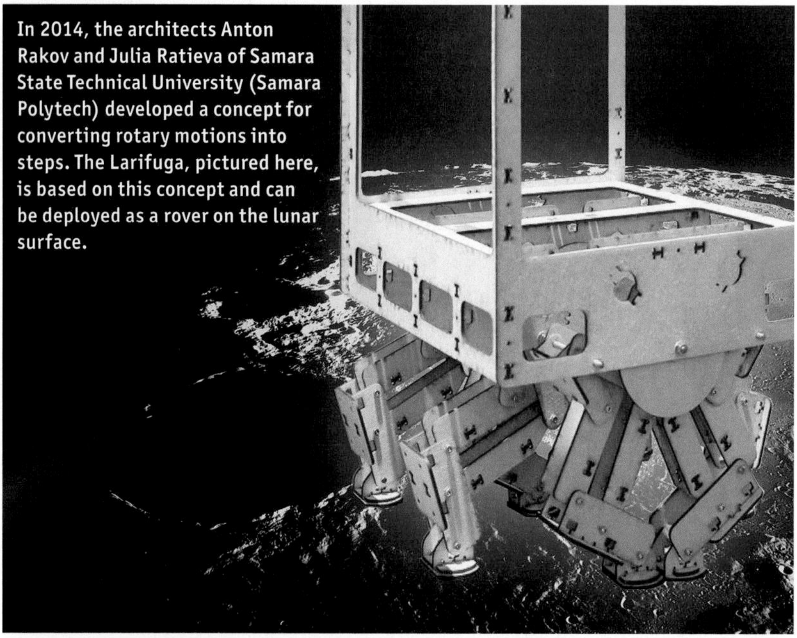
Samara Polytech

In 2014, the architects Anton Rakov and Julia Ratieva of Samara State Technical University (Samara Polytech) developed a concept for converting rotary motions into steps. The Larifuga, pictured here, is based on this concept and can be deployed as a rover on the lunar surface.

Samara Polytech

LIN Industrial

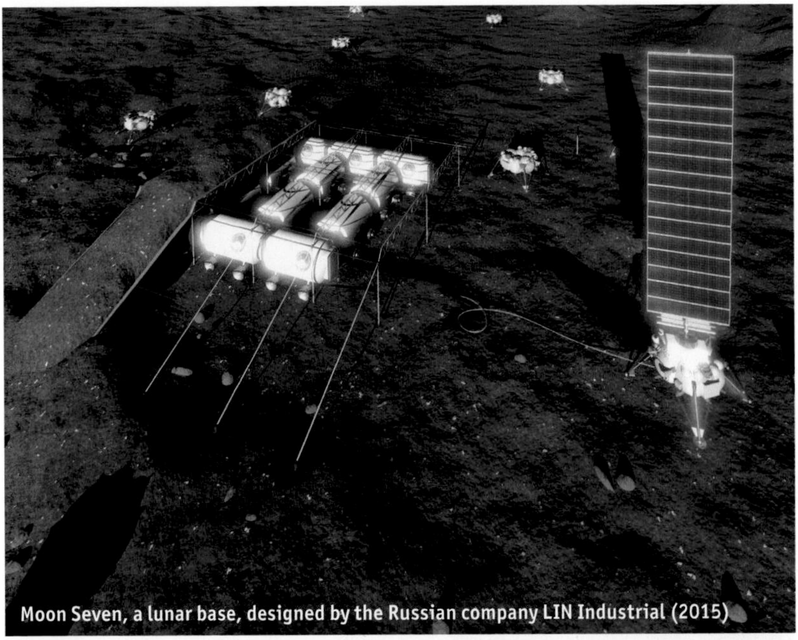

Moon Seven, a lunar base, designed by the Russian company LIN Industrial (2015)

Lunar base with ancient and neo-classical forms, planned by Anton Rakov of Samara Polytech (2018)

NASA

Artist's concept, commissioned by NASA, of an Orbiting Lunar Station (OLS), designed to study the lunar surface and serve as a layover for moon landings (1976)

European Space Agency, Foster + Partners

State-run lunar programmes are currently pursuing two strategies. On the one hand, they are considering a fixed lunar station dedicated to mining natural resources ...

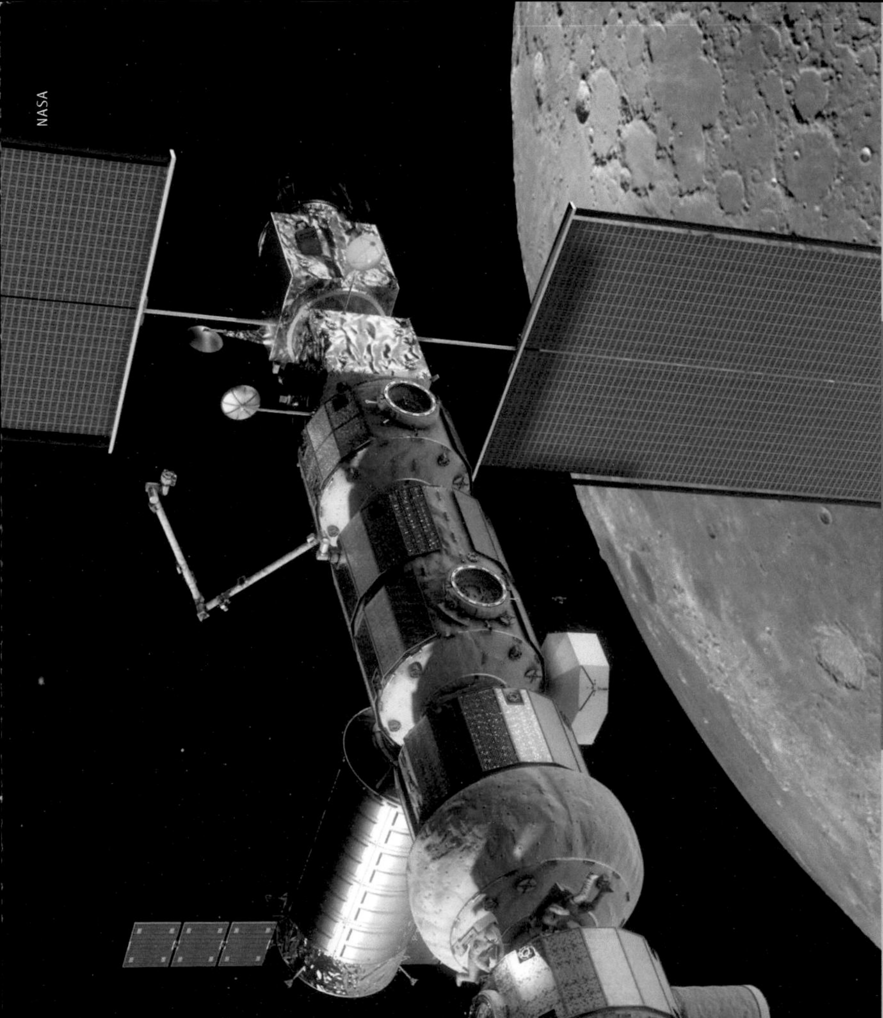
NASA

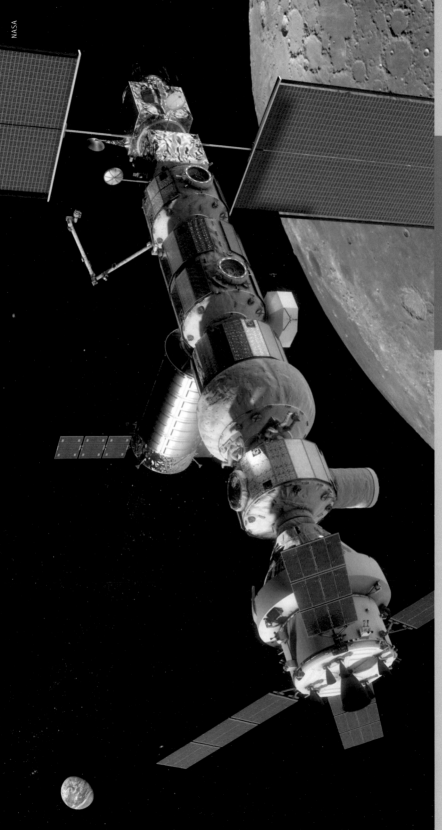

... on the other, they are planning orbiting lunar stations with which to prepare flights to the moon and, in the future, also to Mars.

SICSA

The Future of Lunar Architecture

Olga Bannova

The barren landscape of our Moon may scare or invite, intimidate or inspire. In order to create sustainable settlements for continuous human presence, architects need to combine functionality, safety, robustness, and aesthetics. In the previous century, lunar surface architectures relied on complex and large-scale structures such as Project Horizon, a US study for establishing a lunar military outpost (1959), and the Soviet lunar base Zvezda, also known as Barmingrad (1962). Historical examples of lunar exploration strategies show detailed and thoughtful approaches in many aspects. However, they required the delivery of massive structures and intensive labour. Nonetheless, during this dawn of space exploration, visionaries were at least able to address the first three main requirements for supporting human life on a celestial body: functionality, safety, and robustness. Since then, we have learned much more about space flight, the moon, and ourselves as space explorers. We want and need to establish a permanent settlement on the Moon – for ourselves in the present day, and for our descendants in the future. Some might ask: 'can we afford it?'. But we should rather ask ourselves a different question: what will we miss, and at what cost, if we choose not to pursue lunar exploration and development?

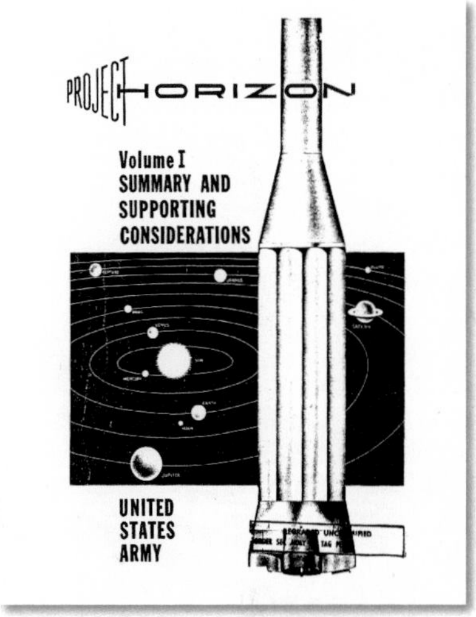

The first Project Horizon report (1959)

Space agencies and private companies that seek to explore the moon and create lunar settlements are pursuing a hierarchical architectural strategy, centred around the establishment of a minimalist beginning, followed by the use of local materials to build infrastructure, and finally the development of a settlement capable of continuous growth. The minimalistic beginning entails missions with multi-use landers and small settlements with minimal living areas for the crew. The Apollo missions can be considered an example of such an approach, although the surface missions did not include any plans to establish surface capabilities after the landing. To this day, the lunar module deployed during the Apollo missions is the only off-Earth habitable space that has ever been built and used. It provided space for two astronauts

Station in a lunar crater, designed by Sasakawa International Center for Space Architecture

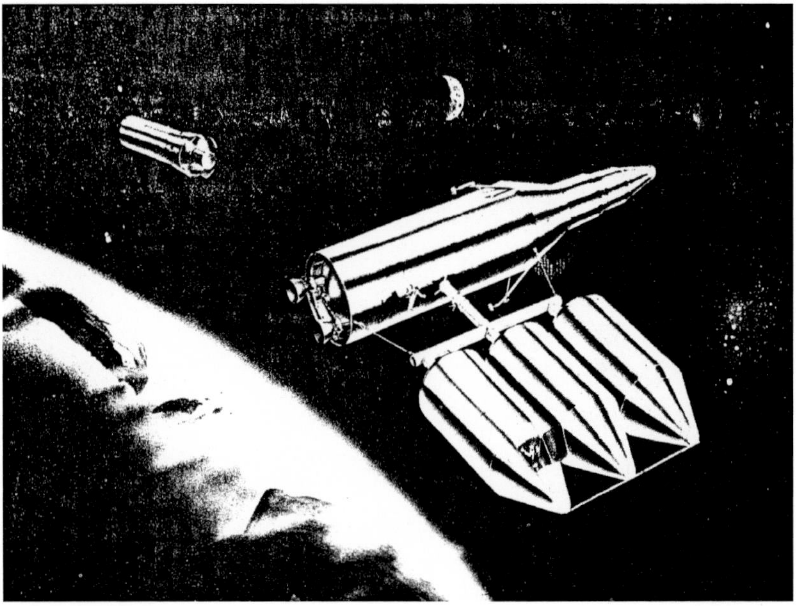

United States Army

Project Horizon: geostationary satellite (1959)

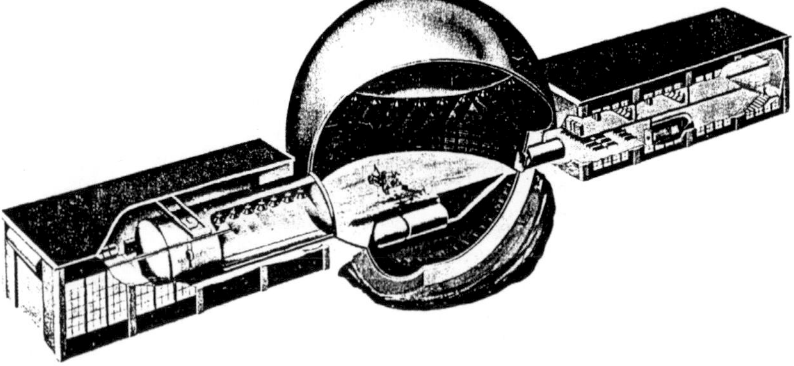

Project Horizon: cross section of the main facility LERUT

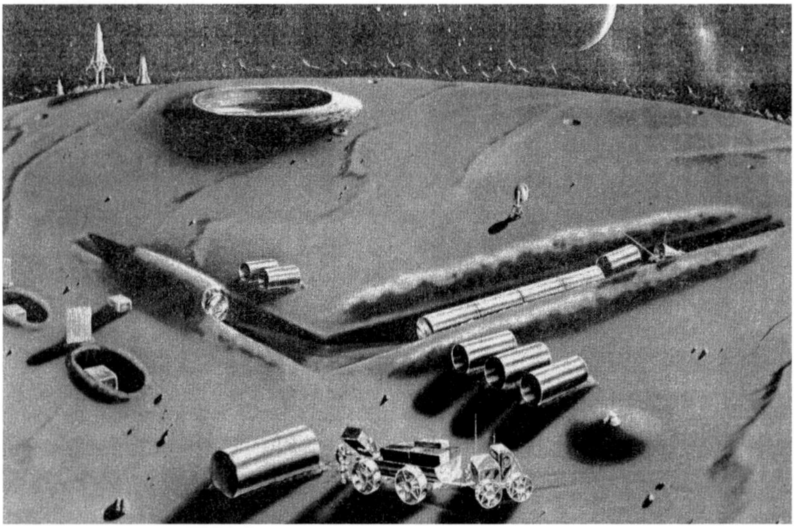

United States Army

Project Horizon: outpost, planned in 1965

Lockheed Martin

Lockheed Martin's concept for the Mars Base Camp. The lunar lander shown above was designed as a prototype.

and had a habitable volume of about 6.6 m^3. The Saturn IB and Saturn V rockets were used as launch vehicles. The longest mission included a surface stay of 12 days and 17 hours (Apollo 15).

Current approaches build upon that strategy and follow a principle of step-by-step exploration. This begins with determining landing sites that would be most interesting for scientific research and may lead to in-situ resource utilisation (ISRU) potentials. The surface architectures that are initially created aim for functionality and robustness and seek to ensure the crew's safety and minimise risk. The use of prefabricated parts or of locally available resources – or a combination of both – offers additional possibilities of meeting safety and robustness requirements. NASA's plans for sustainable lunar exploration with the space station LOP-G (Lunar Orbital Platform-Gateway) entail the lightest and most compact concepts, designed by companies participating in the NextSTEP partnership. These concepts, developed by Boeing Company and Lockheed Martin Corporation, among others, focus on the basic requirements for the crew members' survival. Aesthetics do not play a role.

Other surface architectures aim to sustain long-term human presence on the moon from the very outset, that is, already during the delivery and assembly of the earliest structures. Those architectures also employ ISRU strategies and aim to provide long-term safety and robustness. But they consider both comfort as well as the aesthetics of the interior and exterior as equally important elements for long-term human habitation.

How new is the new vision?

Space architecture addresses the challenges associated with all these stages using a human-centred design approach. 'Human-centred design' has perhaps become a hackneyed postulate today, but what does it actually mean? It means that the design of the living area as well as the planning of the base and settlement incorporate all factors that affect the human users, such as comfort, adaptability, and aesthetics, from the very first stages of the preparations for mission architecture. New lunar surface architectures must offer a viable compromise between efficiency, affordability, and the comfort and well-being of the crew. Architecture is transdisciplinary by nature, and an architect's main role is to ensure that all elements of the respective project receive an appropriate amount of attention.

The Boeing Company

The Boeing Human Lander on the lunar surface (2018)

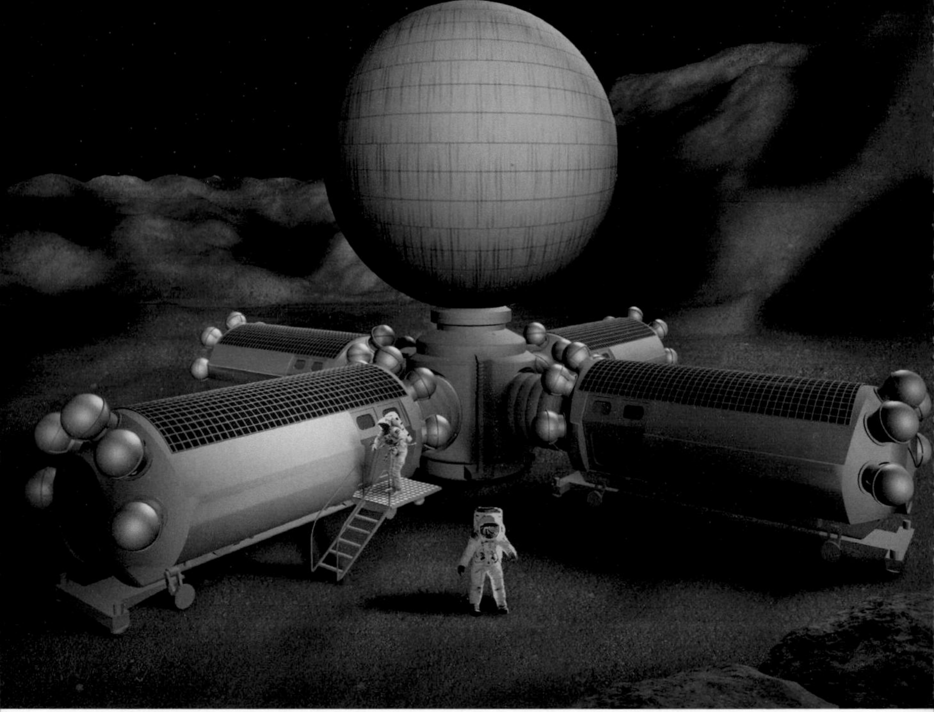

Lunar base: conventional modules have been joined to form a knot-like structure with an inflatable module in the centre. Concept by Sasakawa International Center for Space Architecture, Houston

This means that an architect should be able to synthesise multidisciplinary knowledge, experience, and expertise, and apply all of this to the design process. Whether on Earth, in space, or on the moon, architects should always provide designs that offer practical advantages and benefits. In other words, they must ensure that the systems and interiors function optimally while being cost effective and aesthetically pleasing. The final product, whether it is the design of a habitable area or of other facilities, cannot be considered successful if any one of these facets of architecture does not fulfil human needs.

Three key factors influence architectural planning in such an extreme environment as the moon. First, the outcome must offer a safe environment for the crew; second, it must provide optimal working conditions so that the crew can conduct their research comfortably; and third, it must maximise the 'quality of life'.

As on Earth, the process of designing a structure, especially a habitat, in outer space requires finding a good balance between the factors outlined above and the limitations posed by the harsh conditions outside. For example, when planning architectures in outer space, it is important to remember that all habitable structures must be capable of withstanding internal pressure loads of between 0.6 and 1.0 atmospheres without leaking. The laws of physics have not changed since the first space flights took place and naturally apply to the lunar environment as well. Habitable structures on the moon should therefore meet the following design requirements:

• All pressurised structures are primarily vessels with circular cross sections that include spherical, tubular, or toroidal geometries (regardless of the materials that are used).

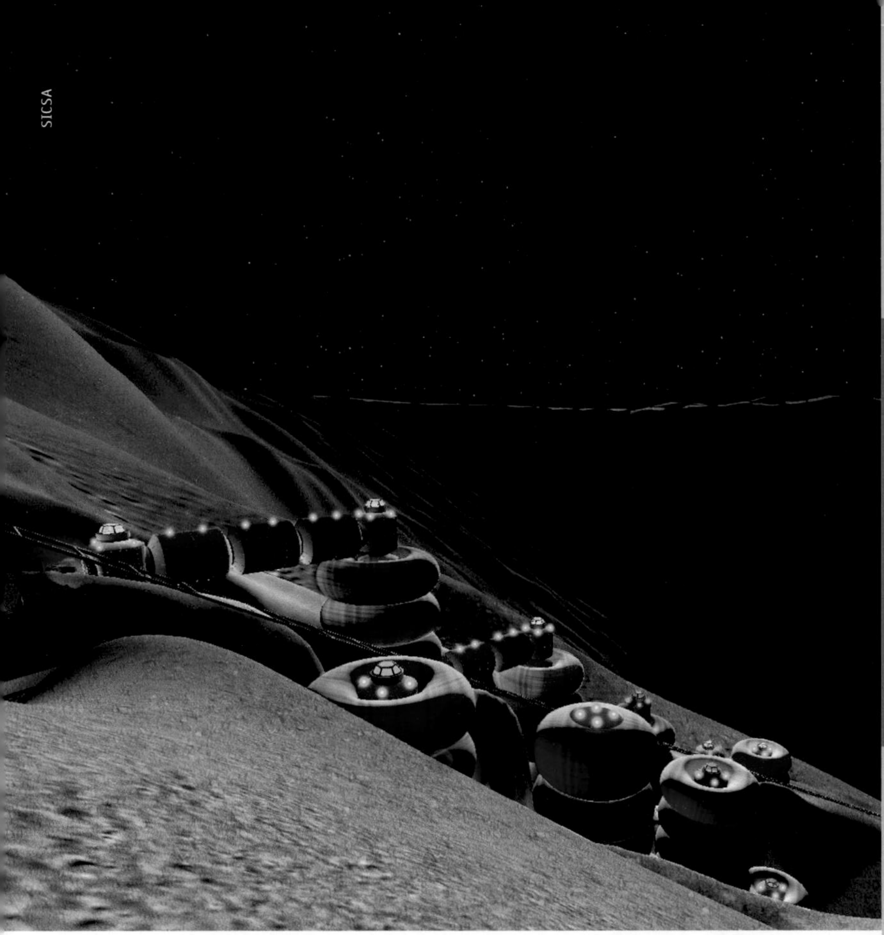

Lunar settlement for 80 people on the slope of the Shackleton crater. It comprises a combination of conventional and inflatable modules. Concept by Sasakawa International Center for Space Architecture, Houston

• Openings for windows, hatches between modules, orbital docking ports, utility passages, and other interfaces raise potential leakage concerns.

In addition to these parameters that define the form, dimensions, and orientation of habitable modules, the following factors must also be considered during the planning stage: the selected launch vehicles, the options for assembly and transfer in orbit, the landing strategy, as well as transport and deployment on the lunar surface. These are the strictest and most critical requirements space architects need to meet. And those architects must find the most optimal solutions within the transport payload and mass limitations.

Lunar surface modules have to provide interfaces for connecting with transport and landing vehicles. They also need to provide EVA access and have dual ingress/egress capabilities. Modules need a stable footprint and a centre of gravity suitable for landings and relocations. The size and configuration of the modules will determine multiple technical and design parameters: the interior habitable volume and its spatial and functional optimisation, for example.

What has changed?

In the future, large settlements on the moon will consist of many different structural types. These can either be: (i) prefabricated, delivered to the moon, and assembled there, or (ii) be built directly on the moon using local resources. Another, more likely, option is to combine the two kinds of structures together. However, all structural elements will still be subject to the same geometric, functional, and aesthetic considerations outlined above.

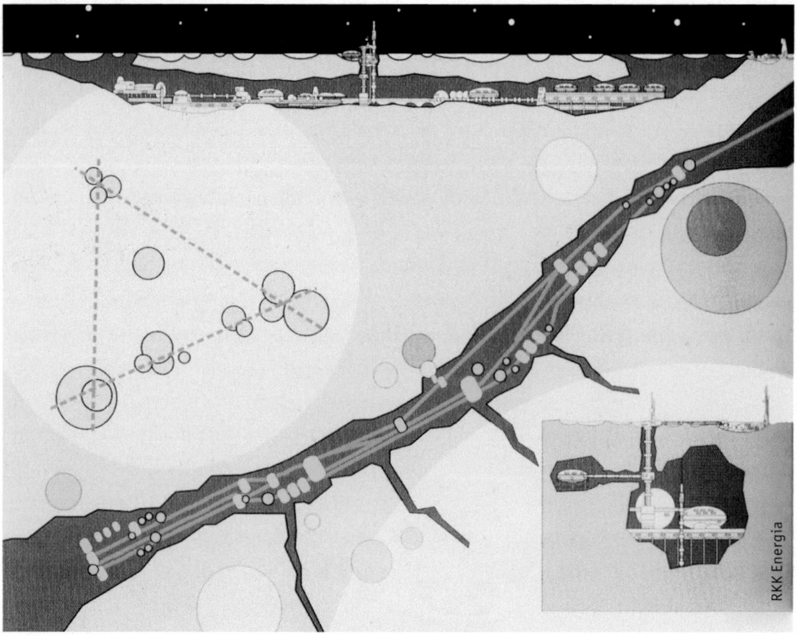

Establishing a lunar settlement in a crater and in a lava tube

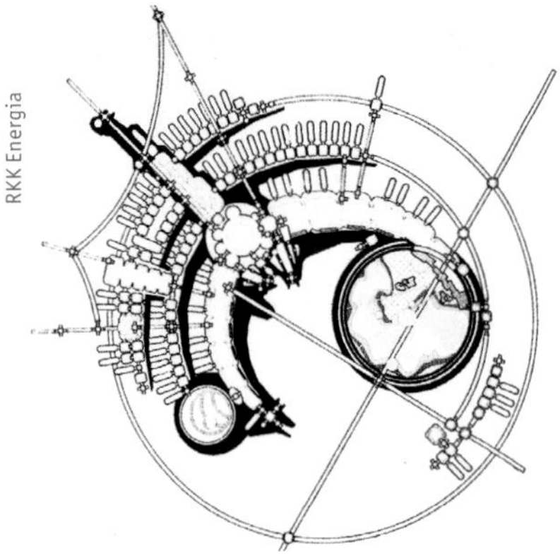

Top view of a lunar settlement in a crater

Current examples of lunar architecture propose settlement layouts arranged in a variety of ways. Based on the needs and desires of the future inhabitants, architects will design a variety of additional modules which can be added to the existing ones, or develop concepts for expanding existing settlements using in-situ resources. Modules that are pre-assembled on Earth are ready for habitation as soon as they are placed within the respective lunar site, but they do not offer much interior freedom for the inhabitants. Meanwhile, expandable structures, such as inflatable modules, do not have built-in fixtures, but they offer much freedom for a variety of uses once they have been inflated to their full size. A combination of both types of modules therefore leads to a symbiotic effect and makes the resulting architectures more humane (see images on previous page).

Objectives

Space architects designing lunar settlements therefore need to incorporate a large number of procedural and technical considerations into their development strategy. These include: the transportability of the modules in the low-gravity environment; access/egress for logistical operations and the crew's extravehicular activities; the maintenance and repair of the crew's habitation modules and other facilities; the supply of power to all elements of the settlement; the site's proximity to the research sites; availability of local resources; and potential for the settlement's future expansion.

It is crucial to plan for the settlement's growth already during the design stage in order to ensure its long-term viability. The long-term or permanent use of a site depends on the correct arrangement of the right number of interfaces between individual elements and between internal and external airlocks. Lunar rovers (with

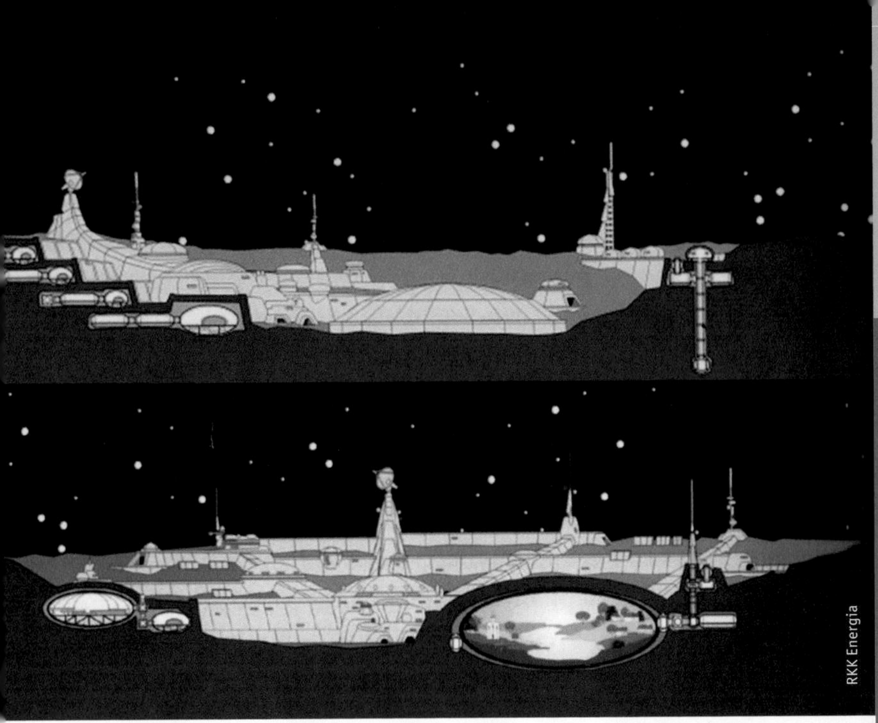

Cross section of a lunar settlement in a crater

a pressurised cabin) can also offer a viable mobility solution. These interfaces determine the geometry of a lunar settlement, whether it is on the surface or inside a crater. The settlement's geometry in turn influences how the site is further developed and defines the dual egress options for the crew in case of an emergency.

Roadmap

Planners developing future architectures and settlements on the moon need to consider expansion and reconfiguration options from the very outset. In other words, they need to incorporate provisions for additional habitation units into the design. Other elements of a lunar settlement include modules for providing medical care and technical supply units for extravehicular operations and for the settlement itself. The location and scale of each element or module will depend on the specific site conditions, operational time frame, and in the case of elements prefabricated on earth, the weight and size constraints defined by the launch and landing procedures.

No less important are the infrastructures that are required to support any large-scale human settlement, whether on Earth or in outer space. Such infrastructures are key elements that ensure not only the survival of humans on the lunar surface but also the operation and maintenance of equipment and robots. Permanent power supply and energy storage systems will be required to support long-term operations. The location of the associated infrastructural elements will depend on various factors. They can either be situated close to the habitation elements, or at a safe distance if nuclear sources are used to generate power, for example.

Equally important are the communications, control, and remotely controlled operating systems. These require a high degree of redundancy and will need to take the form of several modules spread across multiple locations. Although the concept of a permanent lunar settlement assumes closed-loop systems and self-sustaining operations, logistical resupply missions will still be needed. The volume of follow-up supplies may be reduced with ISRU procedures and mature closed-loop systems, but they will be critical for the survival of the lunar base or settlement in the initial stages of development. ISRU-enabling infrastructure will have to continuously expand as the settlement grows, so room for expansion must be incorporated into its design.

The European Space Agency (ESA) and the British architecture firm Foster+Partners ...

Scale and technology

Engineers and architects have extensively explored the use of lunar regolith for on-site construction. Earlier moonbase projects, such as Zvezda, proposed lunar regolith as a cover material to protect surface elements against radiation.

Nowadays, sintering techniques are used to create regolith layers, or building parts are made using a 3D printer with lunar regolith (also referred to as lunar concrete) as material. These processes are recommended to cover the habitat and to shield it from micrometeorites and radiation. When it comes to processing regolith, the lunar poles are the most convenient places for construction activities, due to the moderate temperatures required for the printing process. This may be a decisive factor for determining the location of a permanent lunar settlement in the future. Any lunar settlement will need to use regolith for multiple construction purposes, including 3D printing and radiation protection, in order to survive. The ultimate goal of any planetary surface exploration is to establish a permanent settlement that is completely self-sufficient and does not rely on follow-up supplies. The use of regolith would represent the ideal solution. But is it the only resource available on the moon?

The moon has a unique and beautiful landscape that we can adopt to protect our settlements but also to give their design aesthetic qualities. Using naturally created shelters such as lava tubes and craters may help to minimise the time and effort required to protect the first outposts and provide a natural transition to settlement expansion.

... have planned the astronautical lunar base The Moon Bridge (2013)

Pure colonisation or enrichment of humanity?

We may think that we have learnt a great deal about the moon since the first Apollo landing in July 1969, and we probably have. However, our plans for returning to the moon are still focused on robustness, functionality, and safety. This will likely guarantee a safe return but will not enable long-term presence. Architectural vision, combined with the required engineering expertise, can enable the human-centred design of our future cities on the moon. Space architects, together with engineers, design large settlements not only in moon craters and lava tubes but also on the lunar surface, protected by 3D-printed structures and blocks made on site with regolith. These ideas will address important human factors that go

Lunar Bridge, competition entry by Craig McCormack and Tristan Morgan (2016)

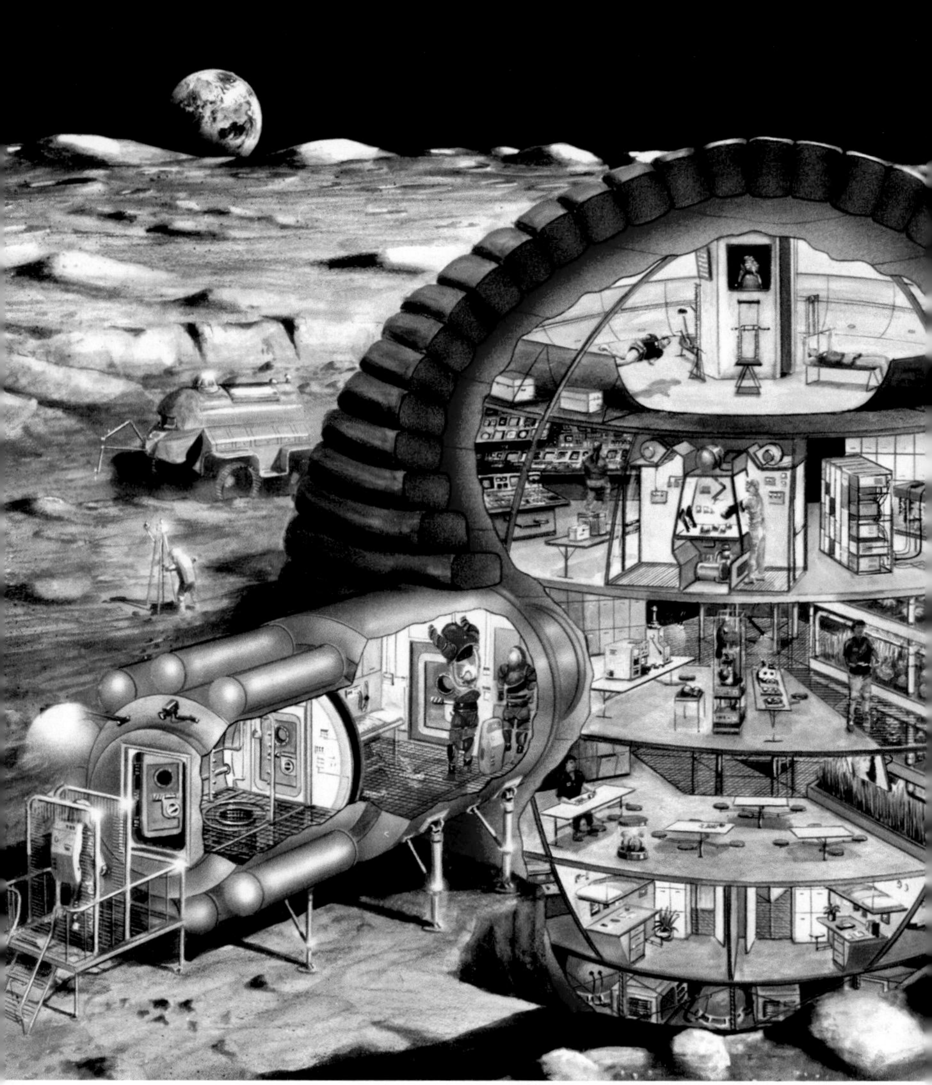

NASA study for a four-storey lunar base with a roof made of moon rock (1986)

beyond mere survival and are necessary for long-term habitation: convenience, comfort, and aesthetics.
Visionary projects such as The Moon Bridge go beyond conventional understandings of space architecture. Perhaps some day, such a structure will be built and inhabited. Constructed from lunar regolith, its roadway provides the inhabitants, who live beneath it, with protection against solar particle events and also spectacular views of the lunar crater Plato. The project seeks to become an infrastructural milestone for human presence on the moon. In the distant future, it may become a monument to the early reach of humankind. It will become a destination for those who travel off the Earth. And perhaps they will ask their friends: 'have you ever walked across the Moon Bridge?'

European Space Agency, Foster + Partners

The European Space Agency's lunar base, designed by Foster + Partners (2013)

DLR Cologne

Regolith samples from an ESA project (2018)

NASA

RegoLight / Liquifer Systems Group

Future lunar base that will be produced with a 3D printing process

The Russian space architect Galina Balashova in front of the space capsule Vostok 1 and a portrait of Yuri Gagarin (2016)

Designing for a Zero-Gravity Environment

Interview with the Russian space architect Galina Balashova

Galina Balashova was born in Kolomna, near Moscow, in 1931. She studied architecture in Moscow and began in 1957 to work as an architect for OKB-1, the Soviet Experimental Design Bureau in Korolyov. In 1963, she designed the first habitation module for the Soviet spacecraft Soyuz. During her career, Balashova not only developed habitation modules for spacecraft and space stations, she also worked as an architect for the Soviet Union's top-secret crewed lunar programme.

How did you end up working for the Soviet space programme? What were your tasks?

At the Design Bureau, I worked under the so-called head architect, who was actually a plumber. She often requested changes, but I rarely agreed to make them. After all, I was the actual architect there. It was not such an easy time for me, as you might imagine. My task was to restore buildings. I needed to plan things for the city of Korolyov, dealing with the everyday problems an architect usually deals with. One day, after I had designed the monument to Friedrich Arturovich Tsander, I was approached by Sergei Korolev. He asked if I could create a design for the Soyuz space capsule. All of this was done in secret. He spoke to me in a stairwell so that no one in my department would find out. To design the interior of the Soyuz space capsules, I had to work from home for over a year. I couldn't do the work from my workplace; my supervisor, the head architect, had forbidden me from participating in this work. But when Korolev established the lunar programme, he brought me into his department as the head architect. From this point onwards, I was officially a space architect.

What were the challenges of designing habitation modules for outer space? How did you approach these challenges as a 'regular' architect?

It turns out that when you're designing for a zero-gravity environment, you need to design the same way you would when there is gravity. It is impossible to live without a floor and a ceiling. Without them, you simply would not be able to orient yourself. All of my designs for the Soyuz programme were guided by a very simple principle: give different colours to the walls, ceiling, and floor.
But I pursued a different concept for the space capsules of the lunar programme. Before I say more though, I should emphasise that it was a good thing that my designs were never realised. The colour concept was designed to reflect the different functions that are performed during a space flight. It was beautiful, but ultimately uncomfortable. Gravity is required to provide comfort, so you need to at least create the illusion of it. I designed the Soyuz capsule's ceiling with bright colours, and the floor with dark ones, to enable the cosmonauts to tell where is above and where is below. You need two things to orient yourself in a zero-gravity space: a sense of direction, and the feeling of being fixed in place. The cosmonauts used Velcro straps to steady themselves, and the interior was clad with pile fabrics for the Velcro to cling to.

Tell us about your design for the lunar spacecraft (see also pp.136–7). Were there particular colours for the ceiling and floor?

In designing the lunar orbiter, I was creating something that had never existed before. And you could not train on Earth for life in a zero-gravity environment.

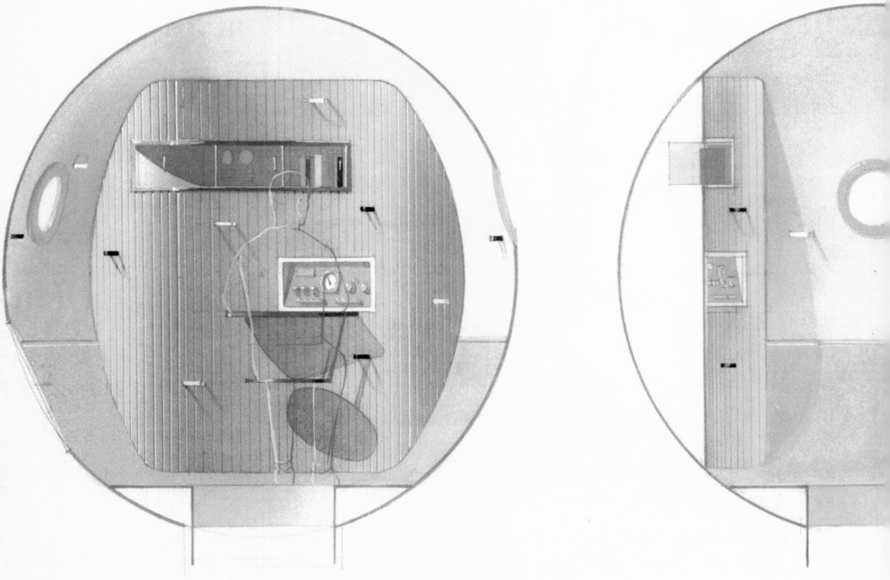

Initial sketches for the interior of the LOK lunar orbiter (1966)

Moreover, the training sessions on Earth were spread out across over three years, and you could go back home after the day's work. But in space, you are in flight for several months at a time. The two experiences are not comparable.
This is why I tried to design a space in which cosmonauts would be able to orient themselves, with a clear above and below. The more the interior resembled an environment on Earth, the more it would be suitable for zero gravity. Zero gravity does not simplify life; it complicates it. An early design of the Soyuz capsule, which I designed for the first time at home during a weekend in 1963, made it clear where one could sleep, where the toilet was, where the book shelves were, and where one could eat and drink.

Are there design differences between the habitation module intended for a short trip to the moon and a space capsule for a stay lasting several months in space, such as the International Space Station (ISS)? What are these differences?

In my view, the interiors must be very different. It would be wrong to consider zero gravity exclusively, while ignoring human habits and needs during the design phase. In any case, associations with life on Earth are desirable.
Even though the two kinds of spacecraft have different functions – one is a multi-unit space station designed for long stays lasting several months, the other is a conventional Soyuz capsule designed to be inhabited for several days – the design objective is always the same: to create a comfortable and habitable space for people. If you manage to create a successful design for a space on Earth, this design is also suitable for life in outer space. In fact, you perhaps need to make the design for outer space even more Earth-like than a design for a place on Earth.

The lunar spacecraft is composed of two parts: the orbiting module and the landing module. What is the reason for this division?

There are both technical and organisational reasons for the division. The landing module and orbiting module have completely different purposes. The landing module flies to the lunar surface and needs to be launched from there again later. A different department developed it, and I was not involved in its interior. I only worked on the lunar orbiting

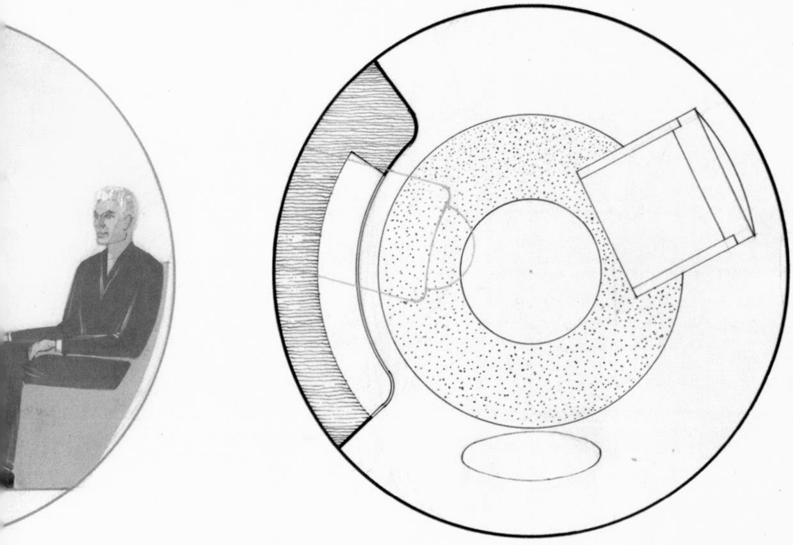

Archive Galina Balashova

module, which was designed to stay in orbit around the moon during the landing. The landing module was to be coupled with the orbiting module, and together they were to fly from Earth to the moon. But unlike the orbiting module, the landing capsule was to stay on the moon instead of returning to Earth. I can't say more. I did not study this part of the programme, and we did not have access to its development at the time. Everything was done in secret.

What differentiates a self-contained spacecraft from another that is designed to dock at an existing structure in outer space? What details must one consider? Which additions are required for the cosmonauts to stay in outer space for a long time?

It is crucial to divide up the space sensibly. For the space station, I created individual cabins for the different crew members. I also designed a lounge and various work areas with devices and control systems. I used different colours to differentiate the functions of the different spaces. The work area was blue, the ceiling was bright, and the floor was green throughout the spaceship. The cosmonauts were very pleased with this design.

Were there any particularities you needed to consider, from a design perspective, for the lunar mission?

No, there weren't. The thing is, there had been no architect at the OKB-1 before I joined, and no other architect has been hired there since. Nobody gave me any specific instructions. I only received one instruction for the design of the first Soyuz capsule, and this came from Korolev himself. He had given me the task of creating a more comfortable space, because initially, the engineers had simply built two boxes that were painted red. He asked me to present him with a habitable model within a week. Moreover, I was to design a living area over the weekend.
And so I did that. For the very first design, I decided that the cosmonauts should sleep on a sofa. This room was also designed for the cosmonauts to take off and dry their space suits overnight. It also needed to have a toilet. When closed, the toilet could also be used as a seat. There was also a control panel. It was a rather old-fashioned design, and Korolev asked me to give it a more 'modern' appearance. So I designed two variations. I also drew all the details for the engineers.

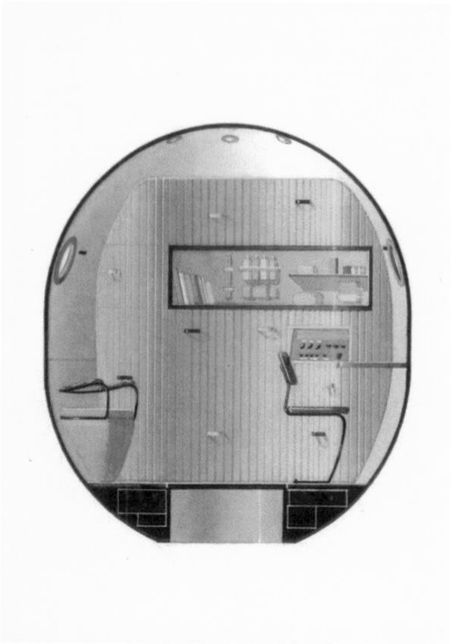

Archive Galina Balashova

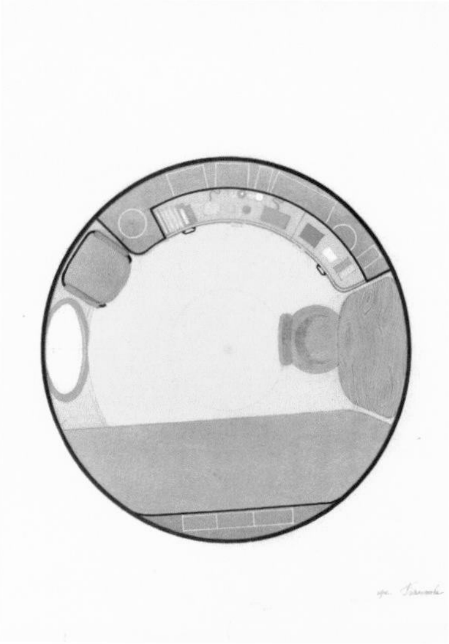

Archive Galina Balashova

Initial sketches for the interior of the LOK lunar orbiter (1966)

And how was the functionality of all these details tested? Did the engineers build the module and then carry out tests on Earth?

Yes, of course! I also accompanied the tests for a week. We worked inside the models. Until then, I had designed all interiors at home. It wasn't until a year after this first test that I was finally transferred to the design department.

At what stage was the model built in its full size so that training sessions could take place inside them?

Engineers were brought in right away once a design was finished. They procured the fittings, arranged the devices, and thought about how to combine the individual parts into a coherent whole.

What became of the space architecture that you developed and tested back then?

The thing is, I was the only one working on the interior architecture, and since Korolev's death in 1966, no one has taken an interest in it to this day. After he died, no one understood my work anymore. After Korolev, the bosses gave me mundane tasks, like painting a weather vane for example. But they were not interested in the architecture of the spacecraft. This is a major cultural difference between the Russians and the Americans, who for example designed the Skylab space station and thereby developed a specific understanding of architecture and space. The Russians on the other hand usually failed to understand the need to involve architects. We used Velcro straps on clothes and shoes to fix the cosmonauts in a zero-gravity environment; the Americans used niches in which the astronauts could anchor their feet with specially designed apparatuses. In other words, they developed a spatial solution!
But when the Americans were with us for the Soyuz-Apollo joint flight, they were lost for words when they saw the interior, particularly its colours. They had only known interiors that were metal all around and compared our capsules to a good hotel. We developed everything with materials scientists, who had large folders full of material samples. We selected everything together. The production took place in Kiev.

When you designed the Soyuz capsules, were you aiming to create capsules dedicated to specific missions? Or did you envision the design as a universal Soyuz capsule?

Every design was actually considered on its own terms and developed for the respective task at hand. Sometimes I worked on three designs at the same time,

Archive Galina Balashova

Design for the interior of the LOK, port side (1967–1968)

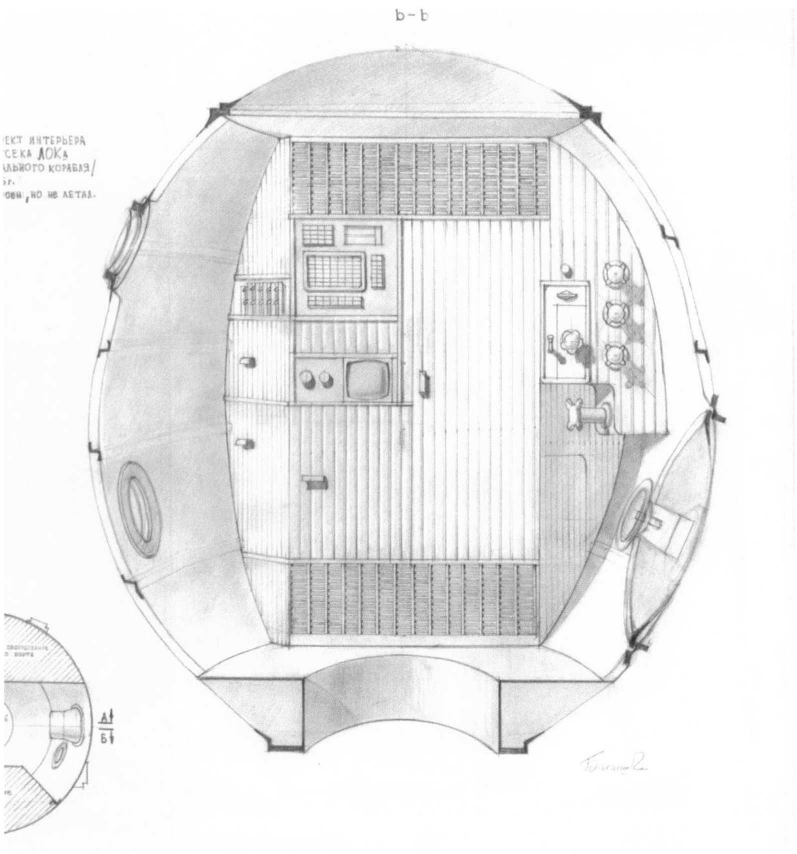

Archive Galina Balashova

Final design for the interior of the LOK lunar orbiter (1969)

while also drawing the technical engineering details for the arrangement of the devices as well as designing the emblems and pennants.

When I joined the design department, the Soyuz-T spacecraft was being developed based on the first Soyuz spacecraft as a model. There was a new combination of devices, so the interior needed to be modified as well. I was called when the individual parts had already been placed. My task was to create a visual representation of the spatial design.

It was decided relatively early on that the colour green would play a key role in the interior design. I'd initially chosen red for the sofa, but this turned out not to be such a great solution. Red looks black in outer space, while green does not change. So we decided to stick to green.

Once a prototype of the Soyuz capsule had been designed, it was revised and modified for alternative versions. It helped that I had worked on the arrangement of spaces and devices during the planning phase until then. This allowed me to achieve better results during the revision stage. If I had worked together with the engineers to find solutions, I don't think they would have listened to me at all. Women in this field usually took care of the supplementary and menial work. They were instructed to do something and needed to carry out these instructions. But my work was creative. For a woman to perform such work was not typical. It was interesting work.

The engineers did not understand the task of the architect – the task of thinking in spatial terms. I was surprised to find that not everyone can do this, since I had been accustomed as an architect to finding spatial solutions. For me, this was as natural as breathing air. Until I took on my position at the department, I was not aware that this was a particular ability.

Interview by Paul Meuser in Korolyov in August 2017.

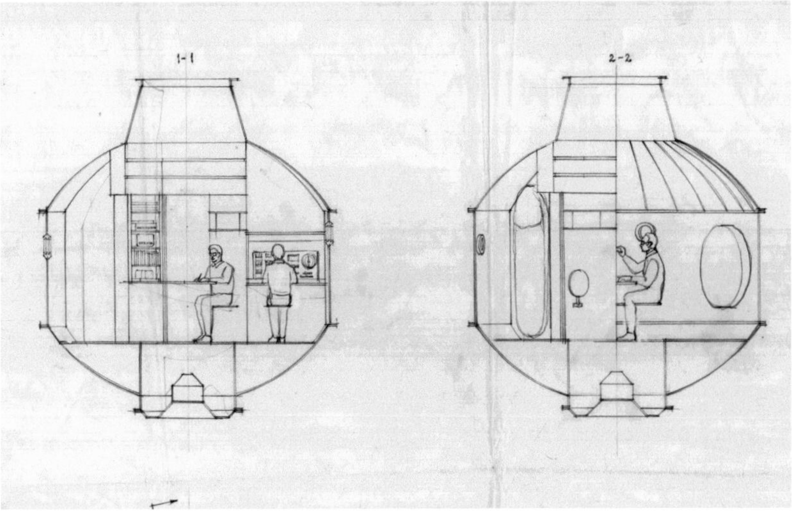

LOK lunar orbiter, section (1964–1968)

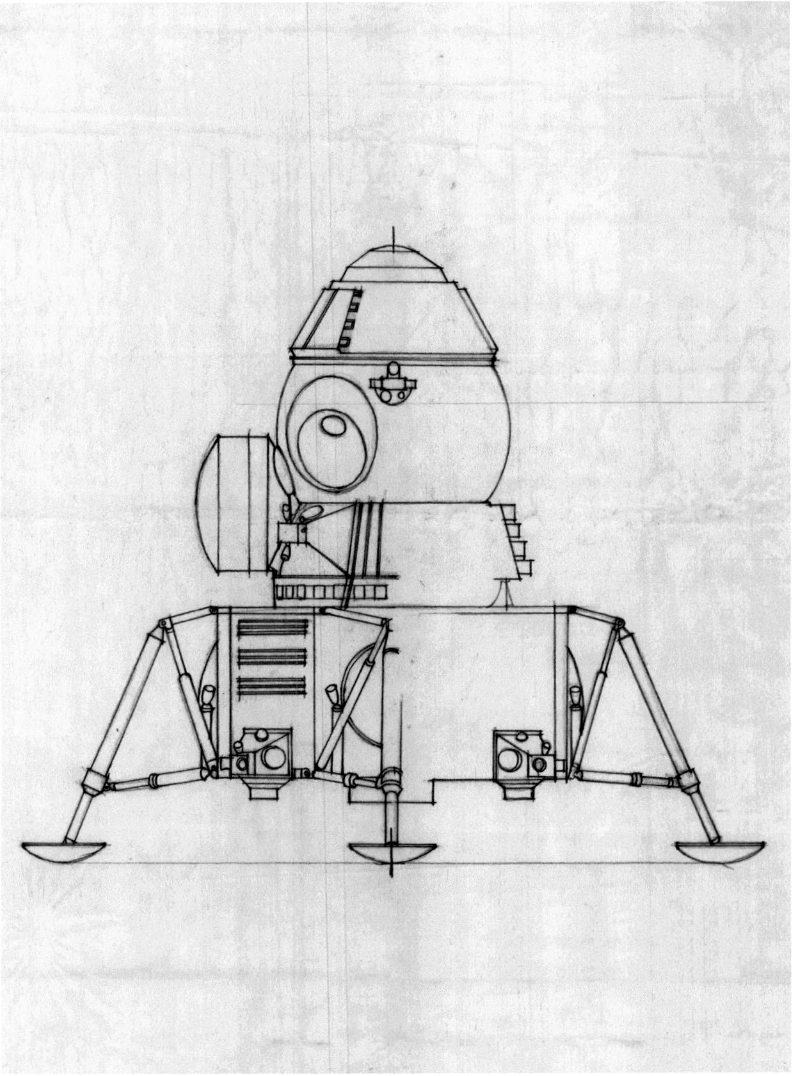

Lunar lander Lunniy Korabl (LK), elevation (1964–1968)

The USSR and US: Competing Ideologies Race to the Moon

1959 to 1969

Humanity's relationship with the moon in the twentieth century was defined by the success of the Apollo programme and its heroic astronauts. The moon landing in July 1969 represented an epochal milestone in the struggle for technological supremacy over the moon and overshadowed the Soviet space programme's long-standing lead in the early stages of the space race.

But it was the Soviet space organisation that sent the first human into outer space, launched the first artificial satellite into Earth orbit, established the first crewed space station (also in Earth orbit), and landed the first space probe on the moon. The launch of Sputnik in 1957 broadened the arena of the Cold War from Earth to outer space. And what followed during the subsequent two decades was a veritable showdown between two protagonists and their proxy organisations: the US with NASA on the one hand, and the Soviet Union with Roscosmos on the other.

The first chapter of this book documents that fierce technological battle. It encompasses the ten years of spaceflights to the moon between 1959 and 1969. During this time, the two rivals developed their rockets not only as launch vehicles for satellites and probes but also as carriers of intercontinental nuclear warheads in preparation for the possible escalation of the Cold War.

The Luna, Ranger, Surveyor, and Orbiter missions laid the foundation for the space-exploration projects taking place today. Although some of the missions resembled one another, both aesthetically and technologically, each mission represented a new milestone.

Moscow Polytech

Luna 2
Soviet Union

Launch: 12 September 1959, 6:39 am
Landing: 13 September 1959, 9:02 pm
Last contact: 14 Sept. 1959, 10:02 pm
Launch site: Baikonur
Landing site: 29.1°N 0°0

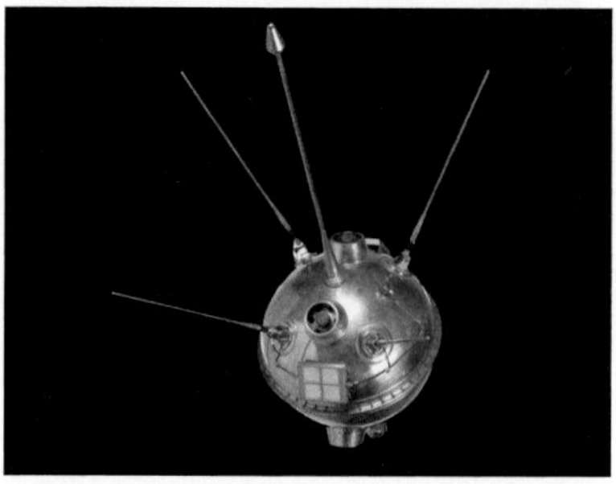

Polytech Moscow

Two years after the launch of Sputnik, the first-ever artificial Earth satellite, the Soviet space programme made its sixth attempt to send a probe from Baikonur to the moon, 384,400 km away. The rocket engineers had worked on this project in vain for almost 12 months. Luna 2 represented the breakthrough, which arrived after four explosions and the failed navigation of its structurally identical predecessor, Luna 1. In autumn 1959, Luna 2 became the first impactor probe to reach the moon, crashing into the targeted destination between the craters Autolycus and Archimedes. The Soviet Union thereby took a decisive lead in the race to the moon – the US made an unsuccessful attempt 12 days later. This success opened up a new chapter of the Cold War, since the Luna 8K72 carrier rocket could also be used as an intercontinental missile, capable of hitting US territories. The lunar probe's outer form was identical to that of the first Soviet satellite, Sputnik 1. It was the size of a football and weighed 390 kg, with five antennae mounted on the upper half of its spherical body. Four extended radially outwards, while a larger one marked the central axis of the round body. The body's joint, perpendicular to the central antenna, kept the two halves of the sphere together. Its rivets accentuated the diameter of the spherical structure like a kind of cornice. Various technical parts protruded from the metal surface. Most of the measuring devices were mounted on the side opposite the antennae, away from the cornice. During its flight through space, Luna 2 created a gas cloud with a diameter of up to 650 km, which was clearly visible from Earth. Once the probe impacted the moon, it dispersed a number of metal pennants, decorated with the hammer and sickle, which became the first architectural ruins on the moon.

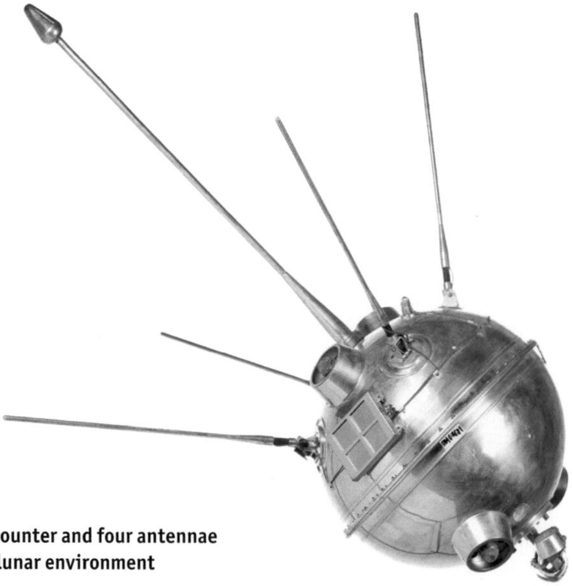

Moscow Polytech

Micrometeorite counter and four antennae for studying the lunar environment

High-performance technology with a zoomorphic design: two extendable antennae between two ion traps make up the fish-like face of Luna 2

Moscow Polytech

NASA

Wood construction in outer space:
Pat McKibben from NASA's Systems Design department, presenting the balsa wood sphere

Ranger 4
US

Launch: 23 April 1962, 8:50 pm
Landing: 26 April 1962, 12:49 pm
Last contact: 26 April 1962
Launch site: Cape Canaveral
Crash location: 15.5°S 130.7°W

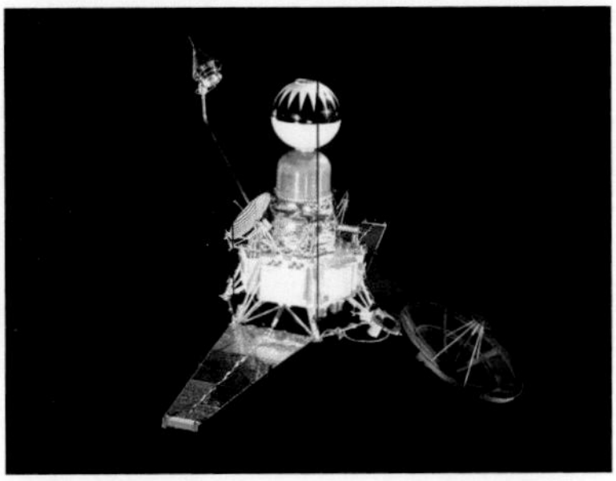
NASA

In April 1962, the US launched Ranger 4, unguided and uncontrolled, into space. However, the onboard computer lost its navigation function around seven minutes after the launch. The probe flung through zero-gravity space at a speed of over 24,000 miles per hour and began to spin on its own axis. The scientists could do little more than follow the weak signal on their monitors. However, against all expectations, the seemingly rogue spacecraft impacted the moon after its 66-hour odyssey.

The probe, a Block II spacecraft, featured a striking element: a balsa-wood ball with a diameter of 65 cm. One half was painted white, while the other half was emblazoned with a black-and-white zigzag pattern. It appeared to be balancing on a cylindrical body, tapering towards the top with a rounded edge. The cylindrical body itself sat on a hexagonal base, made of magnesium, with a diameter of 1.5 m. Delicate instruments were attached to the base. One of them was a high-gain antenna with a diameter of over 1 m, which enabled communication with ground control on Earth. Two trapezoidal solar panels extended in opposite directions from the base, with a span of almost 5 m, and were responsible for the power supply. A thin metal arm held the diverse instruments together. A small, cone-shaped antenna was attached to the tip of the arm, resembling a kind of spire for the three-metre-tall tower.

The impactor probe, weighing 155 kg, was designed to lower its metal arm and release the balsa-wood ball 21 km away from the lunar surface. The ball, weighing 64 kg, was to ignite its retrorocket for the descent, before jettisoning its engine to achieve a soft landing in free fall. It was equipped with a thermometer and seismometer, intended to take measurements on the moon after the landing. The rest of the probe was designed to capture images of the moon's surface during its descent and send these via television back to Earth before the impact. However, the navigation system's failure during the flight hampered the entire mission. But unlike all of its predecessors, which had gone missing in space, Ranger 4 still lies in a grey moon crater to this day, albeit in shattered pieces. The tropical balsa wood represents the first organic material to make it onto the moon.

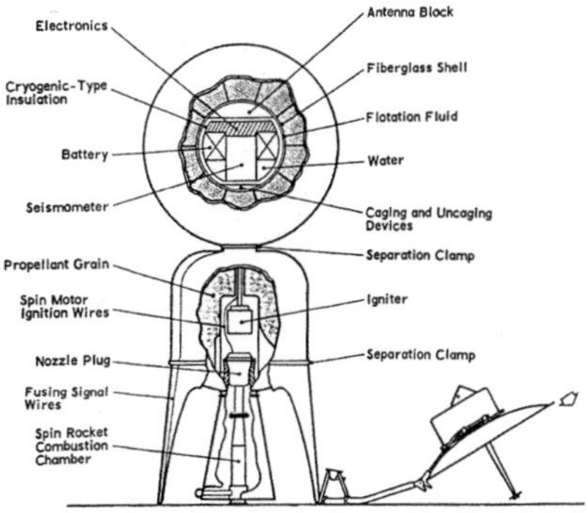

Complex mechanisms inside the probe

Launch of the Atlas-Agena B rocket from Cape Canaveral to the moon with Ranger 4 on board

NASA

National Geographic Magazine, October 1962

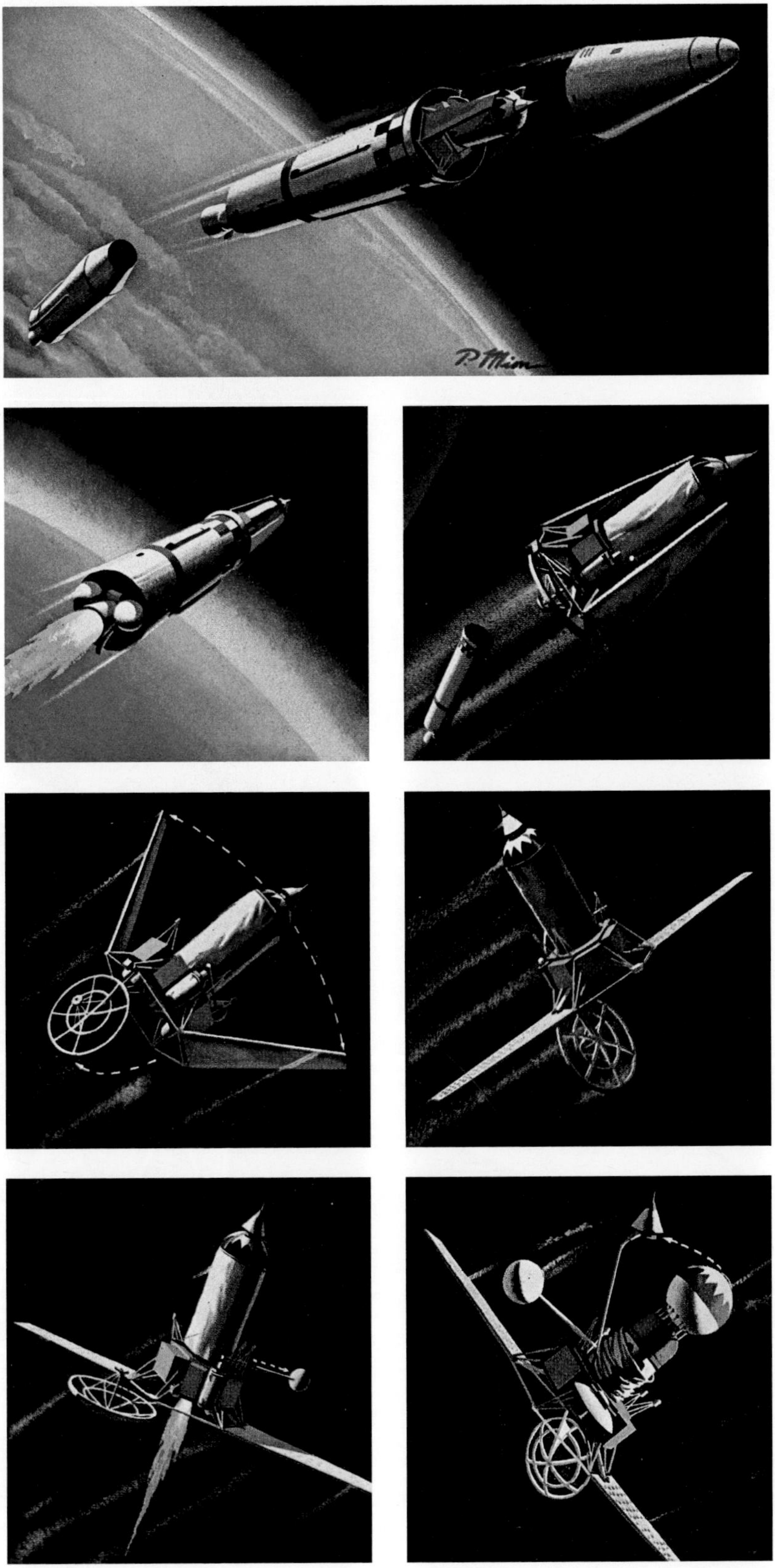

Ranger 4 unfolds itself in space. The individual flight sequences are presented here ...

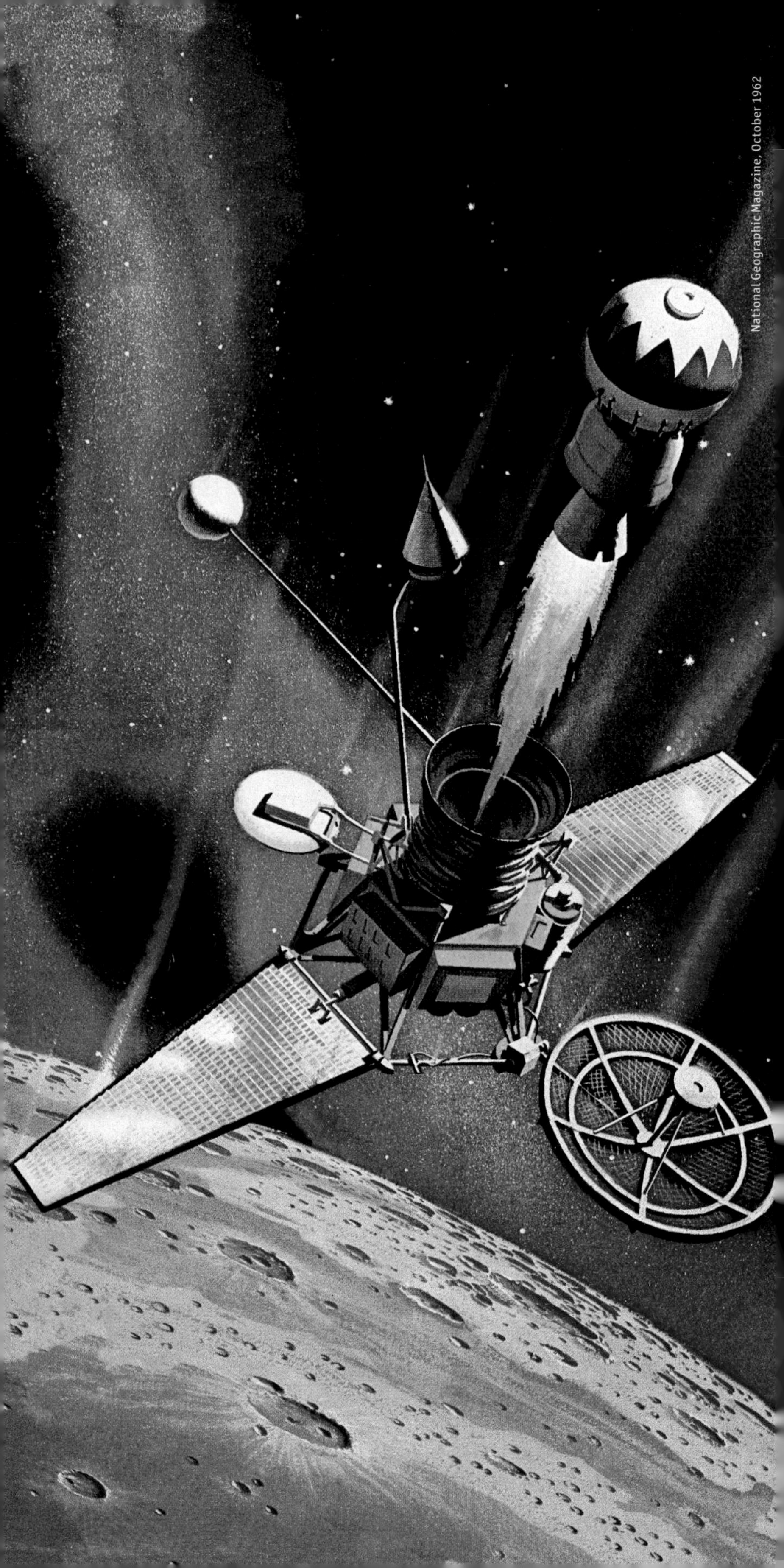

National Geographic Magazine, October 1962

National Geographic Magazine, October 1962

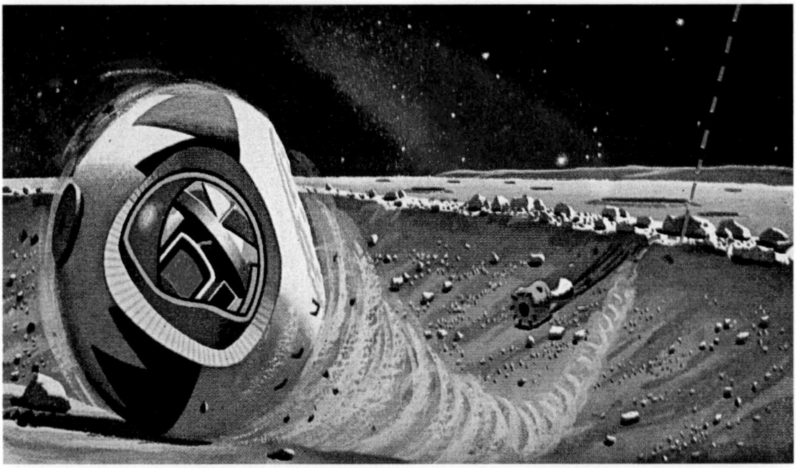

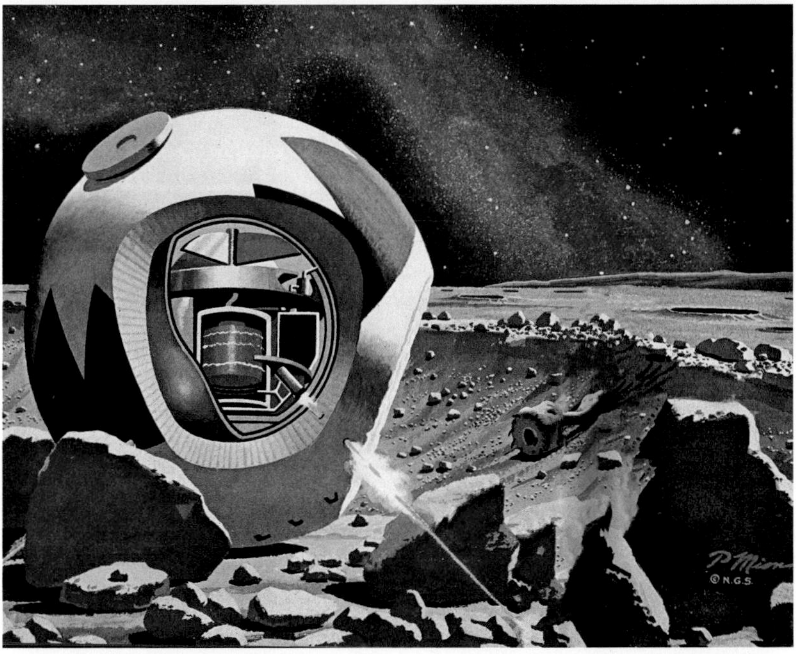

... in illustrations by Pierre Mion. His artistic rendering makes it possible to visually understand all stages of the mission. It also offers a view inside the balsa-wood sphere. These illustrations were published in 1962, in the October issue of the magazine *National Geographic* along with an article on the failed Ranger 5 mission.

An Atlas-Agena B carrier rocket being launched into space

NASA

The Ranger Space Probe Programme

The Ranger probes were the first American impactor probes designed to land on the moon. The US had previously carried out the Pioneer programme, partially aimed at exploring the moon, though the lunar flyby of Pioneer 4 was the programme's only success. The Ranger probes presented in this book can be categorised into three structural types: Block I, II, and III. These types differed not only in their outer forms, but also in their objectives. The first two Ranger probes belonged to the Block I type and were launched to test the technical systems that would be required for later missions. They never landed on the moon, which is why they are not presented in this book.

At first glance, Block I probes look simpler than their successors, though there are considerable similarities between all three blocks. After all, they all shared the same basic structure (for which the technical term is 'bus'). A hexagonal base was supported by a system of 'buttresses'. Two wing-like solar panels and one communications antenna were attached to the base, from which rose a tower-like structure, whose exact form changed from block to block.

Block I's tower, four metres tall, looked rather fragile and resembled the Eiffel Tower due to its lattice structure and pyramidal form. But unlike the Parisian monument, Block I probes featured a cylindrical compartment, tapering towards the bottom, at their peak. This compartment, which housed a magnetometer and ion chamber, was crowned by an omnidirectional antenna, and this composition would reappear in the Block III probes.

The Block II probes were capped by a spherical capsule, which was designed to land softly on the moon. Of these probes, only Ranger 4 was successful, and only to a limited extent. The capsule reached the moon, unlike the capsules of Ranger 3 and 5, but it crashed into the surface, bringing the mission to an abrupt end. Although Ranger 4 already had a camera on board, the Block III probes would be the first to take pictures of the lunar surface as part of an American mission. They were designed to take close-up shots before impacting the moon. To this end, they housed a camera system in a niche in the silver tower, which was also a characteristic feature of the Block III type. However, it wasn't until Ranger 7 that a Block III probe completed a successful mission. The Ranger programme comprised a total of nine missions, which were launched to the moon on Atlas-Agenda rockets. Five of these missions were at least partially successful.

NASA

Ranger Block I

Ranger Block II

Ranger Block III

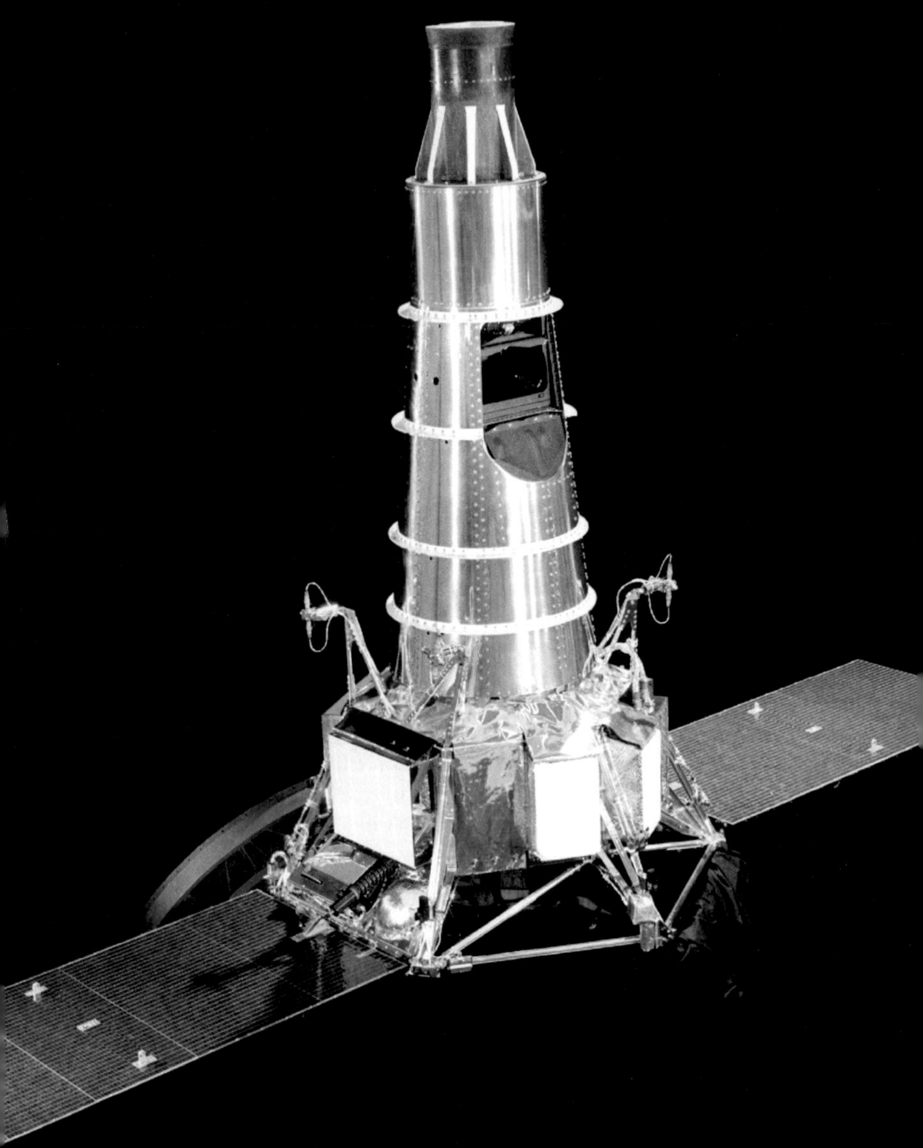

NASA

Standardised design series in outer space: it is hard to spot the differences between the probes from Ranger 6 to Ranger 9, even for space engineers. To this day, NASA uses the same iconic image for all four missions of this series.

Ranger 6

US

Launch: 30 January 1964, 3:49 pm
Landing: 2 February 1964, 9:24 am
Last contact: 2 February 1964
Launch site: Cape Canaveral
Landing site: 9.4°N 21.5°O

The first attempts by the US to fly to the moon failed due to faulty engines and instruments on board. Some probes were unable to resist the gravity of the Earth and fell back into its atmosphere, bursting into flames like fireballs. Others were torn apart during the flight, along with their carrier rockets. Ranger 6 represented the fourteenth attempt by the US to send a functional probe into orbit and to guide it to the lunar surface. The probe, mounted on an Atlas-Agena rocket, was launched in Florida on 30 January 1964. It became the first American probe to capture images of the moon; the Soviet probe Lunik III had photographed the far side of the moon five years earlier. Unlike the Luna impactor probes, Lunik III orbited the moon, though it began to combust once it re-entered the Earth's atmosphere. Ranger 6 resembled its American forerunners in a number of ways. The tower structure rested on the base, called the 'bus', which was a hexagonal structure composed of aluminium and magnesium tubes. A high-gain antenna and rectangular solar panels were attached to the base's open framework. A rocket engine nozzle was located inside. Unlike the heterogeneous bodies of its predecessors, Ranger 6 had a column-like tower structure, shimmering silver, which tapered towards the top and was held together by white, riveted barrel bands. It was crowned by a bronze cylinder, joined to the tower structure by white struts. An omnidirectional antenna was housed inside and was designed to transmit photos to Earth until shortly before the lunar impact. The integrated camera equipment, weighing approximately 160 kilograms and housed in a niche in the tower, was designed to activate automatically, ten minutes before impact, and capture photographs until the collision. The Ranger 6 mission would have been a complete success, except the cameras could not be activated. The Ranger 7 probe was launched half a year later.

Ranger 7

US

Launch: 28 July 1964, 4:50 pm
Landing: 31 July 1964, 1:25 pm
Last contact: 31 July 1964
Launch site: Cape Canaveral
Landing site: 10.6340°S 20.6771°W

It was structurally identical to its predecessor aside from a small number of technical details. The initial stages of the mission went according to plan, and the probe was placed into the correct orbit. But before the landing site's location could be transmitted to the onboard computer, the probe found itself on a collision course and eventually fell under the spell of the moon's gravitational force. It crashed, not so gently, ten kilometres away from the intended landing site. Nonetheless, the mission had enormous consequences for the American lunar programme. Ranger 7 delivered the progress that had been hoped for, transmitting lunar images and testing a range of space-flight technology. The first photos resembled images seen through a telescope. But the image quality continuously improved as the probe approached the moon, and the final photograph showed the moon with the highest resolution to that date of just a few square metres.

NASA

The model served as visual representation in pre-digital times.

Ranger 8
US

Launch: 17 February 1965, 5:05 pm
Landing: 20 February 1965, 9:57 am
Last contact: 20 February 1965, 9:57 am
Launch site: Cape Canaveral
Landing site: 02.6377°N 24.7881°O

As the US lunar programme proceeded from mission to mission, it set ever-higher standards not only for its spacecraft technologies but also for the quality of the photographed images. So when Ranger 8 was launched into lunar orbit, once more with a tower-like body, its objective was to capture high-contrast images of the moon. The NASA scientists were hoping that the clear contrast between the light and dark areas would provide a deeper insight into the conditions on the desolate lunar landscape. However, the telemetry system's power dropped significantly during the planned mid-course trajectory correction. As a result, the camera settings were incorrectly configured, and the travelling speed of over 2.5 km/s led to a stereo effect in the images. The otherwise successful mission thus produced a large number of blurry photographs that fell short of the image quality initially hoped for. The probe's surviving parts remain in ruins on the moon to this day.

Ranger 9
US

Launch: 21 March 1965, 9:37 pm
Landing: 24 March 1965, 2:08 pm
Last contact: 24 March 1965, 2:08 pm
Launch site: Cape Canaveral
Landing site: 12.83°S 02.37°W

The Ranger 9 probe, the last of the Block III type, also completed the initial stages of its mission according to plan. The trajectory correction and orientation of the collision course took place without a hitch. Once the probe began its rapid descent to the target crater, it started to capture images 19 minutes before impact. It transmitted 5,814 photographs during its descent, with the sharpness of the images continuously improving until the point of collision. The last photo had a resolution of 0.3 metres, which was likely the record high at the time. The Ranger 9 mission was also the first lunar flight that could be followed live on American television. The last captured images and the intended collision with the moon were broadcast almost in real time – with a delay of 1.3 seconds – onto screens in regular homes. The last Ranger mission's complete success was therefore witnessed by hundreds of thousands of people.

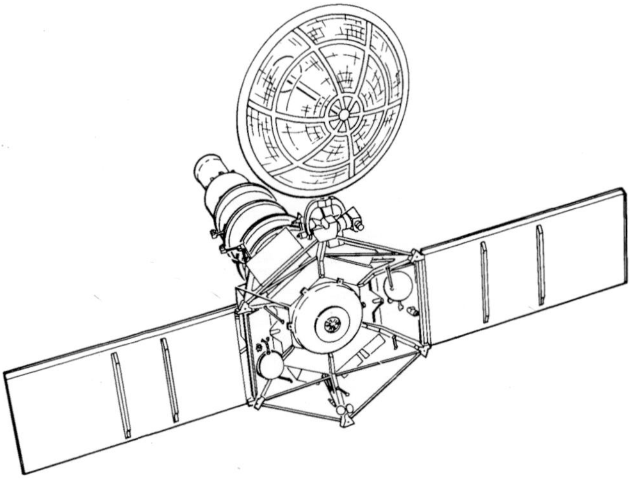

Perspective drawing of a Ranger probe

NASA

Luna 5
Soviet Union

Launch: 9 May 1965, 7:49 am
Landing: 12 May 1965, 7:10 pm
Last contact: 12 May 1965, 7:10 pm
Launch site: Baikonur
Crash location: 8°N 23°W

Luna 5 was launched into space from the Kazakh Steppe almost two years after the successful launch of its predecessor. Luna 4, though launched successfully, had not managed to escape Earth orbit, and the Soviet government re-purposed it retrospectively as a satellite to cover up the failure. (This in fact became the standard procedure for other unsuccessful missions.) The Soviet space programme had gone through many years of failure: three missions had ended with an explosion shortly after the launch, and the Americans had established a comfortable lead in the race to the moon. As such, it was critical for the next Soviet mission to be a success. All hope was placed in Luna 5, which was to represent the Soviet Union's seventh attempt to achieve a soft landing on the moon. It was the second probe of the new E-6 type. Its compact body, weighing almost one and a half tonnes, was composed of three parts: a conical base, tapering towards the bottom, a cylindrical torso, which was almost invisible behind a large number of devices and cables, and an oval head. The head section was covered with a dark piece of fabric – a kind of pillow, which was to inflate like a parachute shortly before touchdown to enable a soft landing. The metal, conical base, housing the propulsion unit, featured a coffered pattern. The majority of the probe's parts served only to guide it during its journey through space. The built-in retrorockets were designed to reduce the speed of the bulky body during the descent. The pillow would then be filled with air to further reduce the kinetic energy, and the head section would detach from the torso, bounce along the lunar surface, and finally come to rest near the planned landing site. Upon

Close-up of Luna 5: open cables, tubes, and instruments on the probe's exterior

Moscow Polytech

Luna 7
Soviet Union

Launch: 4 October 1965, 7:55 am
Landing: 7 October 1965, 10:08 pm
Last contact: 7 October 1965, 10:08 pm
Launch site: Baikonur
Crash location: 9.8°N 47.8°W

Luna 8
Soviet Union

Launch: 3 December 1965, 10:46 am
Landing: 6 December 1965, 9:51 pm
Last contact: 6 December 1965, 9:51 pm
Launch site: Baikonur
Crash location: 9.1°N 63.3°W

landing, the spherical head section would unfold its four leaf-like wings, erect four antennae, and stand up on its own. This, at least, was the plan. In spite of the mission's auspicious launch, Luna 5 began to spin around its axis after one day in space. After a series of malfunctions, it impacted the moon without decelerating. The collision created a cloud of dust, almost 220 km high, which was visible from Earth. The probe's successor, Luna 6, flew past the moon at full speed at a distance of around 160,000 km. It was Luna 7 that became the first Soviet probe to achieve the necessary mid-course trajectory correction, but its mission also failed. Its navigation system had been programmed incorrectly, and the onboard computer activated the retrorockets too early, blocking the engine. As a result, the probe lost its orientation and crashed into the moon, its velocity increasing due to the moon's gravity. The resulting cloud of smoke marked the tenth failed attempt of the E-6 programme. Luna 8 came a small step closer to its objective than its predecessors. Its structure, virtually identical from the outside, was technologically far superior to the others, at least on paper. It perfectly tested the interstellar orientation, guidance, and control systems while hovering through space, carefree, for three days. But while it was preparing to land, a plastic clamp pierced through the head section's pillow and sealed the unhappy fate of the mission.

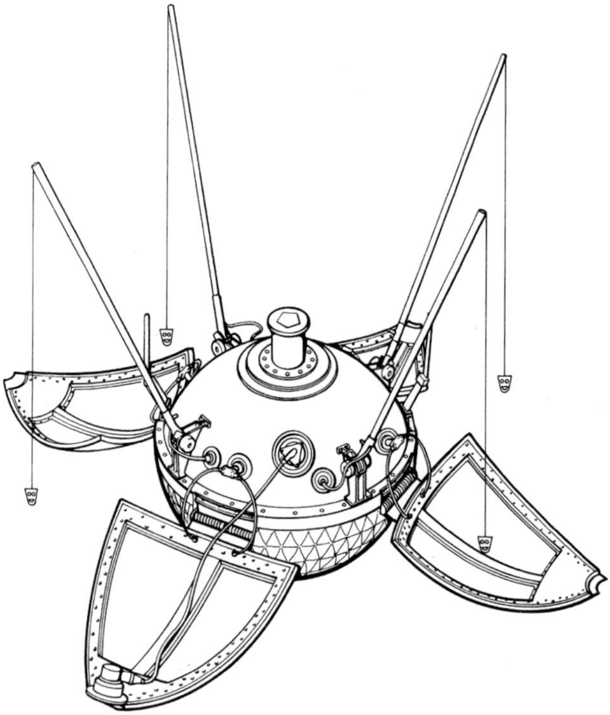

Engel 1988

The metal sphere fully unfurled

Moscow Polytech

The air bag is sewn onto the head of the probe.

Artist's rendering of what a successful mission might have looked like for the probes from Luna 5 to Luna 8

Moscow Polytech

A replica of Luna 9
in the space museum
in Kaluga

Luna 9
Soviet Union

Launch: 31 January 1966, 11:45 am
Landing: 3 February 1966, 6:45 pm
Last contact: 6 February 1966, 10:55 pm
Launch site: Baikonur
Landing site: 7.08°N 64.37°W

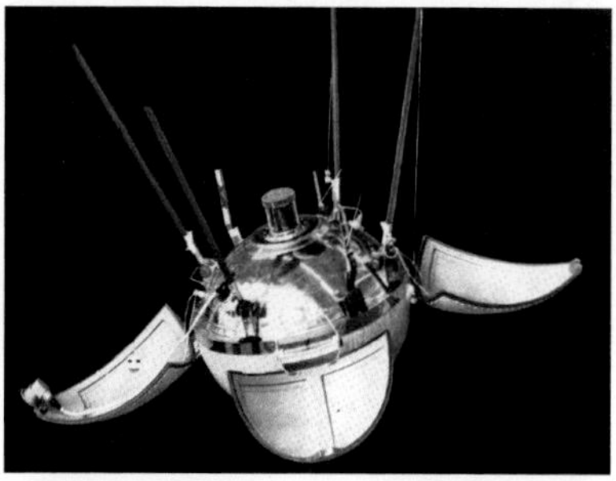

Moscow Polytech

On the outside, Luna 9 very much resembled its predecessors. The only difference, aside from a few minor details, was in the size of its enormous torso, which gave the probe a unified appearance. But Luna 9 featured a number of invisible technical modifications that ultimately led to the mission's success. It had a contact sensor, for example, that touched the moon when it was five metres above the lunar surface. The spherical landing capsule then separated from the torso and began to hover, while the rest of the probe, no longer any use, fell forcefully into the moon and shattered. Meanwhile, an automatically inflated cushion protected the spherical capsule as it bounced along the lunar surface before finally coming to rest, all according to plan.
After the landing, a metamorphosis took place. The protective cocoon deflated, and the metal sphere, weighing 99 kg, unfolded and stretched out its feeling antennae towards its home, the Earth. Unlike its predecessors, which lay in ruins, Luna 9 became the first probe to achieve a soft landing on the moon. This makes it is the oldest artefact still intact on the moon, which means one could argue that the architectural history of the moon begins with Luna 9, in 1966. The event also marked the first soft landing of a probe on a celestial body. Once again, the Soviet engineers, with their thirteenth attempt as part of the E-6 programme, succeeded in making history before their American counterparts. NASA still uses the probe's landing procedure – decelerate using retrorockets and inflate a cushion for protection – for its missions to Mars. Luna 9 transmitted 27 images of the lunar surface for eight hours, spread out over a three-day period, before its battery reserves were depleted.

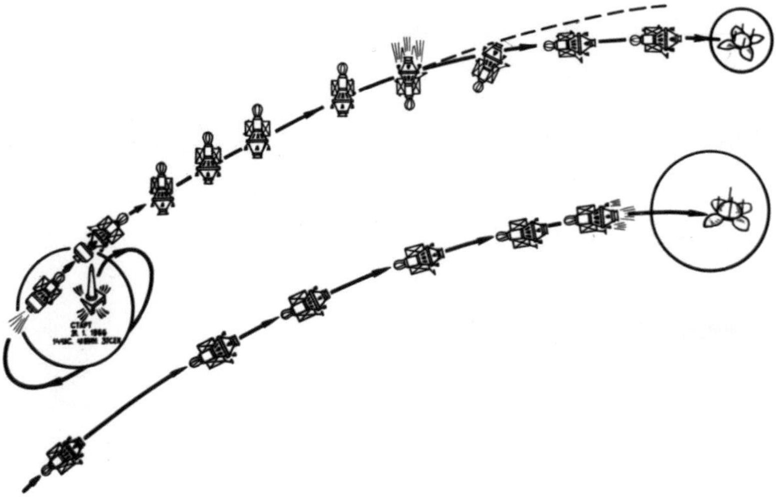

Moscow Polytech

Trajectory and individual stages of the Luna 9 mission

Harvey 2007

Luna 9 carried this device in the centre of its metal cocoon. The panoramic camera made the mission a success, capturing a modest 27 images.

Moscow Polytech

8. Результаты дешифрирования панорамы (вверху), изображение линий дальности (в середине) и горизонталей (внизу) на панораме, переданной станцией «Луна-9».

Technological leap forward: the USSR was the first to capture photos of the lunar surface. Here they are presented in the newspaper *Red Star* and in a book about cosmonautics.

ВЫДАЮЩИЙСЯ ЭКСПЕРИМЕНТ XX ВЕКА

За нашу Советскую Родину!

КРАСНАЯ ЗВЕЗДА

ЦЕНТРАЛЬНЫЙ ОРГАН МИНИСТЕРСТВА ОБОРОНЫ СОЮЗА ССР

ГОД ИЗДАНИЯ 42-й
№ 32 (12863)
…враля 1966 г., вторник
Цена 2 коп.

ЛУННАЯ ПАНОРАМА

СООБЩЕНИЯ ТАСС

ПРОГРАММА ИССЛЕДОВАНИЯ ЛУНЫ С ПОМОЩЬЮ АВТОМАТИЧЕСКОЙ СТАНЦИИ «ЛУНА-9» УСПЕШНО ЗАВЕРШЕНА

5 февраля с 19 часов до 20 часов 41 минуты московского времени с автоматической станцией «Луна-9» был проведен сеанс радиосвязи, которым была завершена намеченная программа исследований Луны с помощью автоматической станции «Луна-9».

Телевизионные изображения, переданные со станции «Луна-9», являются уникальными и, по предварительному заключению ученых, представляют исключительную научную ценность для определения структуры и особенностей поверхности Луны.

* * *

Как уже сообщалось в печати, программа исследования Луны с помощью автоматической станции «Луна-9» успешно завершена.

Вместе с тем, учитывая, что бортовые источники тока имели еще некоторый запас электроэнергии, превышающей расчетное значение, со станцией «Луна-9» был проведен дополнительный двухчасовой сеанс радиосвязи, который начался в 23 часа 37 минут московского времени 6 февраля.

Во время сеанса связи с борта станции были приняты телеметрическая информация, характеризующая работу систем, температурный режим станции и повторные телевизионные изображения некоторых участков лунной панорамы.

В этот сеанс был израсходован практически весь запас электроэнергии бортовых источников тока, поэтому дальнейшая радиосвязь с автоматической станцией «Луна-9» прекращается.

Всего со станцией «Луна-9» было проведено 7 сеансов радиосвязи общей продолжительностью 8 часов 5 минут.

Полученные уникальные телевизионные изображения поверхности Луны и научная информация будут изучаться и исследоваться.

Результаты исследований будут опубликовываться в печати.

Годичное собрание Академии наук СССР

7 февраля в Москве открылось годичное общее собрание Академии наук СССР.

Открывая собрание, президент Академии наук СССР академик М. В. Келдыш отметил, что за последнее время сделан ряд крупных открытий в науке. Наметились новые пути прогресса в ряде важнейших отраслей техники и народного хозяйства.

Доклад, содержащий анализ важнейших достижений в области естественных и общественных наук за 1965 год, сделал главный ученый секретарь президиума АН СССР академик Н. М. Сисакян.

По докладу развернулись прения. Выступили академик Б. Н. Петров, член-корреспондент АН СССР В. М. Чавчанидзе, академик А. А. Имшенецкий, член-корреспондент АН СССР В. Л. Рыжков, академики С. Л. Соболев, В. А. Котельников.

8 февраля годичное общее собрание Академии наук СССР продолжает работу.

(ТАСС).

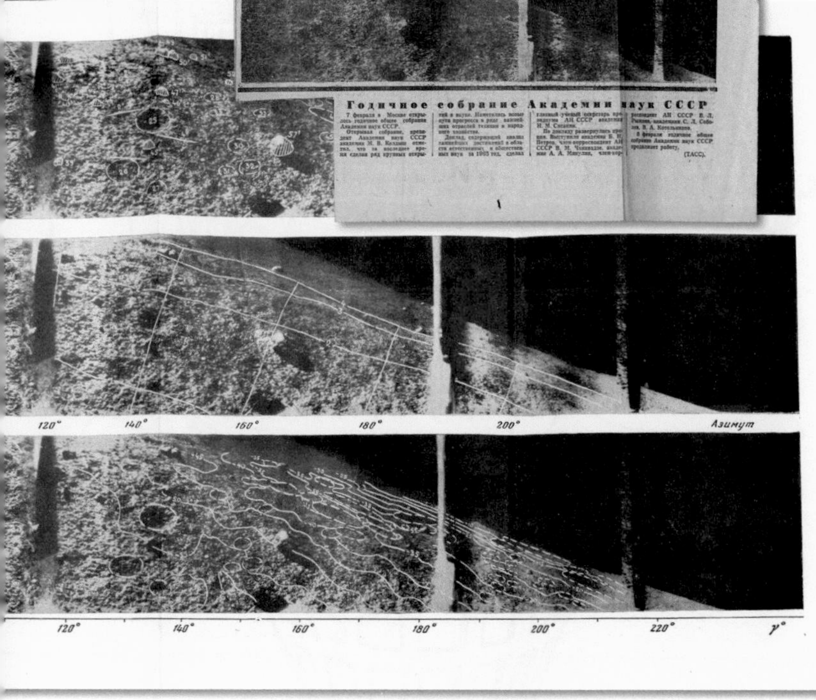

Soviet pragmatism: the missions Luna 9 and 10 were based on a virtually identical concept for the carrier probes. The differences are in the details.

Engel 1988

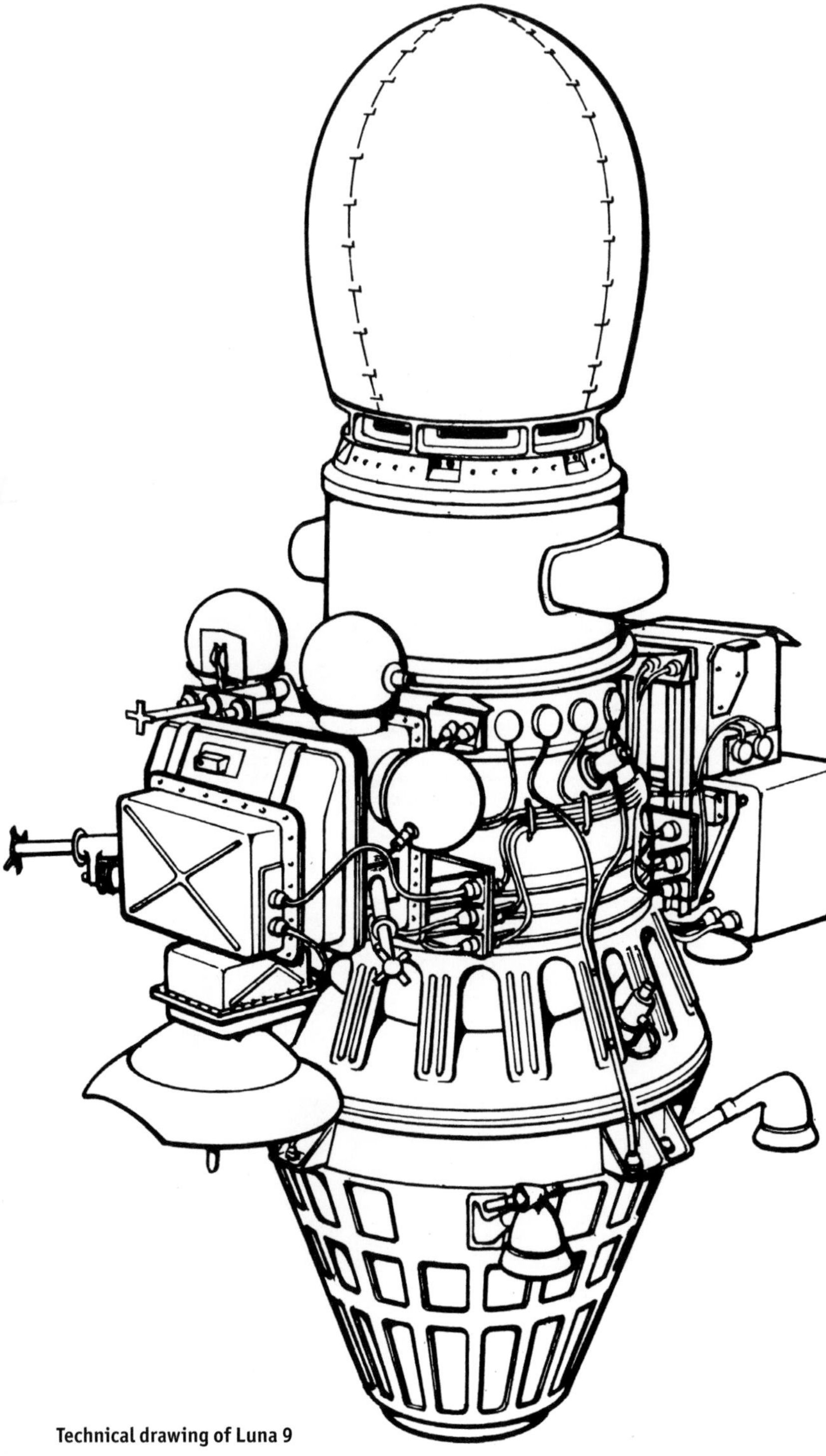

Technical drawing of Luna 9

Engel 1988

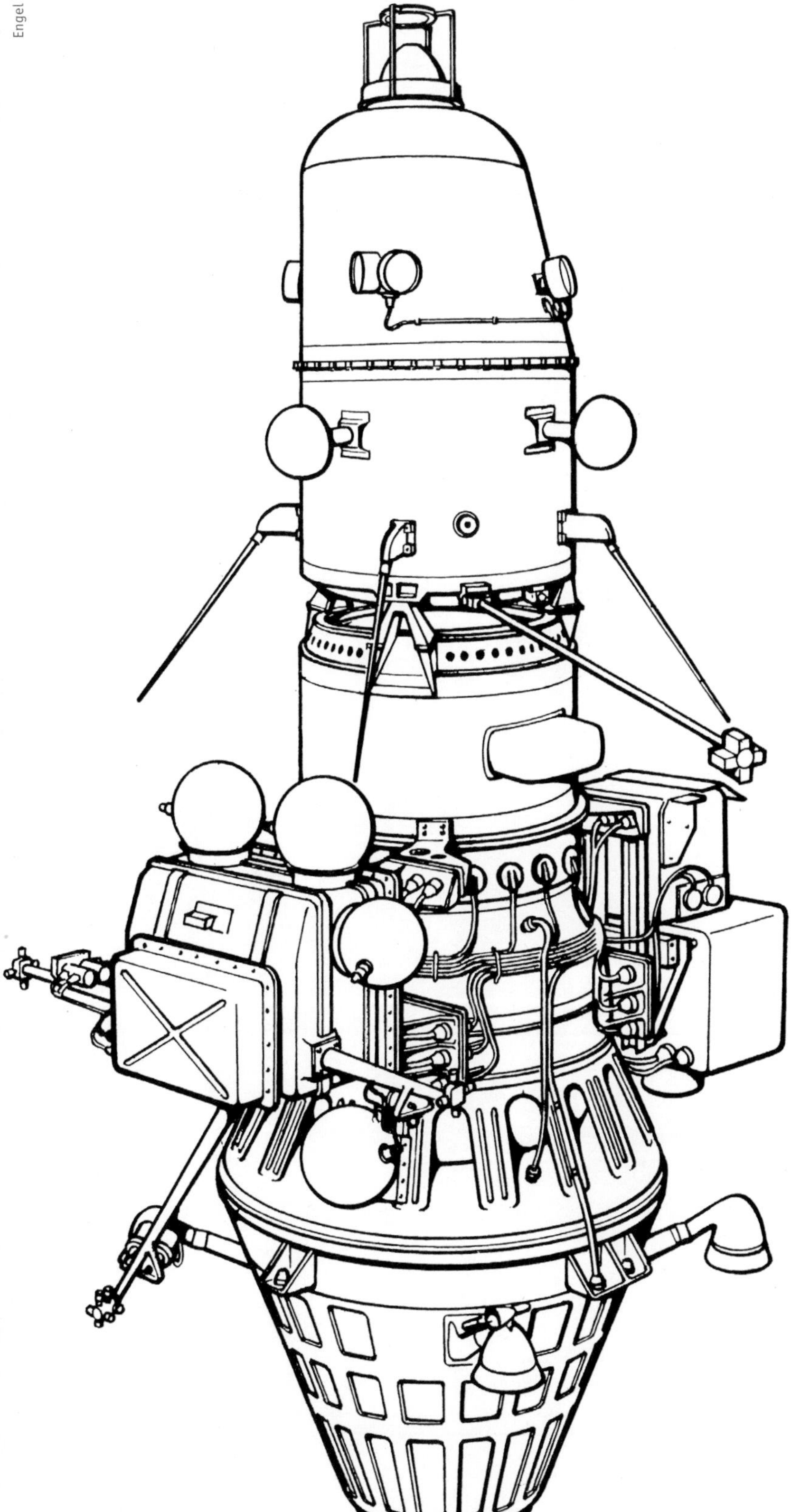

Technical drawing of Luna 10

Moscow Polytech

Luna 10 satellite in lunar orbit, with aesthetic changes added by the illustrator

Luna 10
Soviet Union

Launch: 31 March 1966, 10:48 am
Landing: crashed at unknown time
Last contact: 30 May 1966
Launch site: Baikonur

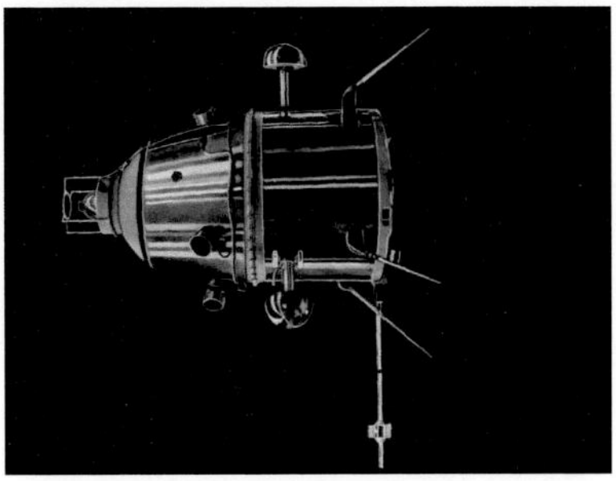

The Soviet Union, having achieved the first soft landing on the moon, set out to reach a new milestone: the launch of a probe into lunar orbit. The Soviet engineers modified the carrier rocket used for the previous Luna mission to achieve their goal as fast as possible. However, the first mission with this rocket failed: the probe was unable to escape Earth orbit due to a technical problem. The Soviet Union attempted to save face by retrospectively describing the probe as a satellite, Kosmos 111. A structurally identical follow-up probe, Luna 10, was launched a month later. It was composed of a cylindrical base and a shiny metal torso in the form of an oblique, shiny, metal cone. A small, dome-shaped radiometer was placed on top. A selection of instruments were arranged around the probe's body, although the equipment was very sparse to enable the soonest possible launch. There was no camera on board. Instead, scientific instruments and measuring devices – such as a magnetometer, whose sensor protruded away from the body, and three mushroom-shaped omnidirectional antennae with silver, polished caps – were mounted around the cylinder. The probe did not have an attitude control system, which meant the scientists on Earth could not orient it during the flight, nor could the probe orient itself. Nonetheless, it was able to study the moon's influence on the celestial bodies in its orbit. After the successful mission, Luna 10, no longer able to withstand the moon's gravitational pull, crashed into it at an unknown time after making last contact with the control centre.

Replica of the Luna 10 mission's satellite in the space museum in Kaluga

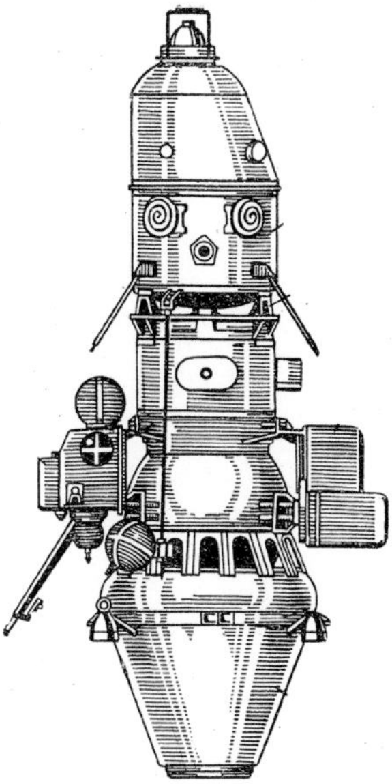

Elevation of Luna 10

**Front, back, above, or below: it is not easy, looking at the probes,
to decipher their orientation, designed as they were for the empty void of outer space.
The above image depicts Luna 12 or 14.**

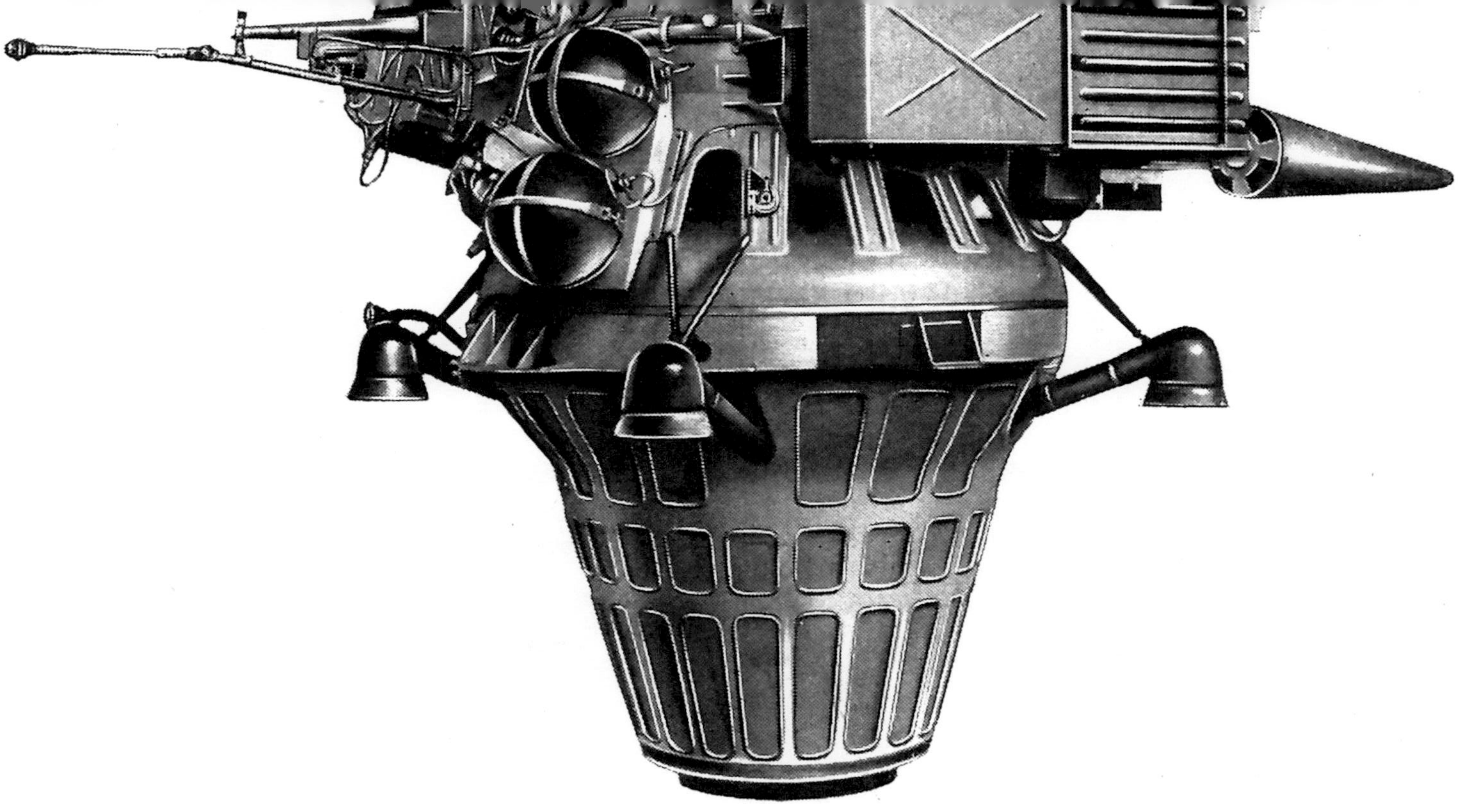

Avduyevski 1979

Badge for the Luna 12 mission, from the front and back

Luna 11
Soviet Union

Launch: 24 August 1966, 8:09 am
Landing: crashed at unknown time
Last contact: 1 October 1966
Launch site: Baikonur

After the success of Luna 9 and 10, the Soviet engineers went back to the drawing board to develop a new design for Luna 11. The outcome was the new E-6LF series, which was more compact than the older designs, although it once again featured a three-part body. The base, in the form of a slightly oblique, upside-down cone, featured the familiar coffered pattern. A series of 'bridges', forming a kind of arcade, connected the base to the central body. This configuration, too, recalled the appearance of older Luna probes. But the metal spheres, fragile cables, and braced antennae were completely new, as were the cameras. The top section also had a much different appearance. Also in the shape of a cone, it was divided into three parts: an undecorated base, a midsection with vertical stripes (emblazoned with the Soviet star), and a pointy radiometer at the top. The mission went seamlessly, almost exactly according to plan. However, a foreign body got caught in the attitude control system and impeded the probe's orientation. So Luna 11, like its predecessor, orbited the moon without orientation. All instruments functioned perfectly and collected data, though the cameras were pointing in random directions as they captured images.

The Soviet space engineers made great efforts throughout 1966 to achieve crewed space flight, undoubtedly motivated by the far greater number of spacecraft and astronauts their Americans had sent into space in the prior year and a half. In the early 1960s, the Soviet Experimental Design Bureau, headed by Sergei Korolev, had begun to develop the Soyuz spacecraft, which was capable of accommodating up to three people. And the Soviet Union devised a plan for two Soyuz capsules to meet in Earth orbit as a new

Luna 12
Soviet Union

Launch: 22 October 1966, 8:38 am
Landing: crashed at unknown time
Last contact: 19 January 1967
Launch site: Baikonur

propaganda spectacle. However, this rendezvous never took place. To compensate for the fruitless plan, the government released a single photo of the moon's surface, with a resolution of 15 to 20 metres per pixel, and celebrated it publicly. This photograph had been captured by Luna 12, the twin sister of Luna 11.

The Soviet lunar programme appeared dormant during the following year and a half, while NASA continued to diligently sent spacecraft to the moon. In 1962,

Luna 14
Soviet Union

Launch: 7 April 1968, 10:09 am
Landing: crashed at unknown time
Launch site: Baikonur

Luna 14 was launched to the moon, just over a year after the death of Yuri Gagarin, the first person to orbit the Earth in outer space. The launch was successful, and the probe was able to test several communications devices for upcoming lunar missions and measure the moon's gravitational fields in order to calculate future trajectories. This marked the last use of this Luna model and of the Molniya carrier rocket, after which the Soviet lunar programme went in a new direction.

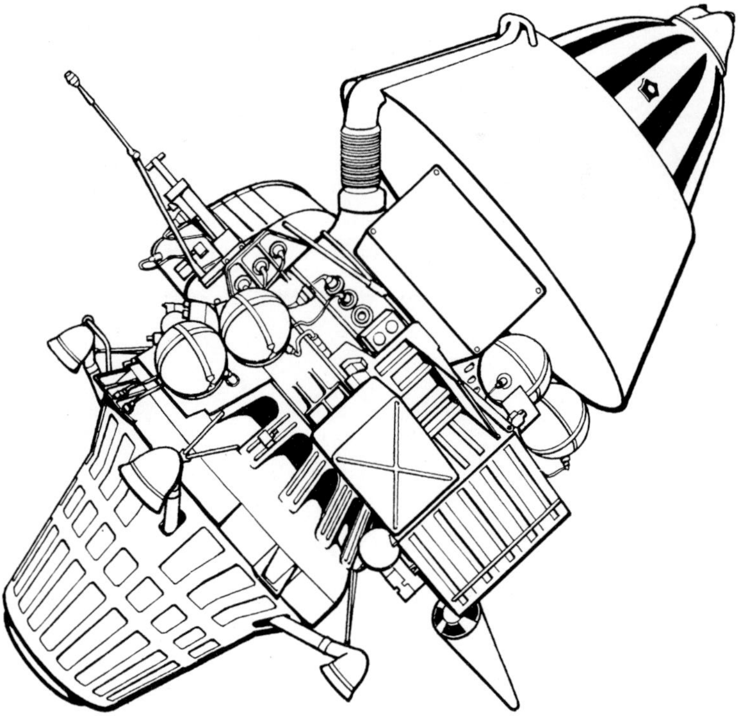

Engel 1988

Technical drawing of the structurally identical probes Luna 12 and 14

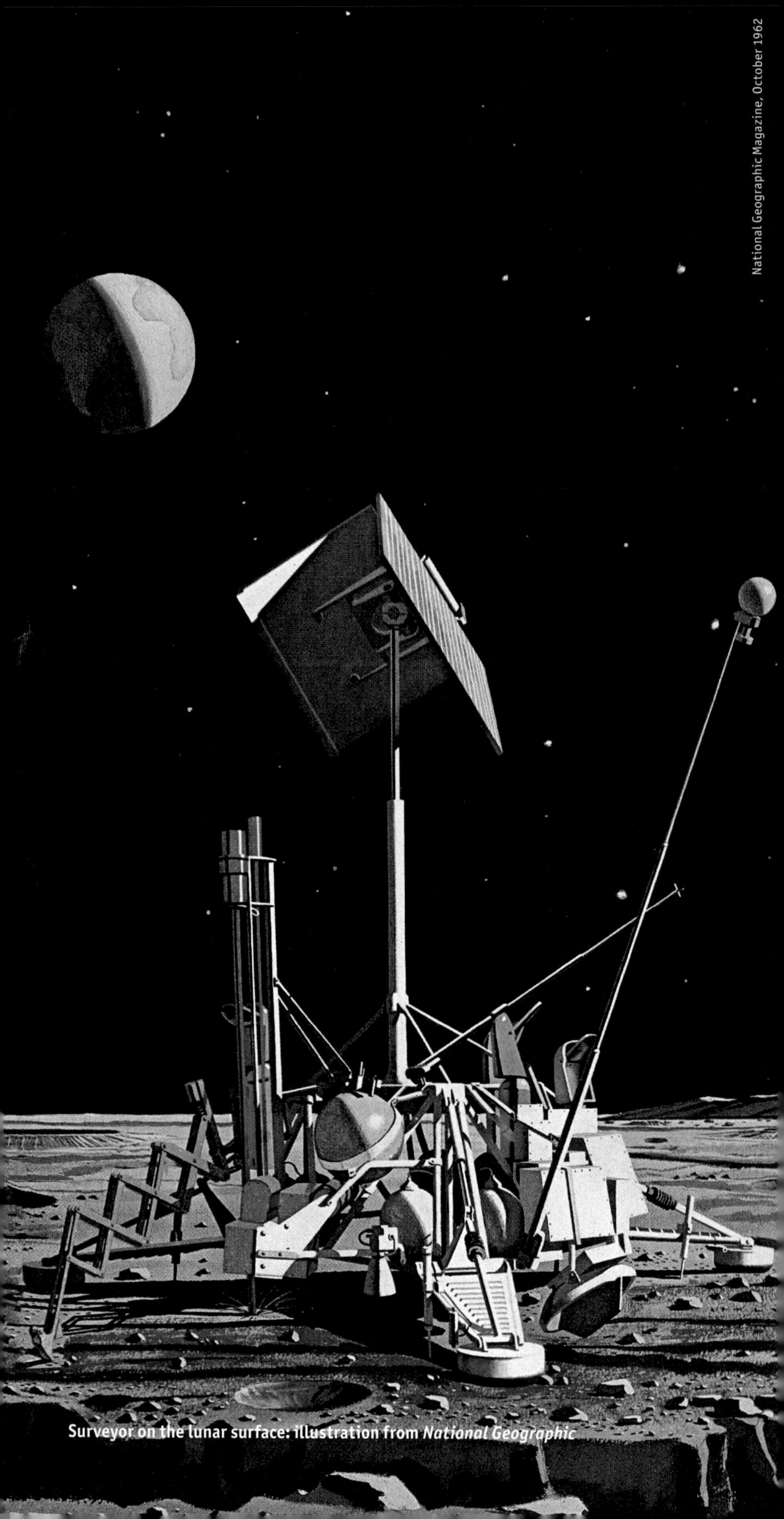

Surveyor on the lunar surface: illustration from *National Geographic*

National Geographic Magazine, October 1962

Surveyor 1
US

Launch: 30 May 1966, 2:41 pm
Landing: 2 June 1966, 6:17 am
Last contact: 7 January 1967
Launch site: Cape Canaveral
Landing site: 2.474°S 43.339°W

While the Soviet space engineers continued to pursue a soft landing, attempting no less than 12 missions, their American counterparts achieved the same goal on their first try, with Surveyor 1. Work on the Surveyor programme had already begun in 1959, but a moon landing remained out of sight for several years, since no existing carrier rocket was powerful enough to transport the probe, weighing several tonnes. NASA conducted several preparatory tests until the mid-1960s, before finally launching Surveyor 1 on an Atlas-Centaur 10 rocket from the Cape Canaveral Air Force Station on 30 May 1966. Surveyor 1, like the earlier probes of NASA, had the form of a tower, but of a completely different kind. In contrast to the solid, column-shape tower of the Ranger probes, Surveyor 1 had a plain, thin mast, braced onto a tetrahedral base. Two rectangular solar panels were mounted on top of the mast. They shaded and supplied power to the probe, rotating in line with the sun's movement for optimal exposure. They also made the probe visible from a distance. The rest of the instruments were mounted on the base, which formed a kind of multi-part plinth. The base, in turn, was carried by three legs, with round feet, extending out of the three corners.

The probe was reoriented during its approach so that it could land on its feet after gently bouncing along the lunar surface. This meant Surveyor 1 also left footprints in the dust – a detail which itself was useful for the scientists, who analysed the prints to study the displacement and scattering of lunar dust. Various propulsion engines guided the approach – which required the probe to rotate – and were jettisoned one by one before the landing. The probe, having

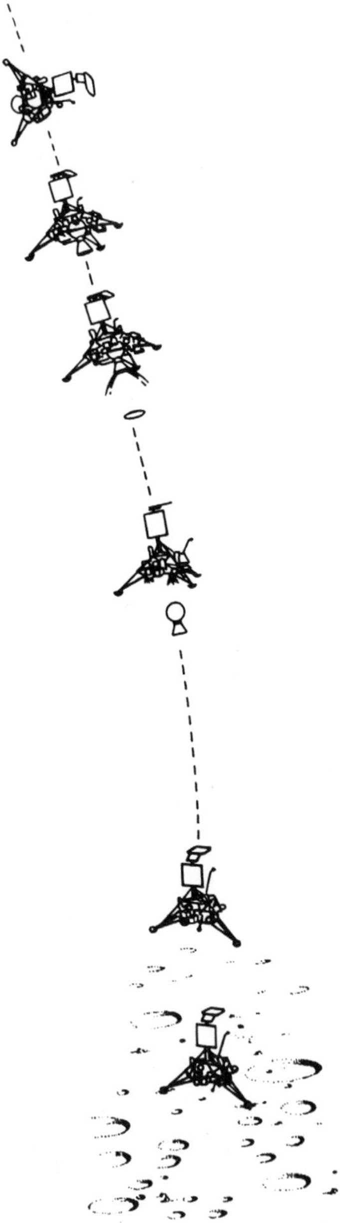

NASA

Surveyor over the lunar surface: a soft landing in five steps

NASA

Oceanus Procellarum – lunar landscape according to Surveyor 1

burnt up its fuel during the flight, was only carrying a quarter of its launch weight by the time it landed on the moon. During its six-week mission, Surveyor 1 transmitted over 11,000 photos of the Earth and of other celestial bodies such as the sun and Jupiter, enabling new insights into outer space. Each image was a million times sharper than what was visible through a telescope on Earth. The probe also used its scientific instruments to gather information about the conditions on the lunar surface. It thereby also served as a scout for later Apollo missions. Surveyor 1, which was actually only a test model of the series, spent 42 days on the moon, withstanding not only the hot days, when temperatures would rise to 120 degrees Celsius, but also the cold nights, when the temperature dropped to minus 160 degrees, until its battery reserves were depleted.

National Geographic Magazine, October 1962

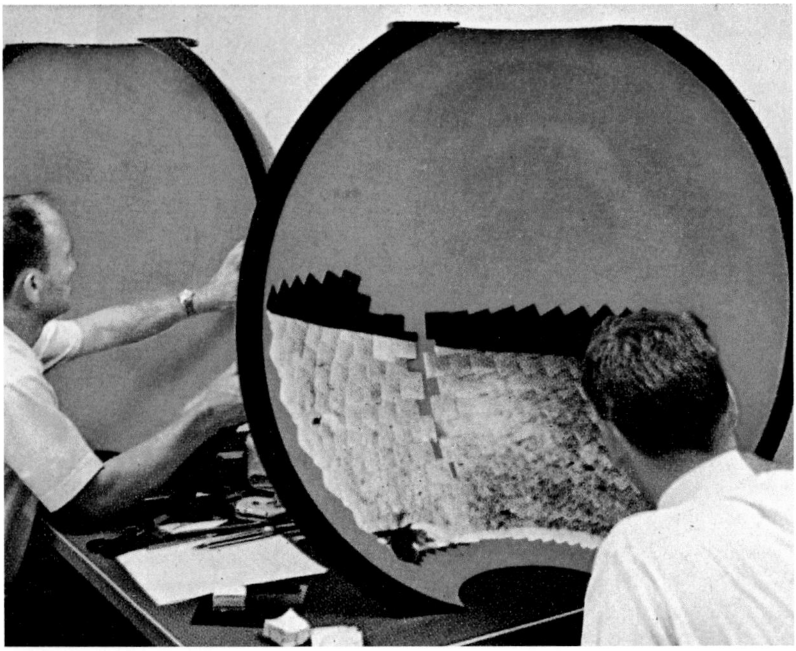

The lunar surface as a puzzle: scientists patching together individual photographs

National Geographic Magazine, October 1962

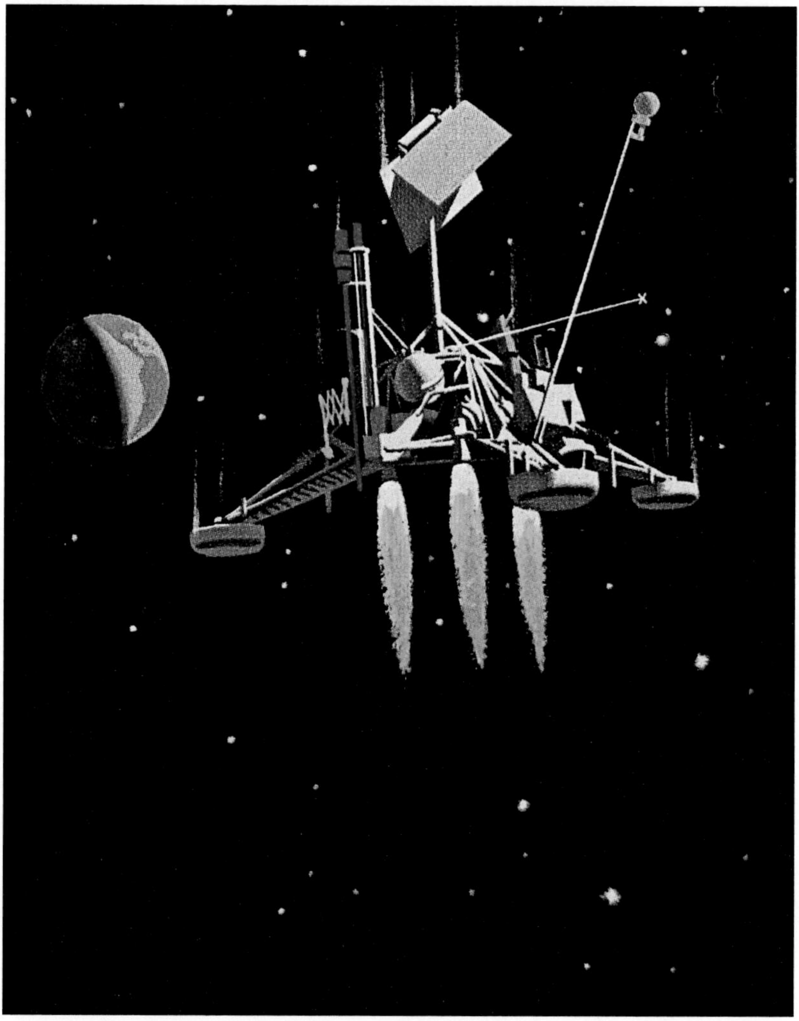

Ignition of the retrorockets shortly before the landing

Engel 1988

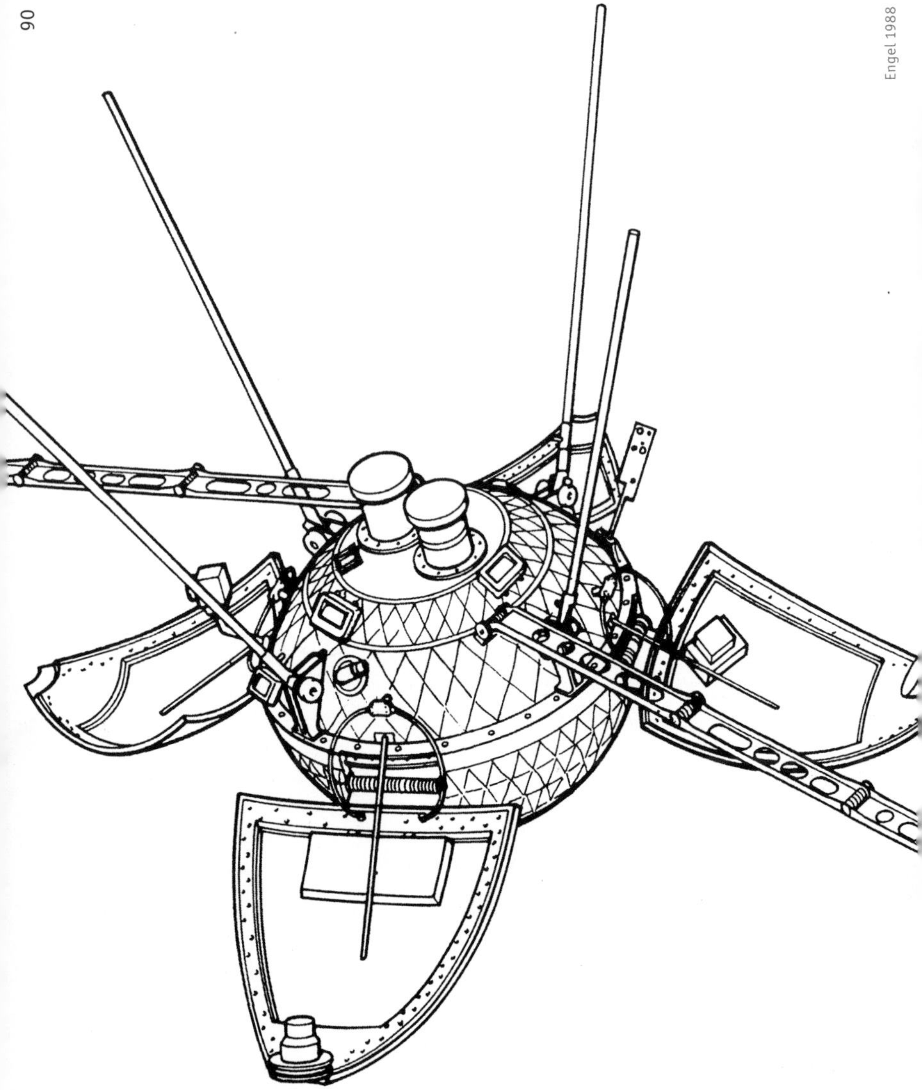

Moscow Polytech

Luna 13 in a technical drawing (top)
and its cephalopodic shadow on the lunar surface (bottom)

Luna 13
Soviet Union

Launch: 21 December 1966, 10:17 am
Landing: 24 December 1966, 6:04 pm
Last contact: 28 Dec. 1966, 6:13 am
Launch site: Baikonur
Landing site: 18.87°N 62.05°W

Moscow Polytech

The first missions of the still uncrewed Soyuz capsules failed in 1966. On 14 December that year, a Soyuz capsule's carrier rocket exploded during the launch, almost completely destroying the launch site, Site 31. And so, the Soyuz and Luna missions came to share Site 1 – also named Gagarin's Start after Yuri Gagarin – and this is where Luna 13 was also launched in the winter of 1966. It flew seamlessly and became the third probe to achieve a soft landing on the moon. Upon arrival, it unfolded its four leaves and extended its antennae upwards. It transmitted the first photographs four minutes after landing. The camera system had been designed to produce stereo images, but it failed to do so due to a malfunction. Nonetheless, the mission was a complete success. The probe transmitted five panoramas of the moon back to Earth during its four-day operation. Luna 13 also documented the force of the impact. It used its penetrometer to measure the force required to penetrate the lunar regolith as well as the density of the lunar surface. The instrument was stored in one of its two cantilevered arms, which automatically extended outwards in opposite directions after the landing in order to carry out various measurements.

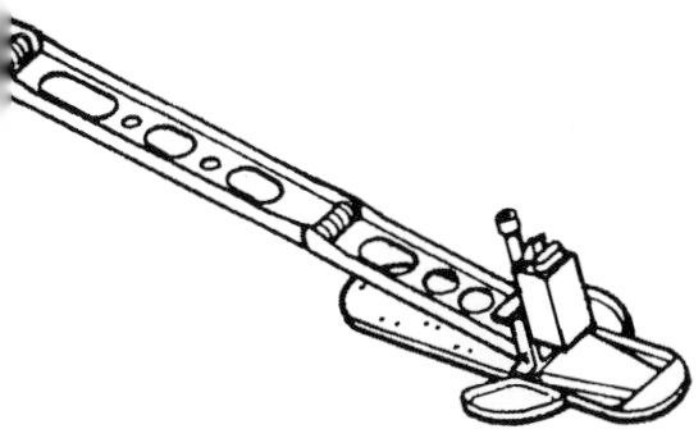

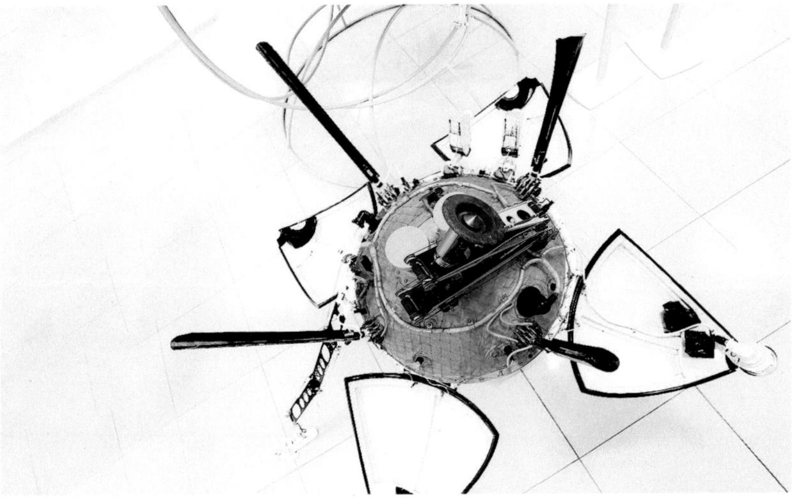
NPO S.A. Lavochkin

Like a beetle with unfurled wings: model of Luna 13, produced in the former Experimental Design Bureau OKB-301 (today: NPO Lavochkin)

NASA

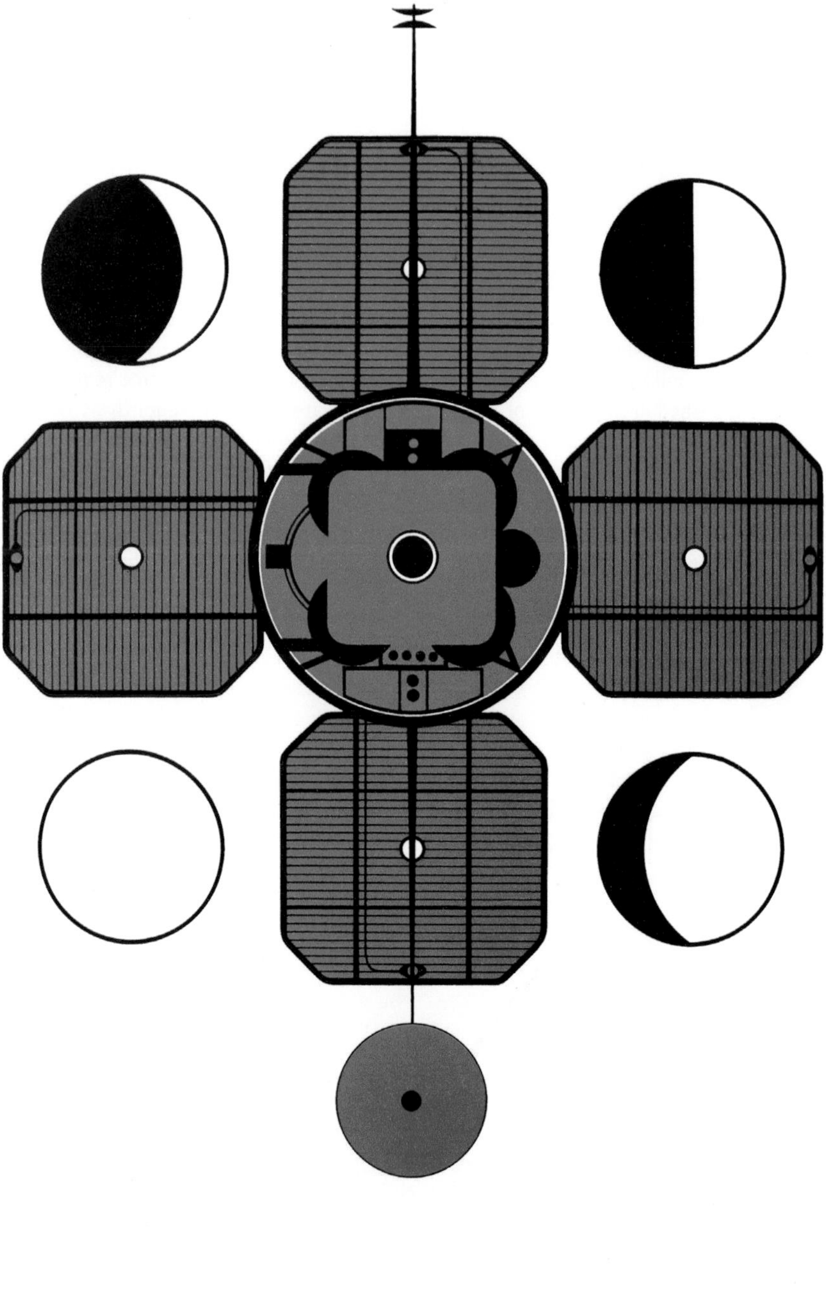

The symmetric layout of Lunar Orbiter 1 with four lunar phases: waxing crescent, first quarter, full moon, and waning crescent

Lunar Orbiter 1
US

Launch: 10 August 1966, 7:31 pm
Landing: 29 October 1966
Last contact: 29 October 1966
Launch site: Cape Canaveral
Crash location: 6.35°N 160.72°O

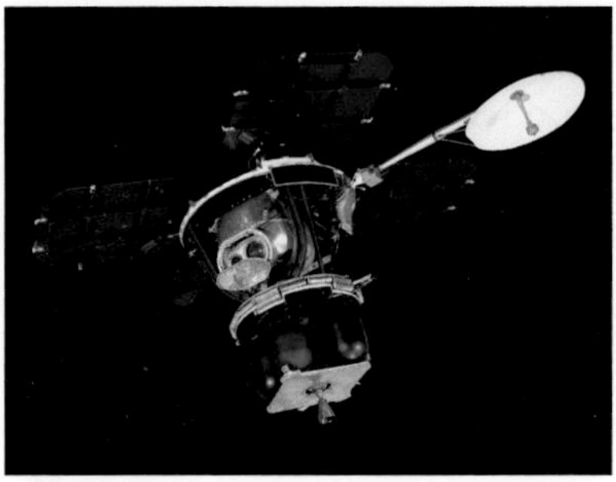
Smithsonian Nat. Air and Space Museum

The American Lunar Orbiter bore some resemblance to the unfurled Luna probes by the Soviet scientists. It featured a round platform, fitted with a camera system, and four axially arranged square solar panels with canted sides. The probe's 'floor plan' – if you ignore the two horizontal antennae pointing in opposite directions – recalled the cross-shaped layout of domed Renaissance churches. A four-legged frame encased a transparent pressure vessel, which in turn housed the camera system. Resting on the frame was an additional round platform with four oval tanks, held together from above by a square plate. Crowning the square plate was a conical element, designed to control the speed of the rocket engine.
The mission's objective was to capture even higher-quality images of the moon and to gather information about the nine proposed landing sites for the Apollo missions. The probe slowly approached the moon's gravitational field and entered a lunar orbit at a distance of 40 kilometres. It transmitted 170 photographs of the lunar surface to ground control over a period of 17 days, using up 60 metres of film that was developed on board. The mission, initially planned as a test flight, was a complete success and gave the scientists greater confidence in the pending Apollo programme. The Ranger, Surveyor, and Orbiter probes together paved the way for the Apollo missions. They successfully forged a path towards the moon, taking close-up photographs, achieving a soft landing, and collecting the first soil samples. Moreover, the completion of the Lunar Orbiter 1 mission enabled a cartographical understanding of the moon.

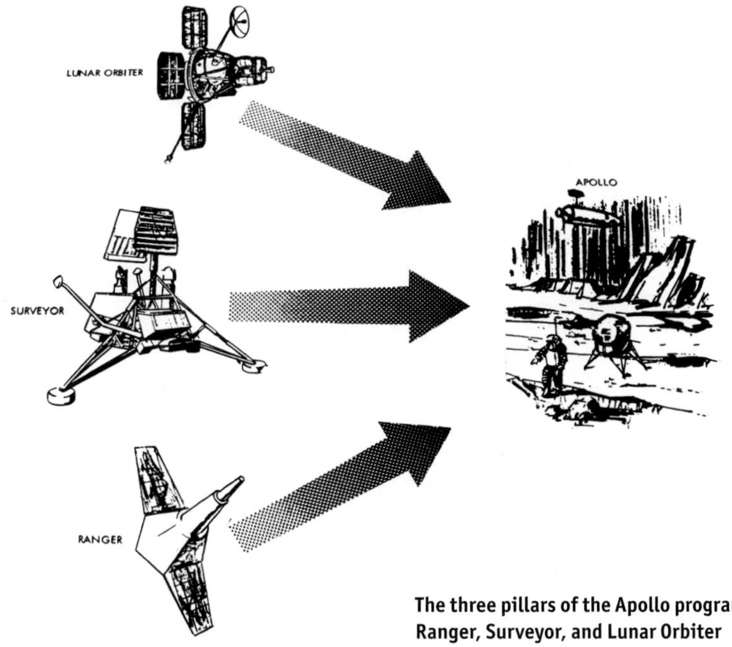

The three pillars of the Apollo programme: Ranger, Surveyor, and Lunar Orbiter

NASA

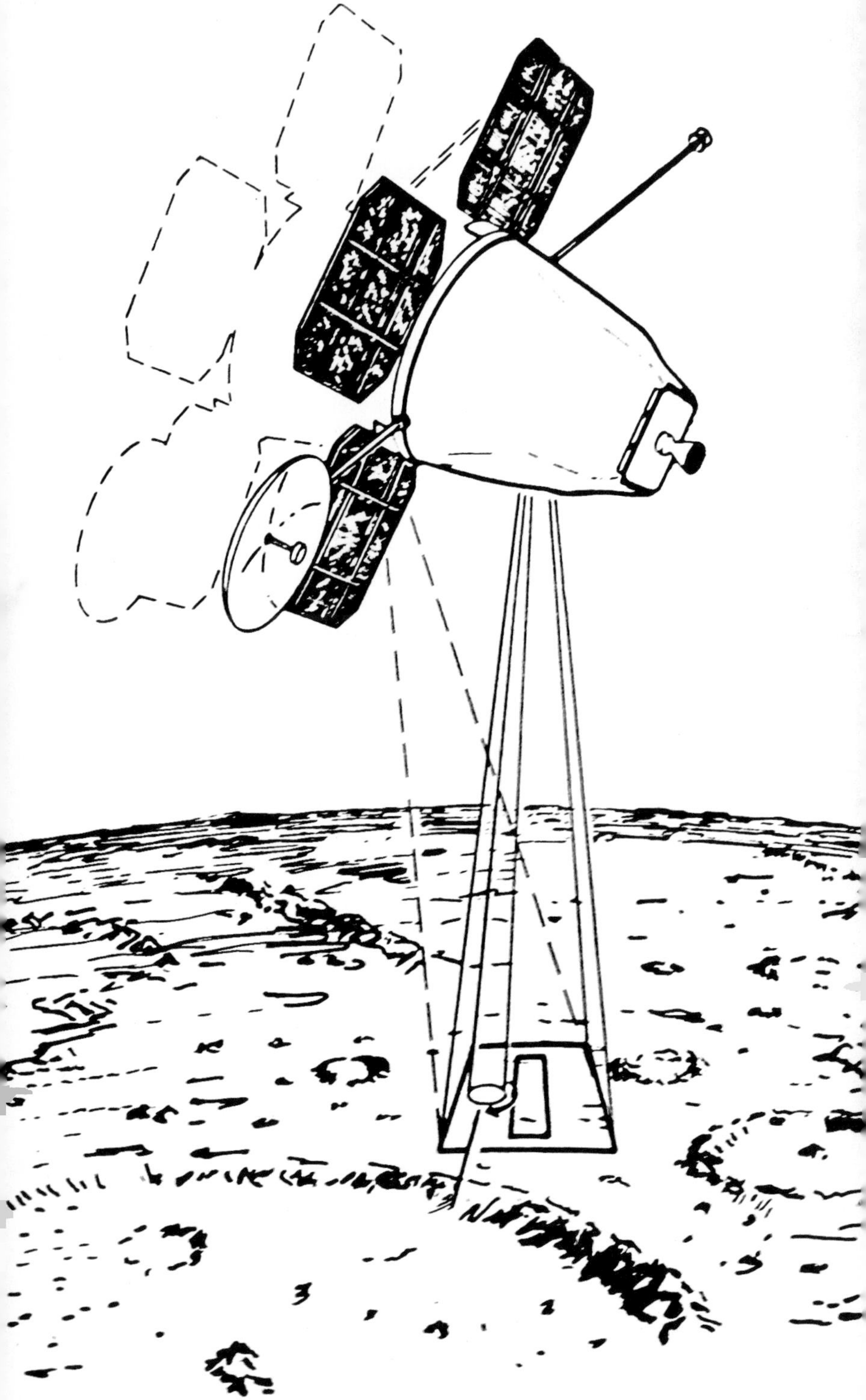

Lunar Orbiter 1: orientation of the cameras and sensors

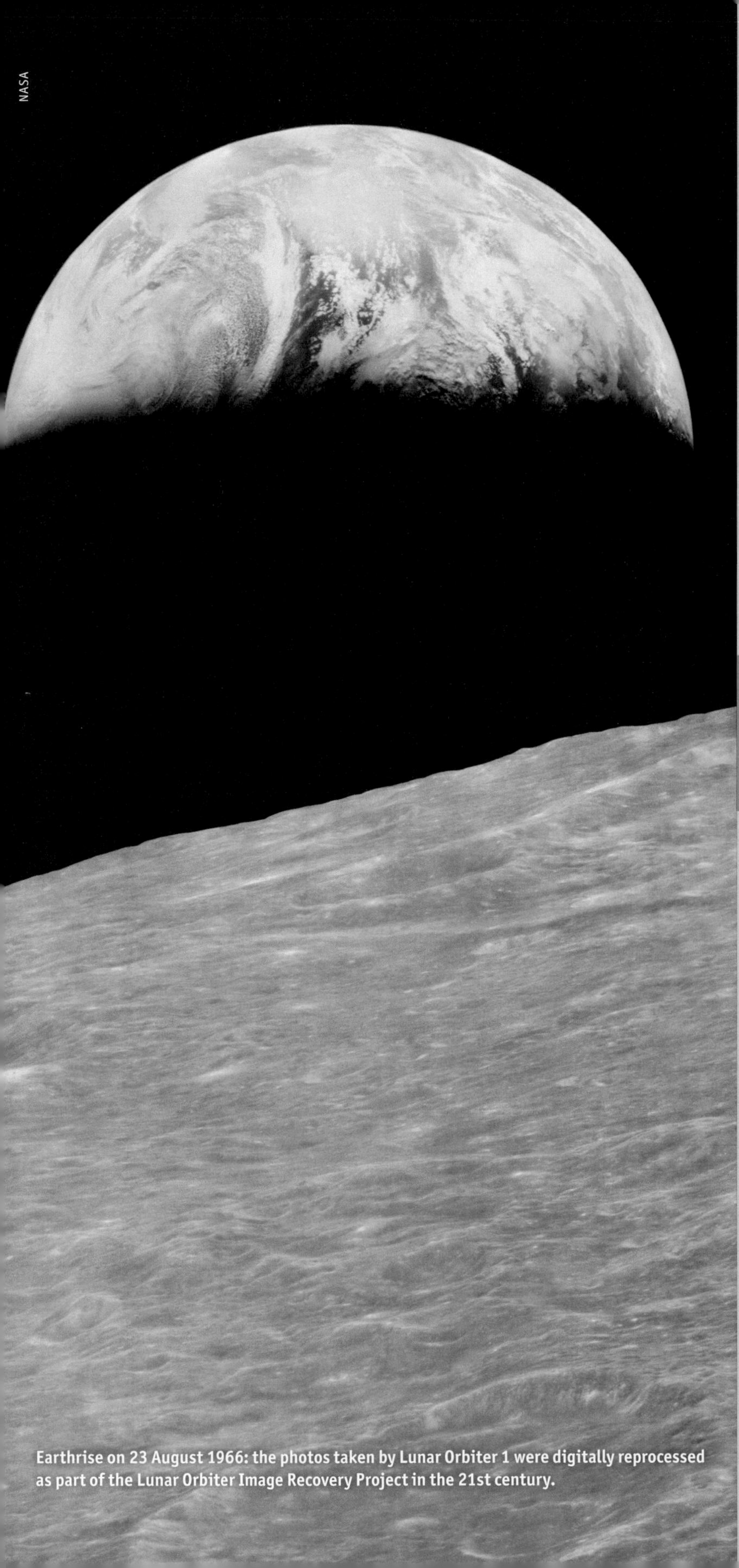

NASA

Earthrise on 23 August 1966: the photos taken by Lunar Orbiter 1 were digitally reprocessed as part of the Lunar Orbiter Image Recovery Project in the 21st century.

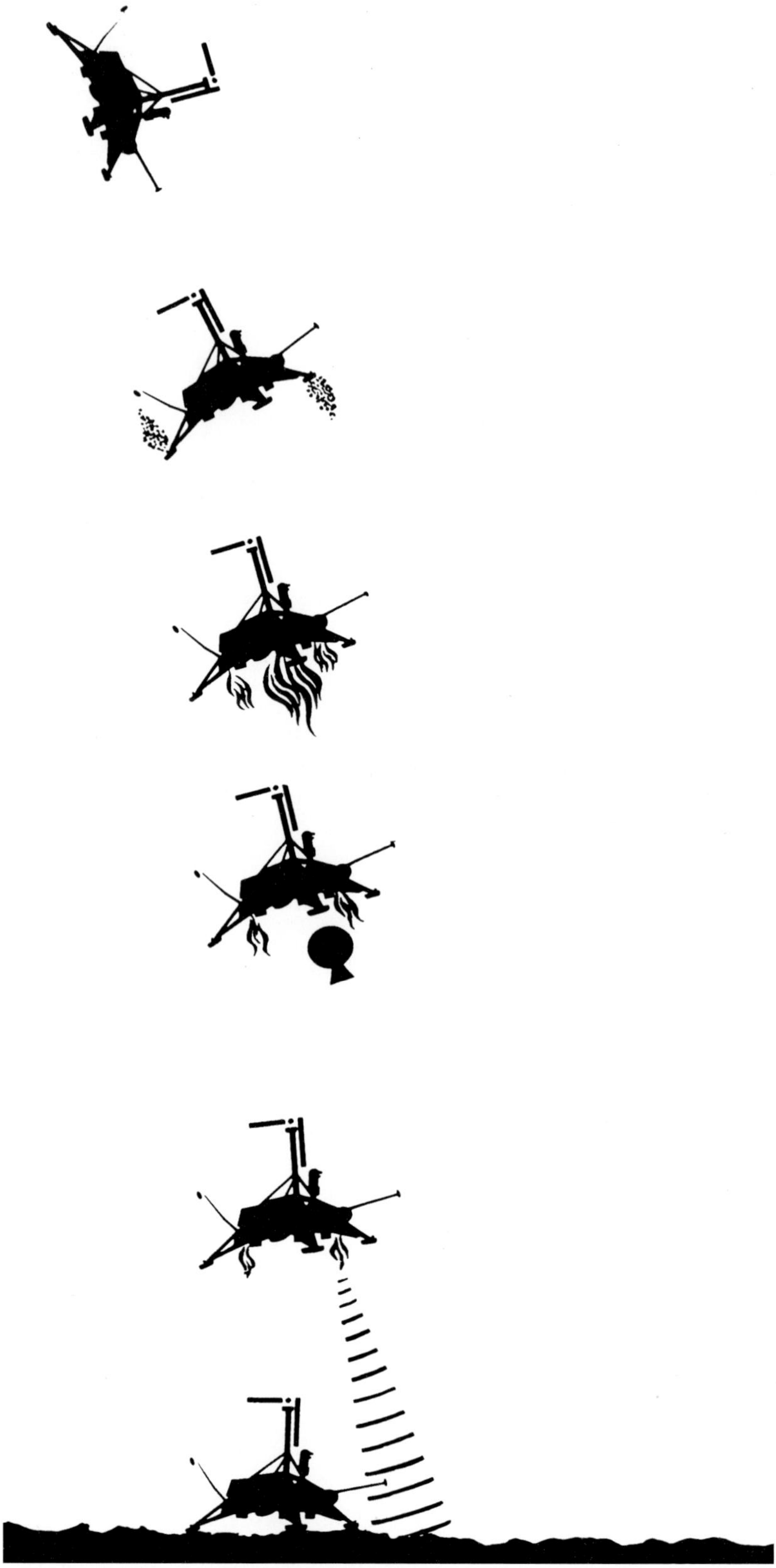

NASA

Landing sequence of Surveyor 2 with automatically activated braking manoeuvre

Surveyor 2
US

Launch: 20 September 1966, 12:32 pm
Landing: 23 September 1966, 3:18 am
Last contact: 22 Sept. 1966, 9:35 am
Launch site: Cape Canaveral
Crash location: 4.0°S 11.0°W

NASA

Each Surveyor probe needed around 250 different commands, which had to be sent from the control centre on Earth during its flight to the moon. The transmitting stations, spread out across the world, from California to South Africa through to Australia, therefore needed to remain in contact with the probe at all times. All of the data was collected and stored at the mission control centre in Pasadena, California. Surveyor 2 was an exact replica of Surveyor 1 and also had the same mission objectives: to achieve a soft landing and thereafter document its surroundings. Only the landing site was different. The scientists expected the mission to run as smoothly as that of the first Surveyor probe. However, one of the three vernier engines failed to ignite during the trajectory correction. The probe lost control as a result of the uneven distribution of rotational forces. While attempting to regain control of the spiralling probe, the scientists collected as much data as possible with the still functional instruments before a crash became unavoidable.

NASA

Back when space research was still largely analogue: an engineer analysing Surveyor's black-and-white photos

Colour filter wheel at the foot of a Surveyor probe

National Geographic Magazine, October 1966

A curious round disc appears, almost inconspicuously, in some of the images captured during the Surveyor missions. The disc's colours – orange, green, and blue – stand out from the black-and-white image in the background. The colours are arranged in a circle and are separated by black-and-white stripes. They are further encased by an outer ring of grey hues. What could be the function of this colour wheel in the grey lunar landscape?

Most images of the moon were monochrome due to technical limitations. But the scientists were still able to produce a number of colour photographs, with which they could obtain a deeper understanding of the materiality of the lunar surface. The onboard camera would photograph the same detail multiple times, each time using a filter with one of the colours in the disc. The scientists on Earth could use these image details to reconstruct the colours of the entire image.

The palette of colours can also be found in a work of micro-architecture that was sent to the moon during the Apollo missions: the gnomon. This is a contraption composed of a vertical rod mounted on three legs. It was taken to the moon with a stripe of colours – grey hues alongside green, orange, and blue – on one of its legs. Its form differed very little from mission to mission. The shadow it cast was analysed to determine the sun's position and direction of movement. Moreover, the mast served as a reference point for analysing the size of the boulders and rocks on the moon. The grey hues, which had a reflection coefficient of between 0.05 and 0.35, and the colour scale made it possible to determine the colours of the seemingly grey moon rocks.

NASA

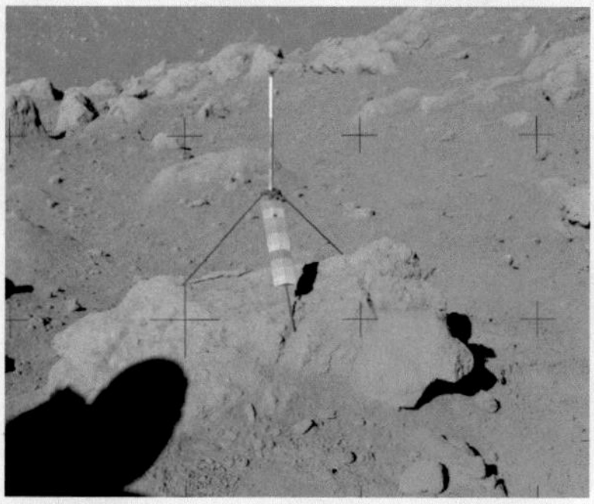

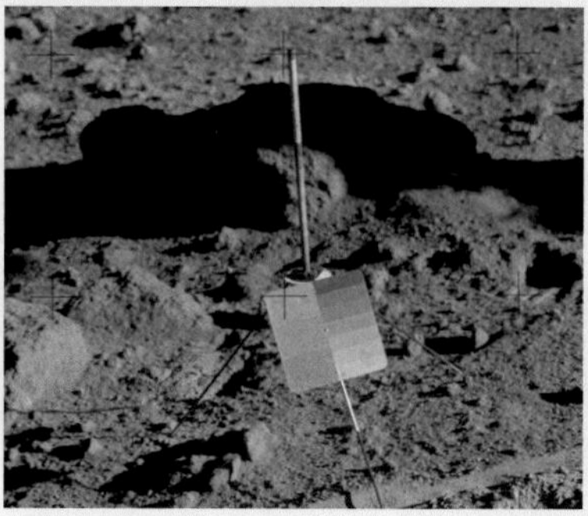

Use of the gnomon during Apollo 14 (left) and 15 (right)

Smithsonian National Air and Space Museum

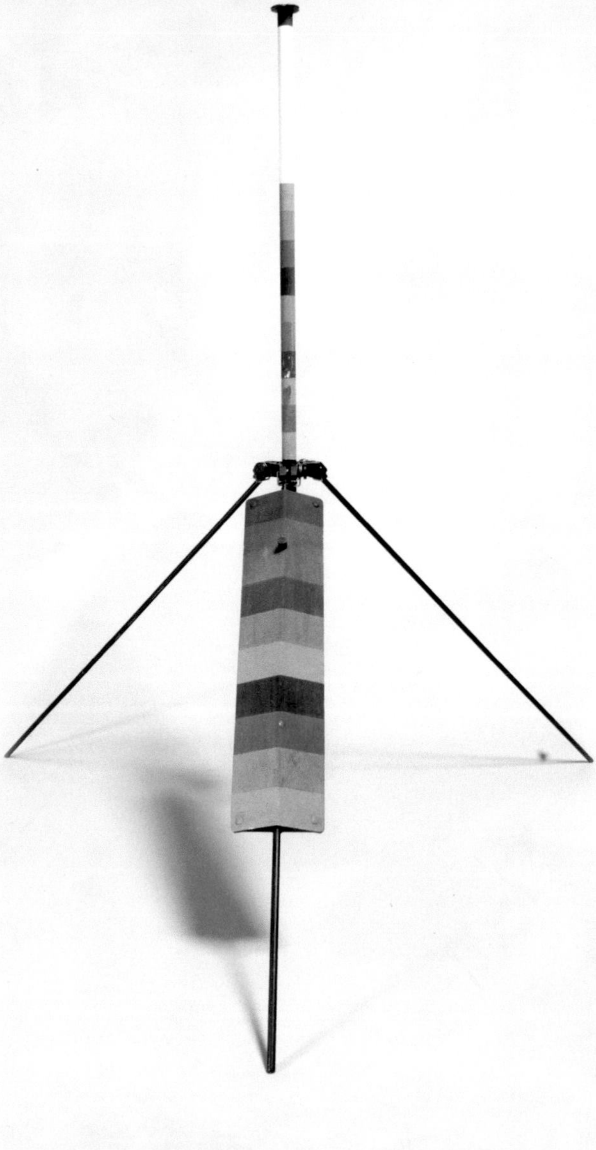

Since 1974, this gnomon has been housed at the Smithsonian National Air and Space Museum.

'Picture of the Century', captured by Lunar Orbiter 2

NASA

Clear for all to see: the landing site of Surveyor 1

Lunar Orbiter 2

US

Launch: 6 November 1966, 11:21 pm
Landing: 11 October 1967
Last contact: 11 October 1967
Launch site: Cape Canaveral
Crash location: 3.0°N 119.1°O

Lunar Orbiter 3

US

Launch: 5 February 1967, 1:17 am
Landing: 9 October 1967
Last contact: 9 October 1967
Launch site: Cape Canaveral
Crash location: 14.3°N 97.7°W

For their second lunar satellite, Lunar Orbiter 2, the Americans chose a different path to the moon from the one taken since Surveyor 1. The new mission's objective was to explore the northern region of the moon. The plan was for the satellite to photograph 13 different locations independently, which would require 40 manoeuvres – without a doubt a challenge for the control centre. The orbiter completed a productive eight-day mission, after which its passive observation mode was activated. Lunar Orbiter 2 orbited the moon for three months, before it was joined by the third orbiter of the same series. The three Lunar Orbiter probes had the same outer form and initially all had the same task of photographing the lunar surface, though from different orbits. All three orbits required the probes to spend a good hour without sunlight in the moon's shadow. However, a longer sunless period was looming: a total lunar eclipse was to take place in April 1967 and would block direct sunlight from reaching the moon. Lunar Orbiter 2 and 3 avoided the problem of low visibility by entering a higher, sunnier orbit, and ended their respective missions successfully. Even after the completion of the main mission, both satellites were kept in lunar orbit for as long as possible, where they served as reference points and destinations for later crewed missions. Moreover, they retained a scientific function even after their film rolls had been used up: both continued to measure radiation intensities and studied the effect of micrometeorites on the lunar environment. In the end, both probes were deactivated at the press of a button and crashed into their resting places on the moon.

Smithsonian National Air and Space Museum

Lunar Orbiter, as a full-colour model, at the Smithsonian National Air and Space Museum

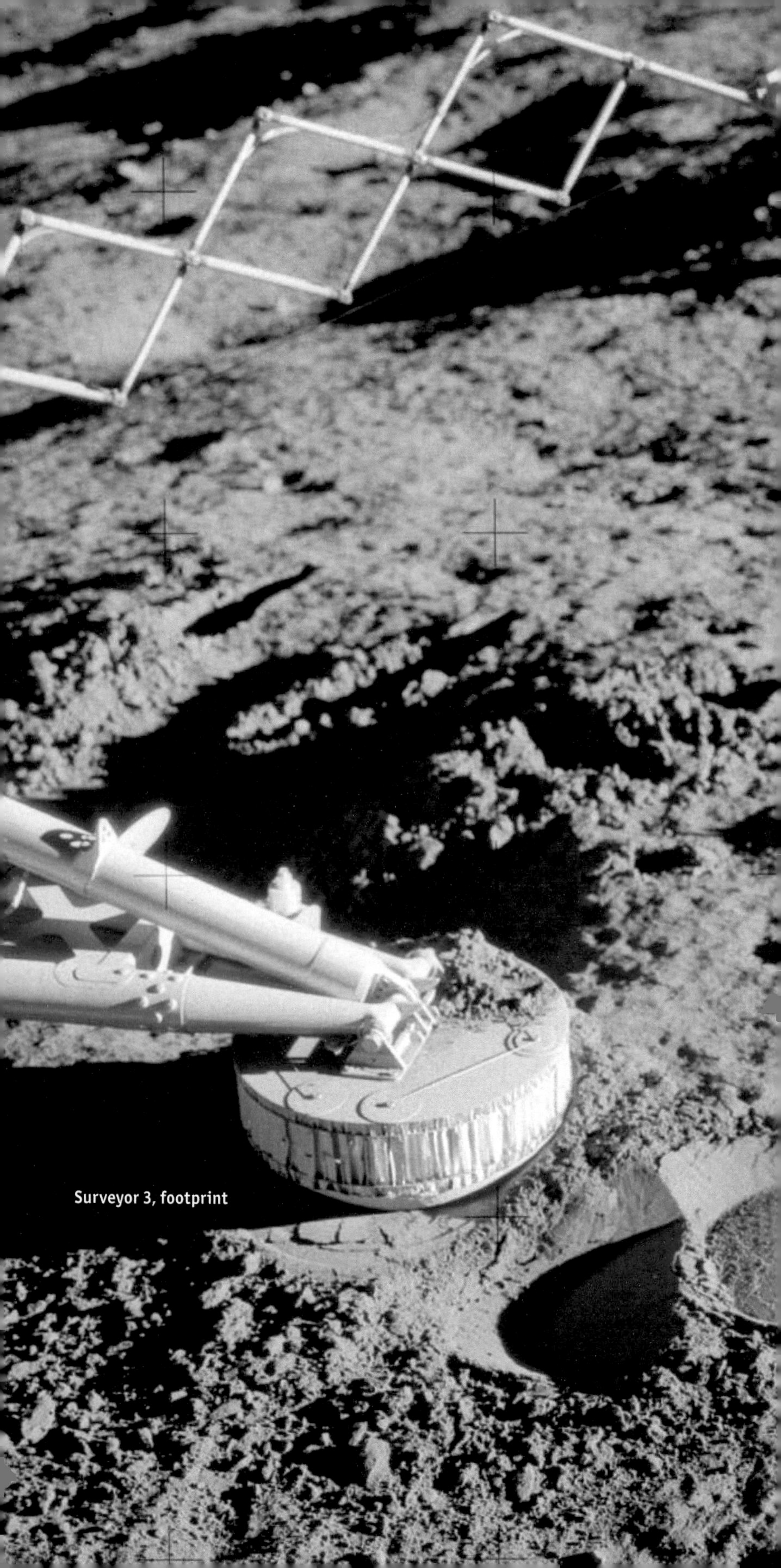
Surveyor 3, footprint

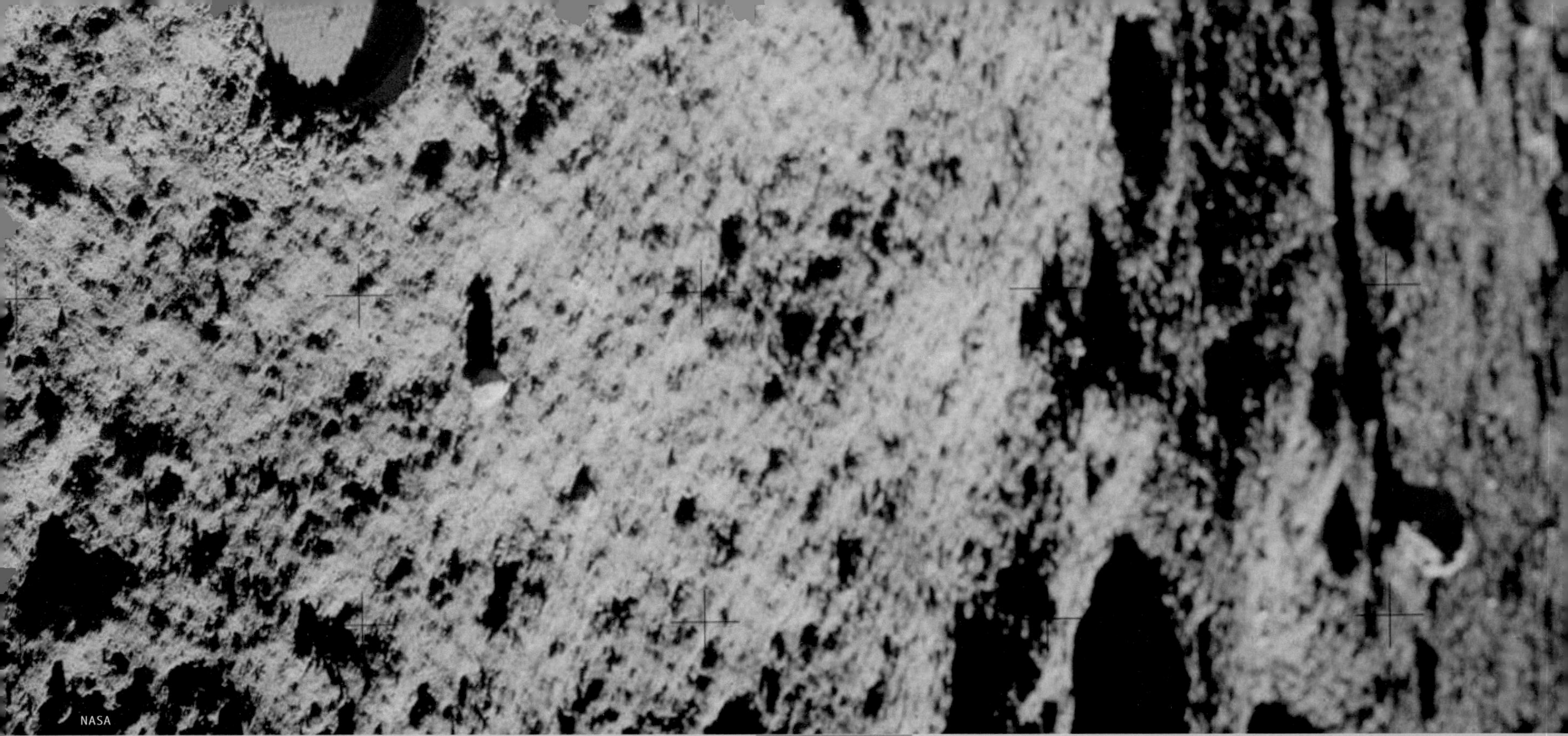
NASA

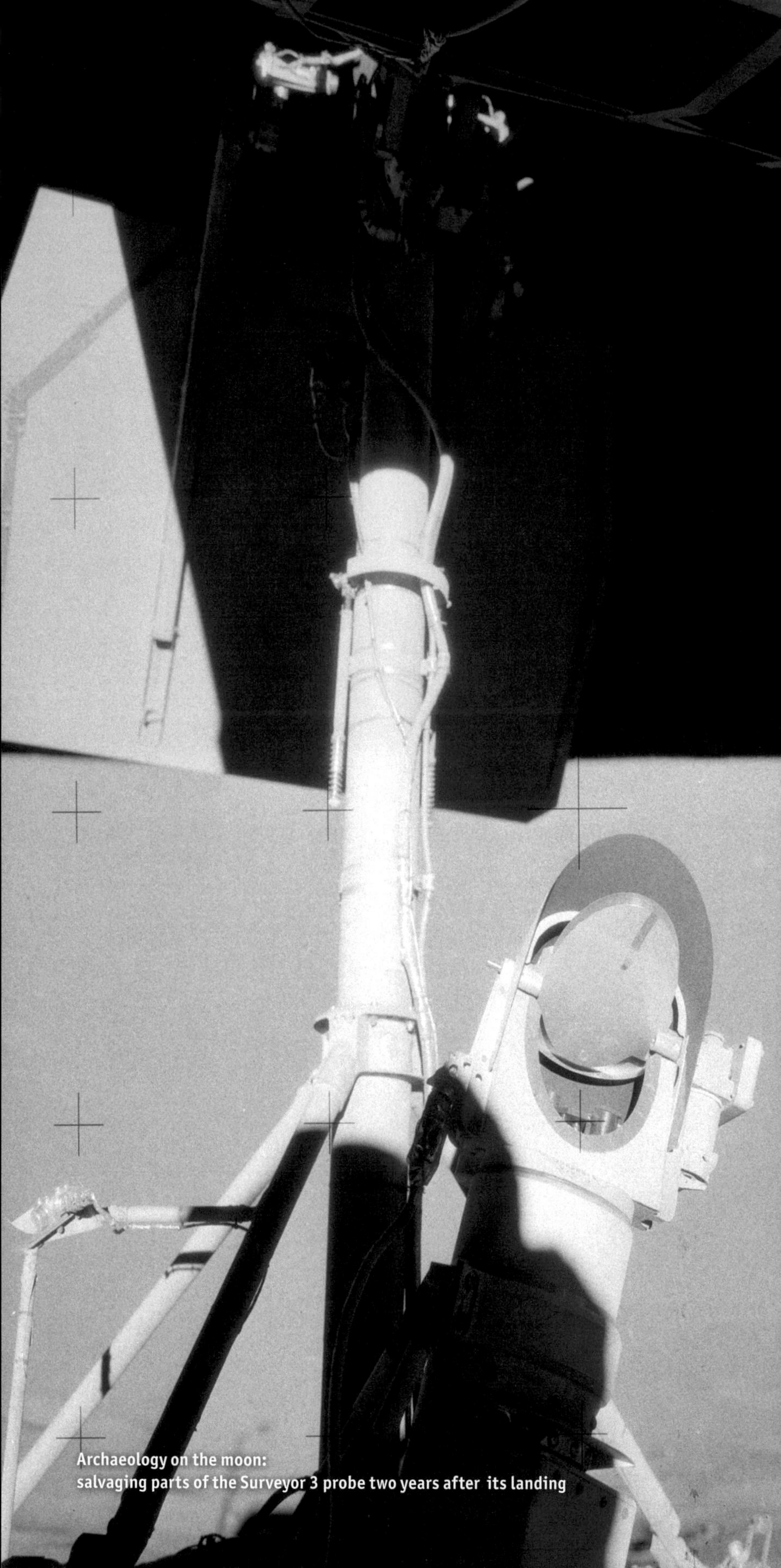

Archaeology on the moon:
salvaging parts of the Surveyor 3 probe two years after its landing

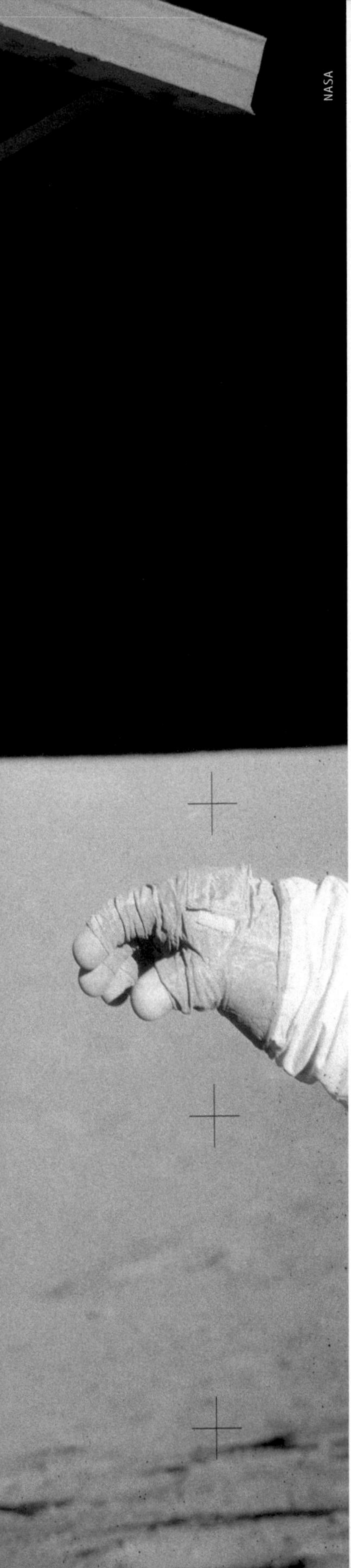

NASA

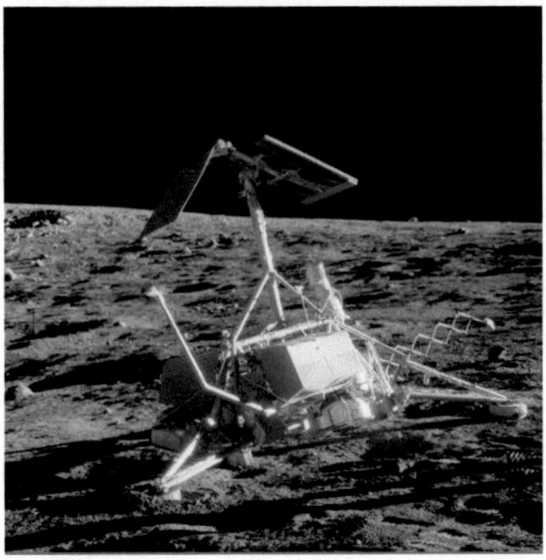

NASA

Surveyor 3
US

Launch: 17 April 1967, 07:05 am
Landing: 20 April 1967, 00:04 am
Last contact: 3 May 1967
Launch site: Cape Canaveral
Landing site: 2.94°S 23.34°W

Surveyor 3 came with a brand new feature: a gripper arm, in the form of a pantograph, mounted on the probe's tetrahedral base next to a range of other instruments. The soil mechanics surface sampler (SMSS), featuring a little scooper at the end, could extend by up to 1.5 metres, scratch the lunar surface, and raise soil from a depth of up to 45 centimetres below the surface. The scientists on Earth observed and analysed the work of the SMSS via video transmission in real time. Two mirrors optimised the soil study. They were positioned such that the camera could photograph the surface below the probe, which in turn provided more information about the excavated material. The investigation led to the finding that the lunar surface would be able to support the Apollo capsules, whose landing had been planned since the early 1960s. At first, Surveyor 3 was merely seen as a stepping stone towards a greater objective. But NASA decided two years later to send Apollo 12 – which would become the second crewed spacecraft to land on the moon – to salvage parts of Surveyor 3 and bring them back to Earth. Of course, there was no guarantee that this plan would succeed. For one thing, sending the

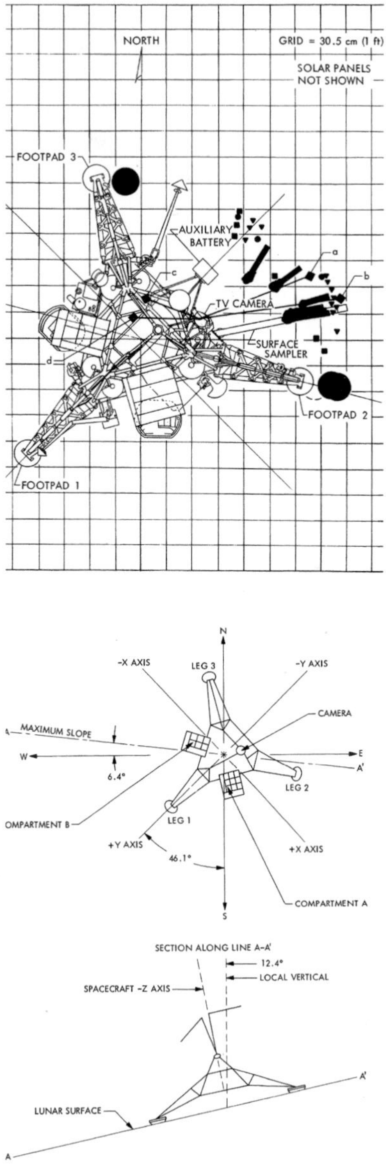

Map of the Surveyor 3 landing site (top); plan view and elevation (bottom)

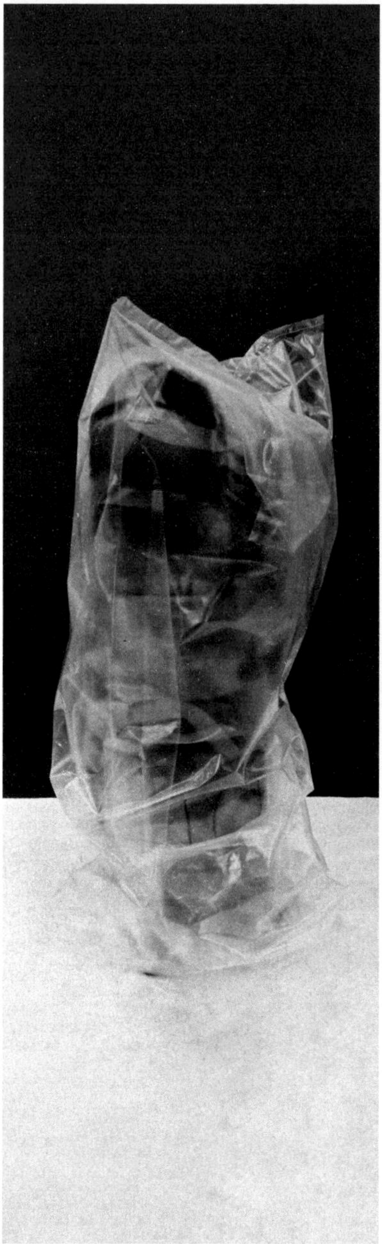

Camera in protective packaging

crewed probe to the correct landing site was no simple task – Apollo 11 had landed a few kilometres away from its actual destination. For another, Surveyor 3's design made it difficult to dismantle, especially for astronauts using bolt cutters with bulky gloves. As such, the final plan was to simply disassemble the camera and a few cables from the probe. These would be used to investigate the wear and tear of materials on the moon. In the end, Apollo 12 achieved a soft landing just 180 metres away from Surveyor 3. This enabled the astronauts Alan Bean and Pete Conrad to remove individual parts from the fallen probe, including the TV camera. They also managed to bring back the gripping arm – a tear in the joint made it easy to remove. The scoop still contained residues of the lunar surface, and the fine particles were taken to the labs for space scientists to study.

Surveyor 3 with Pete Conrad and Apollo 12 on the horizon

Surveyor 4
US

Launch: 14 July 1967, 11:53 am
Landing: 17 July 1967, 2:05 am
Last contact: 17 July 1967, 2:02 am
Launch site: Cape Canaveral
Landing site: 0.45°N 1.39°W

Surveyor 4, identical on the outside to its forerunner, failed its mission, but it nonetheless forms another chapter in the success story of the Surveyor programme. The probe was launched on an Atlas-Centaur rocket from the Cape Canaveral Air Force Station in Florida in July 1967. Although the thrust manoeuvre lasted slightly too long, the probe made it onto a trajectory to the moon, oriented by a star tracker on board. A recalculation of the course-correction manoeuvre compensated for the initial misorientation. After a calm, three-day flight, the landing procedure was also initiated without further complications. The probe transmitted a wide range of data regarding its height and velocity, and the onboard computer began the countdown to the braking manoeuvre. However, the mission suddenly came to an abrupt end. The probe went silent two and a half minutes before the landing. The scientists on Earth initially assumed that the probe had landed as planned, in spite of the radio silence, and began to transmit the first instructions via radio. But in the absence of further responses, they soon had to conclude that it had exploded during its descent. The true reasons for the crash are unclear, due to a lack of data. In spite of the accident, no drastic changes were made to the subsequent Surveyor probes. A commission ruled that the crash was likely the result of one or more unpredictable small errors, and that it would serve no purpose to carry out changes to the already successful Surveyor design.

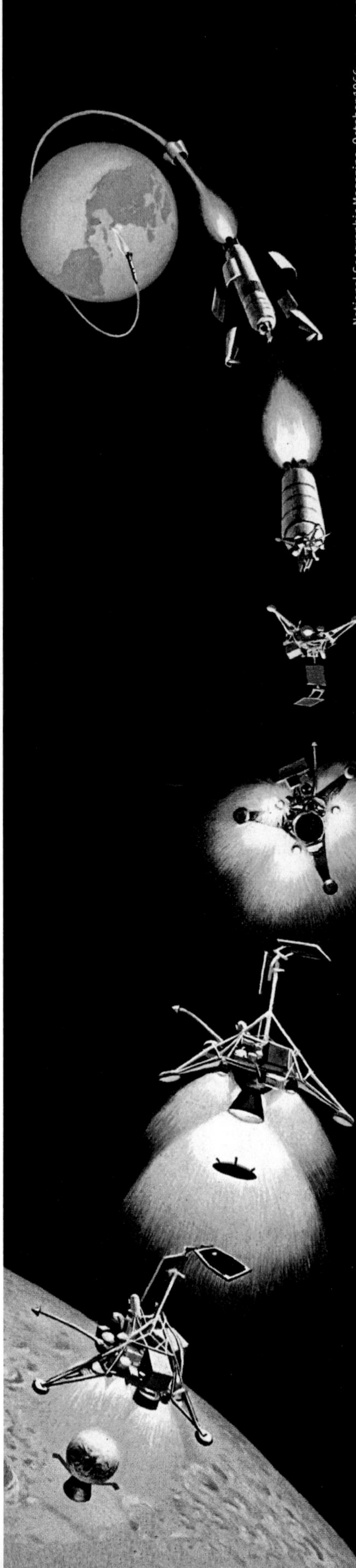

Explorer 35
US

Launch: 19 July 1967, 2:19 pm
Landing: crashed at unknown time
Last contact: 24 June 1973
Launch site: Cape Canaveral

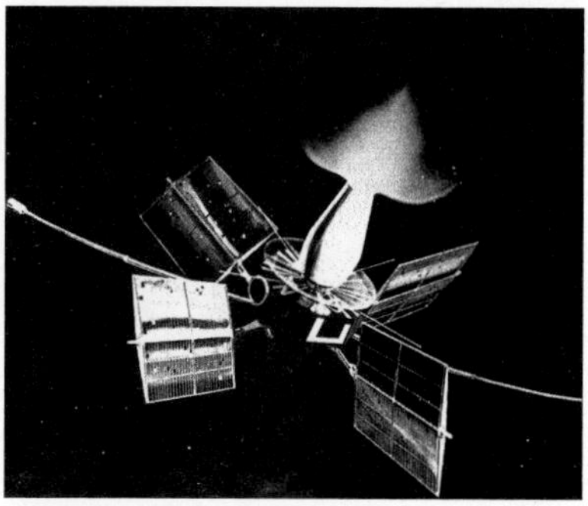
NASA

Explorer 35 was launched to further investigate the conditions on the moon and to expand on the findings of the Lunar Orbiter and Surveyor probes. Previous studies had concluded that the moon is mostly made up of a homogeneous mass, which meant it was difficult to differentiate between different areas of the lunar surface. But surprisingly, strange reflections had also been observed in the surface, and volcanic activity had been identified, which motivated further investigations. Explorer 35 was densely packed with instruments, though these were not visible on the outside. The probe was composed of a central octagonal body, on which a shiny, silver sphere with a cone-shaped propulsion engine was mounted. Four antennae were arranged symmetrically around this engine, while two long magnetometers were mounted on the opposite side of the octagonal body and extended in opposite directions. Like the Lunar Orbiters, rectangular solar panels were attached on four sides of the octagonal centre. Each panel rotated around its own axis, adjusting its position to stay oriented towards the sun. This made the spacecraft resembled a water wheel flying through space. The probe was spin stabilised – it rotated around its own axis – which allowed it to independently maintain a particular direction in space. Unlike its forerunner, Explorer 33, which had failed to withstand Earth's gravitational force, the satellite achieved elliptical lunar orbit. It was deactivated on 24 June 1973, after a long and fruitful mission. It had investigated not only plasma, solar wind, and moon dust; it had also continuously provided NASA scientists and universities with information during its five years in lunar orbit as the only operational lunar orbiter at the time.

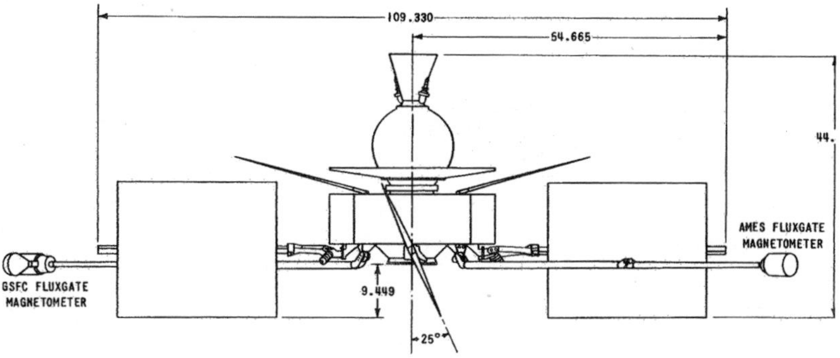

Schematic elevation and dimensions of the Explorer 35 probe

Illustration of the trajectory
of Lunar Orbiter 4

National Geographic Magazine, October 1966

Lunar Orbiter 4
US

Launch: 4 May 1967, 10:25 pm
Landing: 6 October 1967
Last contact: 17 April 1967
Launch site: Cape Canaveral

The first three Lunar Orbiter probes had searched the moon for appropriate landing sites, and the Surveyor probes had achieved two successful landings. This meant the American scientists had already identified eight potential landing sites for the Apollo programme based on the large number of photos that had been taken. Lunar Orbiter 4 entered lunar orbit – shortly before Surveyor 3 – and thereafter proceeded to map the moon. The probe photographed almost all of the visible lunar surface as part of its main mission. Afterwards, it went on to photograph the far side of the moon. 75 per cent of the dark side of the moon had been photographed as part of the previous Lunar Orbiter missions, which resulted in a total of 163 images.

Photograph by Lunar Orbiter 4

Lunar Orbiter 5
US

Launch: 1 August 1967, 10:32 pm
Landing: 31 January 1968
Launch site: Cape Canaveral
Crash location: 2.79°S 83.1°W

A Lunar Orbiter was launched for the last time in August 1967. Its mission was to capture images that would help narrow down the selection of 70 potential landing sites identified by NASA for the upcoming Apollo missions. Around 20 per cent of the photographs served this purpose and were used to select the landing sites for Surveyor 7 and for Apollo 14, 15, and 17. However, the majority of the images were used to map the far side of the moon. At the time, the mission was the most complex photography project conducted as part of the American space-flight programme. It produced over 200 high-resolution photographs and just as many wide-angle images. The mission began with the probe entering an elliptical lunar orbit, at a distance of just over 6,000 kilometres. Over the course of the twelve-day mission, the probe's distance to the moon was reduced to 99 kilometres at the perigee. On 31 January 1968, the satellite spiralled towards the lunar surface and was intentionally crashed into the moon.

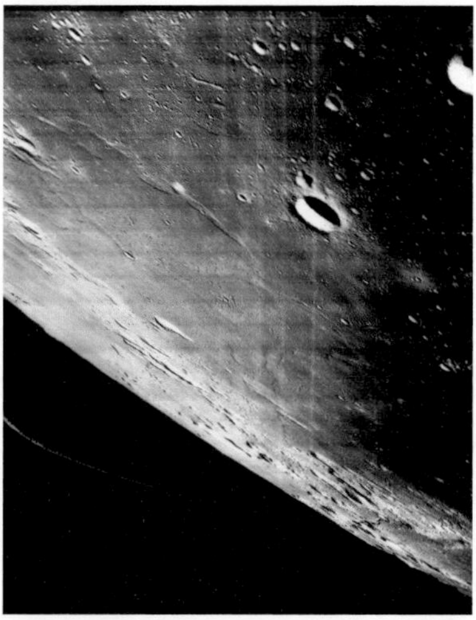

Photograph by Lunar Orbiter 5

National Geographic Magazine, October 1966

Lunar Orbiter 4 on the moon, with the sun rising behind it: conceptual illustration in *National Geographic* (1966)

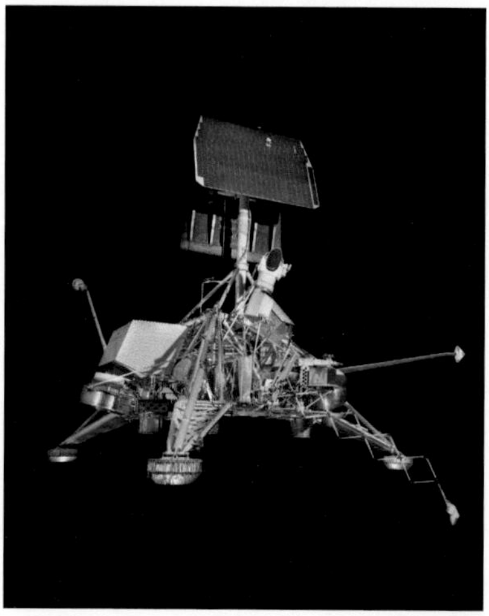

Smithsonian National Air and Space Museum

Surveyor 5
US

Launch: 8 September 1967, 7:57 am
Landing: 11 September 1967, 00:46 am
Last contact: 17 December 1967
Launch site: Cape Canaveral
Landing site: 1.41°N 23.18°O

Surveyor 6
US

Launch: 7 November 1967, 7:39 am
Landing: 10 November 1967, 1:01 am
Last contact: 14 December 1967
Launch site: Cape Canaveral
Landing site: 0.49°N 1.40°W

Surveyor 7
US

Launch: 7 January 1968, 6:30 am
Landing: 10 January 1968, 1:05 am
Last contact: 21 February 1968
Launch site: Cape Canaveral
Landing site: 41.01°S 11.41°W

All probes of the successful Surveyor programme were designed to seek the perfect landing site for the Apollo missions. They pursued various objectives on the moon and tested different landing angles. The main differences between them were in the experiments and soil analysis equipment they carried. The last three Surveyor probes all had magnetic stripes embedded in their landing feet, which were designed to capture ferrous particles. Surveyor 5 landed in Mare Tranquillitatis, where Neil Armstrong and Buzz Aldrin would later arrive and become the first humans to walk on the moon. Equipped with a TV camera, an alpha particle X-ray spectrometer, and a number of other scientific instruments, the probe felt around the lunar soil around it. Surveyor 6 followed suit just two months later, landing in Sinus Medii to carry out the mission Surveyor 4 had failed to complete. The probe, unlike its forerunner, did not bring any devices for analysing the surface structure of the moon. But it captured 30,000 images of stirred-up dust, which allowed the NASA scientists to draw conclusions about the mechanical properties of the moon. Although the American probes generally landed on flat surfaces, Surveyor 7 was sent to the hilly landscape in the southern region of the moon. This last Surveyor probe successfully completed its descent manoeuvre and transmitted over 20,000 images of the surroundings during the first long lunar day. It deployed a wide range of equipment, including a pantograph arm for manipulating lunar soil, to explore the still uncharted surroundings. Contact with the probe was lost on 21 February 1968, and this marked the end of the Surveyor programme. Five of the seven missions had been a success. The probes had transmitted data for a period of over 17 months and captured over 90,000 photographs of different landing sites. This outcome lay a solid foundation for the first crewed mission to the moon, whose launch date was drawing ever closer.

Surveyor 7 captured the photo montage to the right, with a projection of its own shadow.

NASA

U.S. Geological Survey
SEGMENT 5
SECTOR 18
SECTOR 17
CATALOG NO. 7-14 SURVEY NO. 010 054
TITLE Seg 5
GMT START 010 : 07 : 31 : 29
END 010 : 08 : 27 : 59
FILE START END
AZIMUTH BEGIN 63 END 126
IRIS f/8.2 FILTER #1
FOCAL LENGTH 100 mm TAPE 245
0 inches

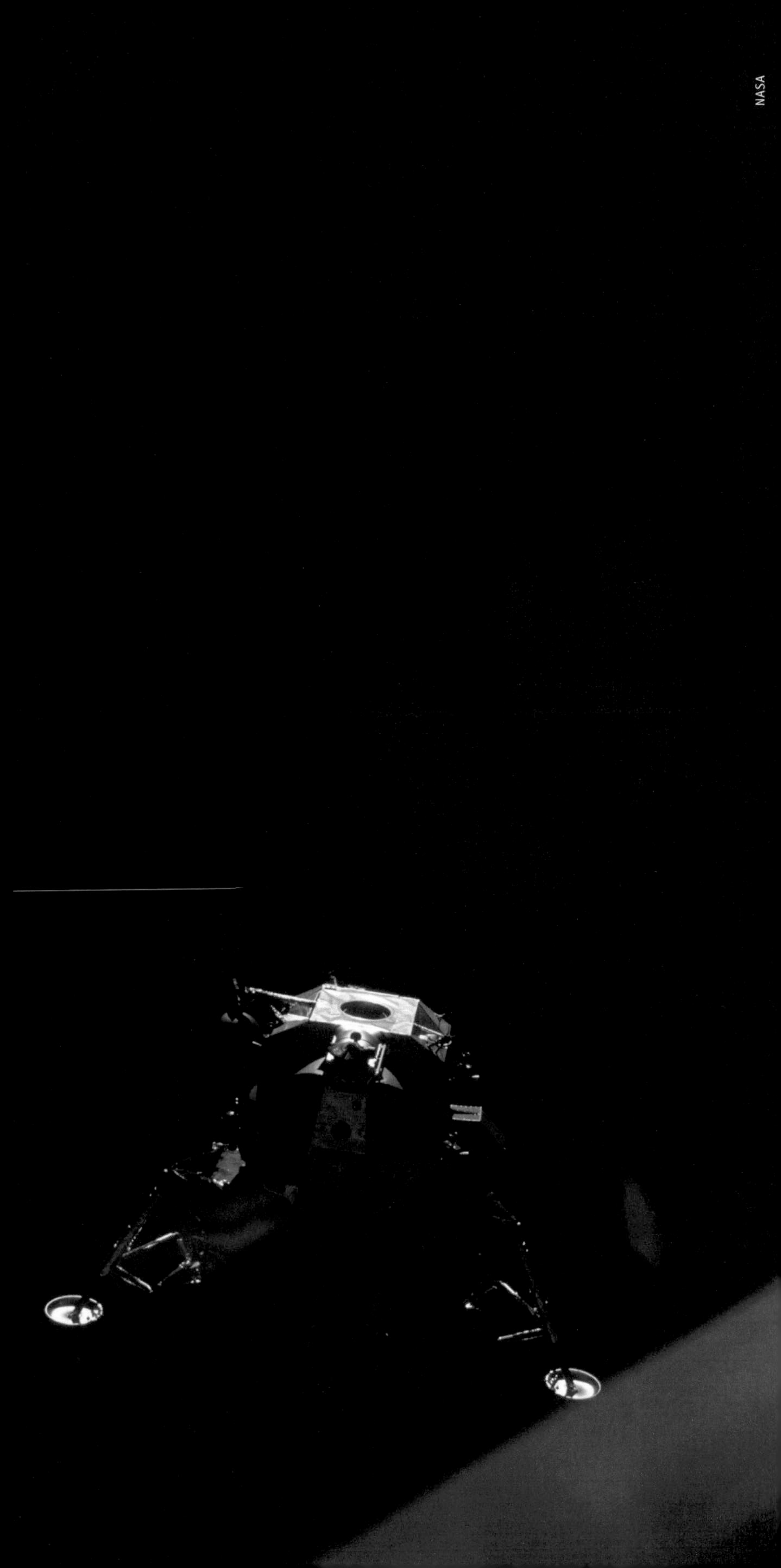
NASA

Apollo 10 Lunar Module
US

Launch: 18 May 1969, 4:49 pm
Landing: 26 May 1969, 4:52 pm
Launch site: Cape Canaveral
Crash location: 15.03°S 164.65°O

NASA

The Apollo programme was the second US crewed space-flight programme, preceded only by the Mercury programme. Various missions had been carried out since 1961 to prepare for the first human landing on the moon. The Apollo 10 lunar module would become the first component from this programme to arrive on the moon. Apollo 10 comprised two space capsules: the command and service module (CSM) and the lunar module, which was in turn divided into a descent stage and ascent stage. The lunar module comprised an octagonal base, wrapped in a gold foil, and a complex upper section above it. It stood on four legs, also coated in a gold foil, forming an airy ground floor below. The upper section, which contained the cabin for the astronauts, looked like a work of brutalist architecture, with a circular disc and octagon standing on their sides next to each other (see p. 188). But a wide range of other polyhedrons were arranged near this core, along with antennae and measuring devices, giving rise to an impenetrable yet compact structure. The mission was seen as a dress rehearsal for the planned Apollo landing. During the mission, the lunar module, Snoopy, separated from the CSM, Charlie Brown, 110 kilometres above the lunar surface. (The crew had named the two modules after characters from the comic strip *Peanuts*.) While Thomas Stafford, the commander, stayed behind in the CSM and orbited the moon, Eugene Cernan and John Young tested the landing routine. People could follow this procedure in real time from around the world, since a television camera had been placed inside the lunar module's cabin. The lunar module descended to a height of 14 kilometres above the moon before jettisoning the descent stage and igniting the ascent stage, which was to fly back to the CSM. During the descent, the astronauts identified several flaws that needed to be corrected before the following Apollo mission. In fact, the mission almost ended in a disaster. Both astronauts, one after the other, flipped the switch for initiating the lunar module's rendezvous with the CSM. But the switch, having been flipped twice, remained off, which caused the ascent stage to spin out of control. Fortunately, the two pilots managed to bring the module back under control manually and were subsequently able to complete the docking manoeuvre. Afterwards, all three astronauts flew back to Earth in the CSM, while the ascent stage was intentionally crashed into the moon. The gold-coated vehicle thus became the first component of an Apollo spacecraft to touch the moon, though it lies in ruins as a result of the hard landing. Meanwhile, the landing capsule re-entered the Earth's atmosphere at a speed of almost 40,000 km/hr before splashing down safely in the Pacific Ocean. The exercise was a success, in spite of a few hitches, and laid the foundation for the following mission, aimed at the first human landing on the Moon.

The Apollo 10 lunar module, separated from the command and service module (CSM), gliding through space. The photo was captured from the CSM.

NASA

Details of the Apollo 10 lunar module

NASA

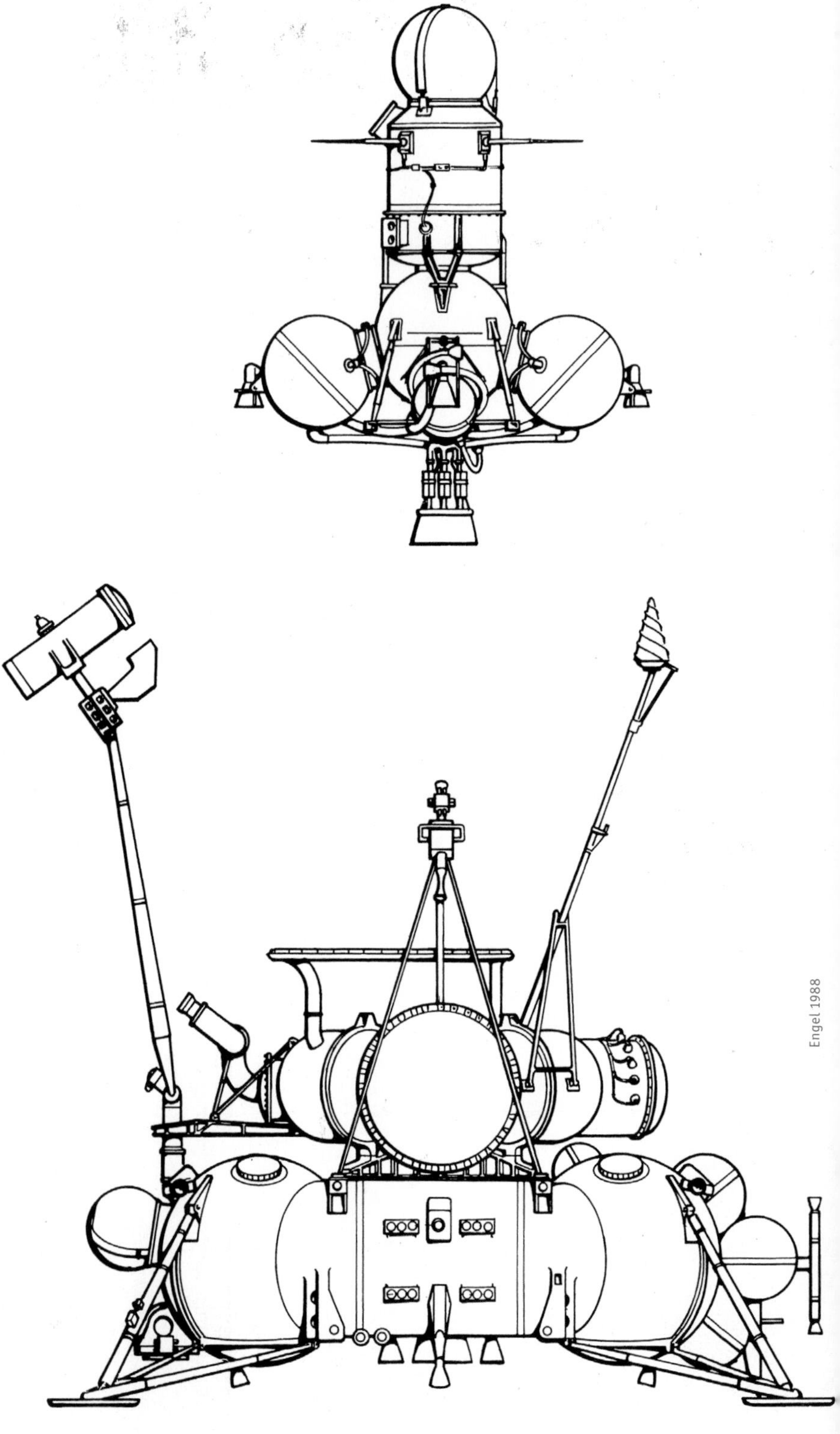

Engel 1988

Schematic drawing of the phallic return probe's launch

Luna 15
Soviet Union

Launch: 13 July 1969, 2:54 am
Landing: 21 July 1969, 3:51 pm
Launch site: Baikonur
Landing site: 17°N 60°O

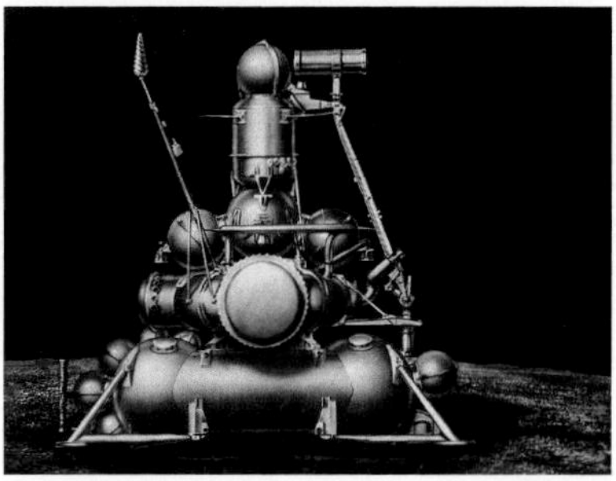

With the mounting successes of NASA's lunar programme, the Soviet Union was falling far behind in the space race by a seemingly insuperable margin, especially since its missions continued to fail, one after the other. Its last hope was Luna 15, a sample-return probe, designed to fly back to Earth after its mission rather than remain on the Moon. The Soviet Union had long since given up on carrying out a crewed mission to the moon before the Americans, though their official reason for this decision was that they did not wish to put any human lives in danger. The Soviet engineers developed a new bus (E-8-5) for the Luna 15 mission. It displayed a number of elements borrowed from American designs: a pyramid-shaped body, a four-legged frame, and round, plate-like feet. Only the shiny, silver spheres harked back to older Soviet spacecraft. The body was composed primarily of spherical forms of different sizes and in some ways anticipated blob architecture. Mounted on the two-storey bus was the return rocket, which, with its spherical head section and spherical fuel tanks placed on both sides at the bottom, can only be described as a veritable phallic symbol. Technical instruments, such as a cylindrical drilling mechanism and a snail-shaped antenna, were mounted on bendable masts. Luna 15 was launched, uncrewed, on a Proton-K rocket to the moon on 13 July 1969. This launch date came three days sooner than that of Apollo 11, though delays during its flight meant the American probe reached the moon first. The Soviet engineers lost contact with Luna 15 during its descent. It most likely collided with a mountain that had not yet been charted. The crash landing therefore cannot be seen as a failure of the probe itself but of the prior missions. No lunar orbiter from the Soviet space programme had delivered information about the conditions of the lunar landscape or scouted for potential landing sites. But even if Luna 15 had been successful, the probe's planned return to Earth would have taken place two hours after the return of the American spacecraft. The success of Apollo 11 would have brought public fame to the American space programme regardless. Soviet victory would have only been possible had the US mission failed.

NASA

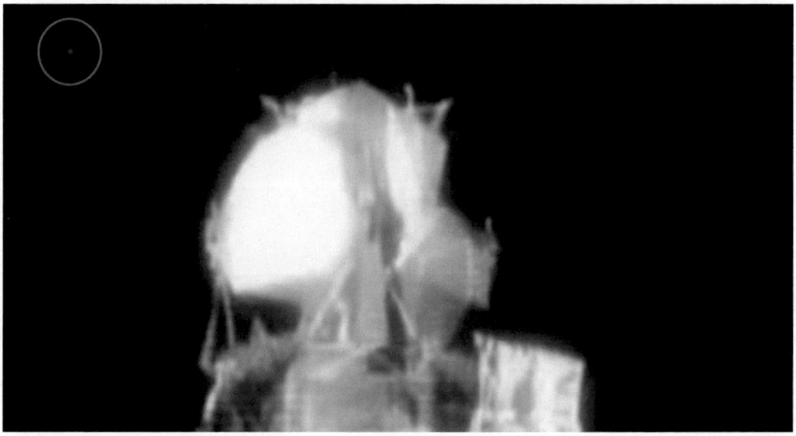

Luna 15 in front of Apollo 11

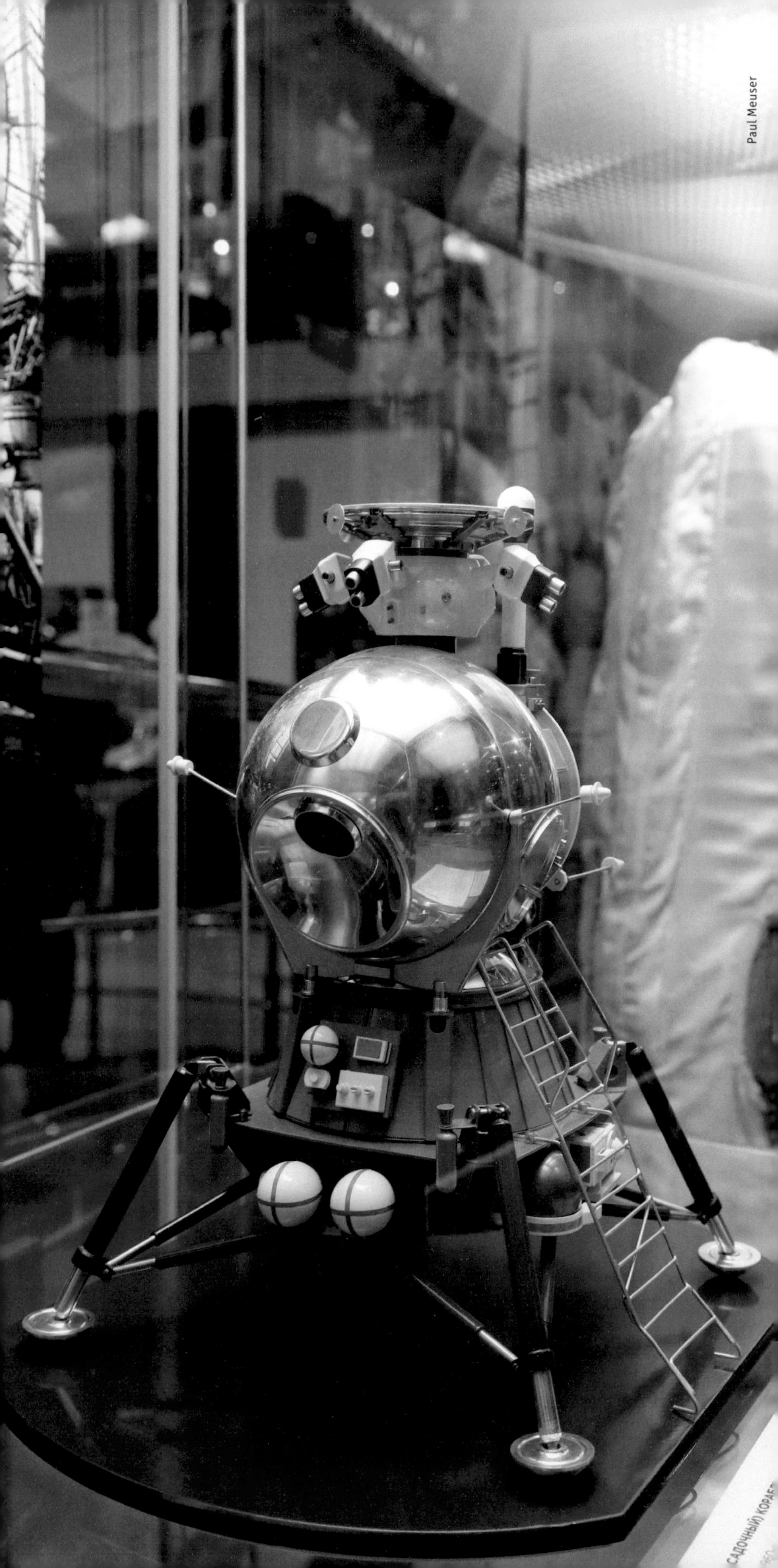

Paul Meuser

The Soviet Lunar Programme

Alexander Glushko

The beginning

The genesis of the Soviet crewed lunar programme can be traced back to 1963. At first, the project seemed to promise little success. But in 1965, the crew was granted official status with Yuri Gagarin as its commander. The rest of the members were selected during the assembly of the Military-Industrial Commission of the USSR in 1966. The group subsequently flew to Somalia, seeking to hone their skills in celestial navigation in the southern sky over the desert. A planetarium was also created there, near a training centre, and special training devices and helicopters for simulating a moon landing were also provided. Unlike the Americans, the Soviets only intended to land one cosmonaut on the moon: another would remain in lunar orbit during the landing procedure. Of course, this led to the question of who would be assigned which role.

When it transpired in 1968 that a team of US astronauts might land on the moon before them, the Soviet crew sent a pleading letter to Leonid Brezhnev, General Secretary of the CPSU, affirming their resolve to arrive on the moon first at all costs and declared their readiness to make whatever sacrifice necessary. Still awaiting an answer, they drove to the site where the space capsule Zond 7 was being prepared for launch. Had Brezhnev granted the Soviet crew's request, the cosmonauts of the Luna mission would have arrived on the moon on 8 December 1968, two weeks before the Americans. But he refused to give his approval, so the Apollo 8 astronauts became the first to fly to the moon, though they did not actually land there. On 20 January 1969, the Soviets launched an uncrewed Zond capsule into space. However, the capsule crashed, and it was clear that a crew would not have survived the flight. After this setback, the L1 (Luna 1) programme was discontinued. And in 1974, the USSR Council of Ministers issued a decree to terminate its follow-up programme L3 (Luna 3). The cosmonauts were re-assigned to different missions, although they were under contract until 1974. The crewed mission to the moon was put on ice.

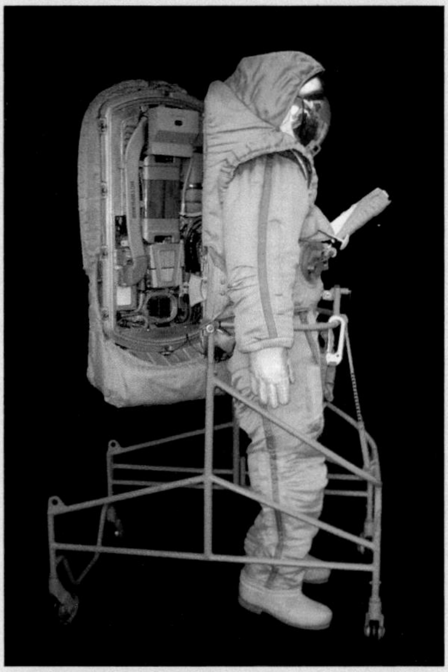

NPP Zvezda

Lunar space suit Krechet-94

However, another lunar mission of a different kind had been taking place at the same time. Any account of this project should begin with the friendship between Valentin Glushko and Vladimir Barmin. Valentin Glushko, the head Soviet designer of rocket engines, had a very close relationship with his engineer colleague Vladimir Barmin. Both figures were interested in colonising the moon for industrial purposes. (In fact, Glushko

Model of the Soviet lunar lander LK-3 at a scale of 1:18

Archive Alexander Glushko

Sergei Korolev, Valentin Glushko, and Vladimir Barmin (from left to right)

had published a set of plans directed to this end in his first and unpublished book *The Conquest of Planets*.) The two figures therefore ran the Zvezda ('Star') project together, seeking to establish a crewed lunar base, between 1961 and 1974. Glushko jokingly dubbed the project Barmingrad (Barmin City).

Zvezda represented the world's first detailed design of a city on a celestial body. The plan was to first land an uncrewed base module on the lunar surface. Various automated modules would follow, one after the other, and bring back to Earth samples of lunar rock from the landing site. A moon rover would then be sent as the second module, and it would analyse the intactness of the lunar surface below the first module. All modules would be mounted on a chassis to form a train that could travel along the lunar surface.

Vladimir Barmin's design bureau had been commissioned to develop a home for lunar residents already in 1962, but his employees would take over ten years to complete this task. Many questions arose during the development process. What requirements should the base module fulfil? What materials should be used to produce the scientific devices and structures? How might the station be used for military purposes?

Eberhard Marx

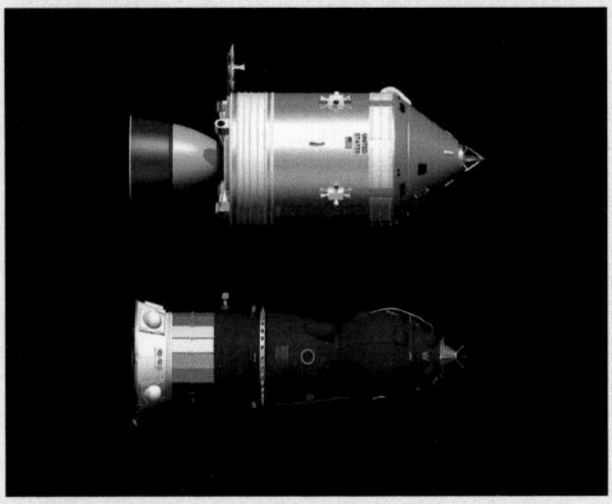

Apollo 10 (US) and LOK (USSR)

The lunar city would comprise nine habitation modules for nine cosmonauts. Each module would have a length of 4.5 metres and be delivered to the moon separately on its own rocket. The uncrewed module would collect soil samples and bring them back to Earth. Afterwards, moon rovers would survey the terrain, and four cosmonauts would go on expeditions in the lunar train to inspect potential building sites before selecting the one most appropriate for their base. The lunar train would also construct a provisional city and go on exploratory trips to the surrounding areas. The train would comprise a tractor, a caravan, a radioisotope generator with a capacity of 10 kW, and a module with a drilling rig. The modules' chassis resembled the one used for the Lunokhod rover. Each wheel was powered by an independent electric motor, so that the failure of one or more of the 22 motors would not incapacitate the entire train.

The habitation modules would consist of three layers of material that would protect the inhabitants against meteorites, heat, and UV radiation. The inner and outer walls would be made of special alloys and connected by a layer of foam. The lunar train's total weight was listed as eight tonnes. The crew's task would be to carry out scientific experiments and geological studies and to select the best site for the residential complex as well as the launch and landing pad. The use of drilling machines would ease the process of procuring soil samples and prevent them from coming in contact with the cosmonauts' space suits. Oxide had already been discovered in previously collected lunar soil samples, which meant it would not be necessary to transport large quantities of water to the moon. Hydrogen could be delivered instead, which would be used to generate the required amounts of water through chemical reduction. Finally, the aerospace company NPO Lavochkin and the engineers from Barmin's team invented a device for obtaining water, although they were ultimately unable to transport it to the moon and test it there.

During the development process, it became clear that the moonbase would be

Maciej Rebisz

The N-1 carrier rocket exploded time and again during four launch attempts, either during or shortly after the launch, until the programme was finally terminated in 1974 due to the continued failures. This also destroyed the dream of the crewed lunar programme.

able to house up to 12 people. It would comprise nine cylindrical blocks, each with a length of 8.6 metres and diameter of 3.3 metres, which would together weigh a total of 18 tonnes. The blocks would be manufactured on Earth, in the form of a metal accordion, 4.5 metres in length. On the moon, compressed air would be fed into the blocks to make them expand and reach their full length of 8.6 metres. Each block was assigned one of several functions, which included: a command post, a science laboratory, a warehouse, a workshop, a first-aid post, a fitness room, a kitchen with a dining room, and three living rooms.

In 1971, Vladimir Barmin finally presented his plans to Dmitry Ustinov, Defence Minister of the USSR, who was also responsible for the Soviet space programme.

Archive M. F. Rebrov

Oleg Makarov and Vasily Lazarev during a theory lesson at the Yuri Gagarin Cosmonaut Training Center, 1970 (first publication)

After a six-hour discussion, the minister gave the green light for the Zvezda plan to go forward. Subsequently, detailed drawings and models were produced of the habitation modules and expedition vehicles. The Zvezda project was part of the Soviet crewed lunar programme, N1-L3, and the project's realisation depended on a key element of that programme: the super-heavy carrier rocket N1. However, all four launch attempts with this rocket failed between 1969 and 1972. In response, the USSR government terminated the N1-L3 programme in 1974.

When Glushko realised that he could not implement the Zvezda project with the N1 rocket, he turned to the UR-700 carrier rocket, designed by his colleague Vladimir Chelomey. However, this rocket was decommissioned even earlier than the N1, due to a bitter rivalry between the rocket engineers Chelomey and Korolev. A complementary programme, N1F-L3M, was initiated before the termination of the Soviet lunar programme. Its purpose was to achieve long-term stays on the moon, or at least to enable cosmonauts to dwell on the moon for a far longer period than their American counterparts had managed. The ultimate goal was to develop new transport options to the lunar base (LEK Lunar Expedition Complex).

Barmin Design Bureau of General Engineering

A moon base module in a pit on the lunar surface (Zvezda, 1962–1970)

Archive A. W. Glushko

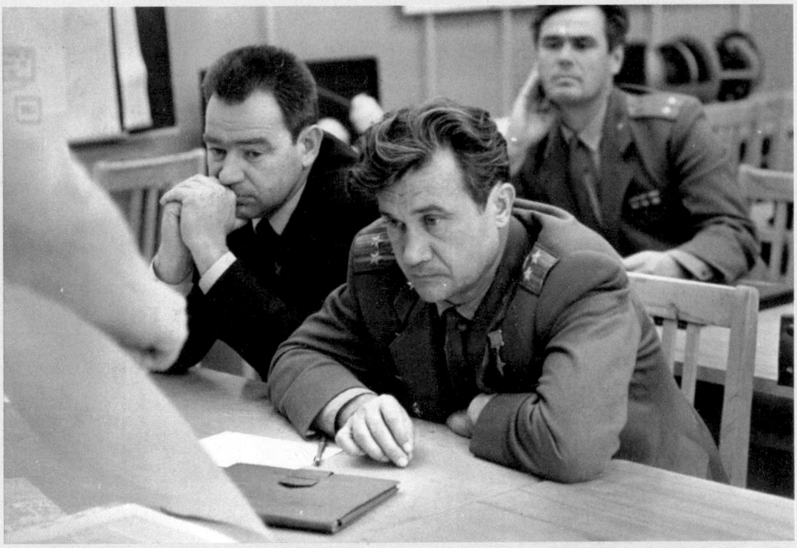

Members of the Contact programme: Georgy Grechko (left) and Anatoly Filipchenko, with Vasily Lazarev behind them, during a theory lesson, 1970 (first publication)

The follow-up

Once Glushko had begun his work on the lunar programme, he began to look back on an idea he had conceived in his younger years: the design of a Helios spacecraft (see p. 133). His idea was for this spaceship, fully clad in solar panels, to first fly towards the moon and Mars before going beyond the limits of the solar system. He developed a unique lunar programme based on this idea, seeking not only to land on and explore the moon but also to construct a moonbase. This was a visionary idea, and even Barmin, the head planner of the Baikonur Cosmodrome, had been guided by Glushko's first deliberations, which had already been published in *The Use of Planets* in 1929.

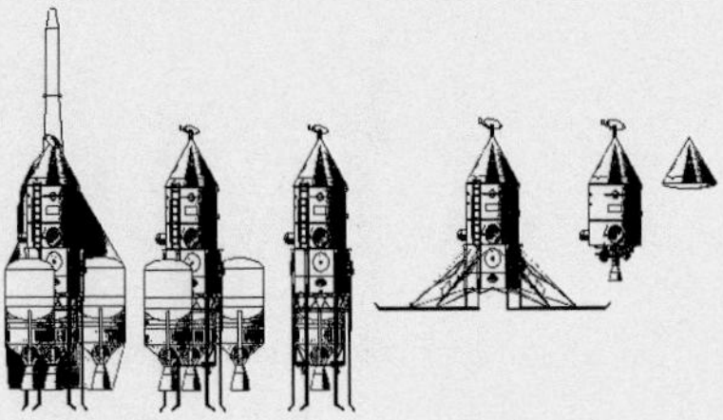

LK-700 moon lander and ascent module

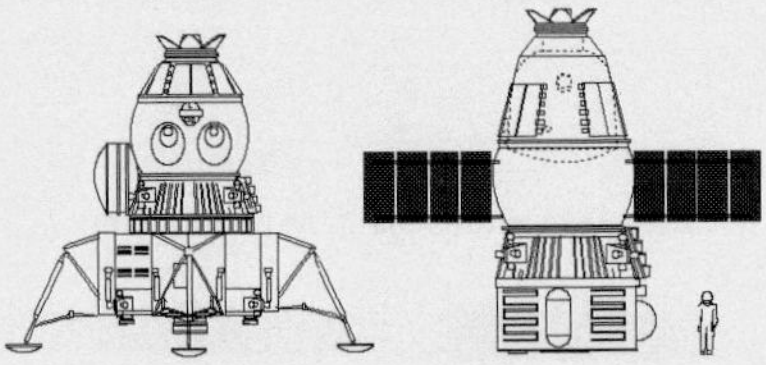

Once the development of Buran was complete, Glushko initiated a new study to develop a crewed lunar base for mining helium-3, a helium isotope rare on Earth.

Barmin Design Bureau of General Engineering

Studies for a hexagonal interior

Paul Meuser

The modules are covered with regolith on site (schematic model, 1970).

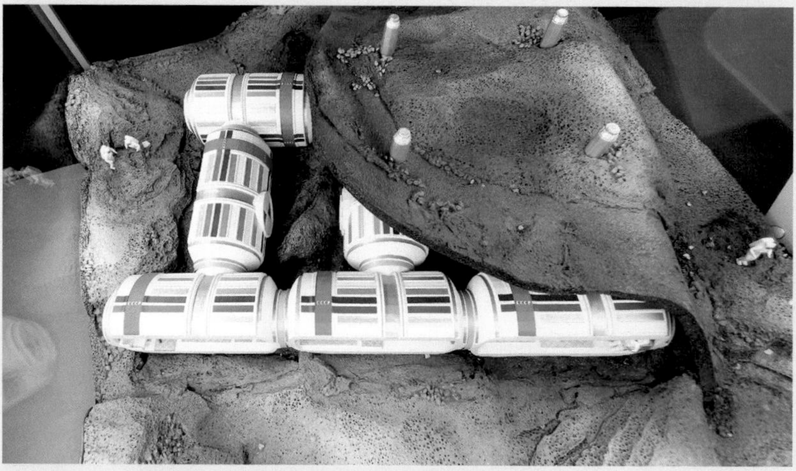

Paul Meuser

The Zvezda modules are connected using the same technology used by the ISS today (1970).

Barmin Design Bureau of General Engineering

True-to-scale model of a Zvezda lunar base module

Barmin Design Bureau of General Engineering

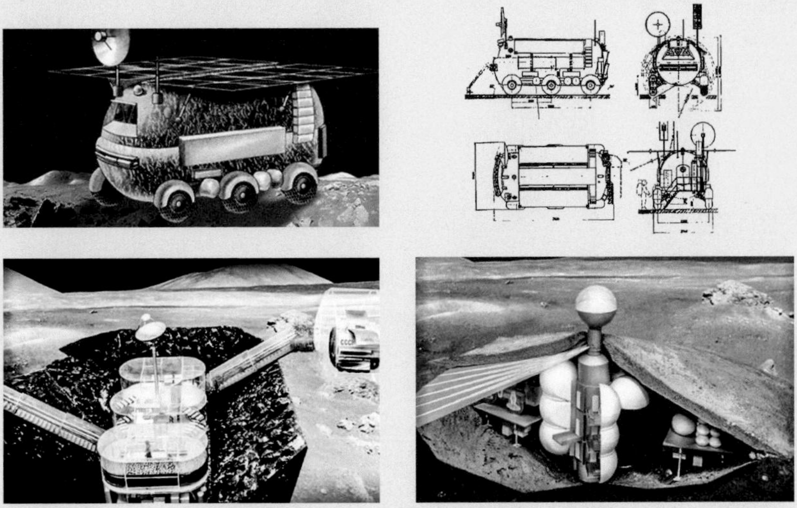

Design variations by Vladimir Barmin and his team of engineers (circa 1972)

Barmin Design Bureau of General Engineering

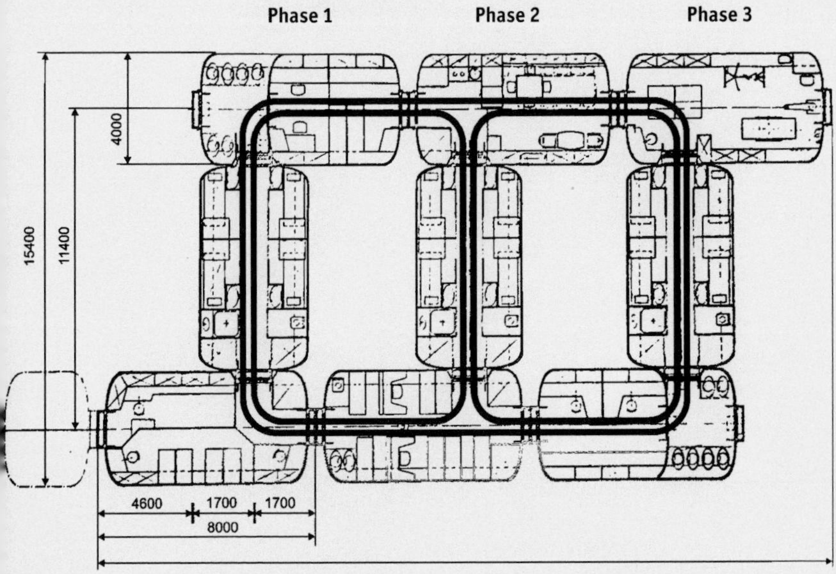

Layout of the three-phase moonbase composed of inflatable modules (circa 1972)

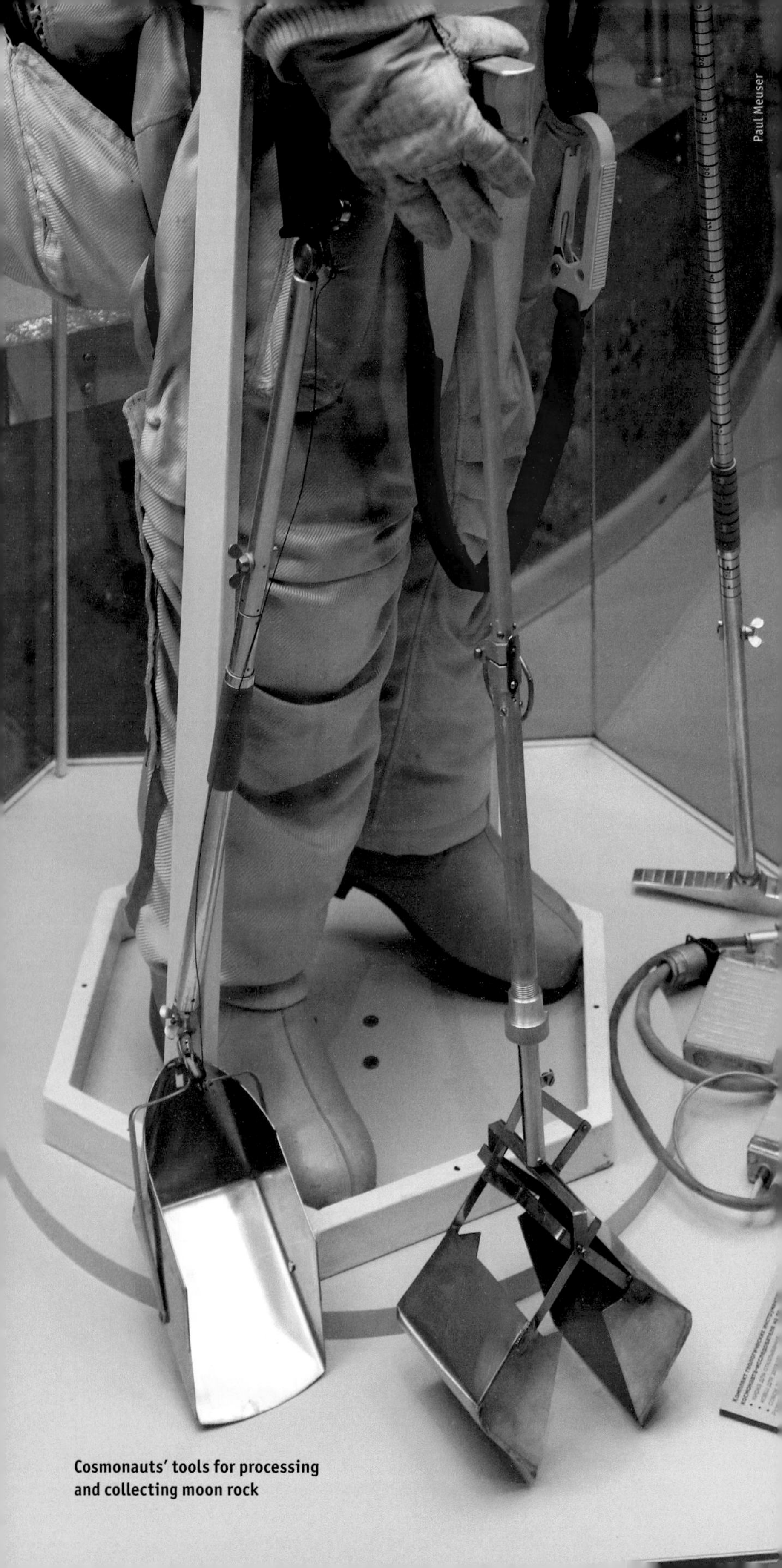

Paul Meuser

Cosmonauts' tools for processing and collecting moon rock

Sputnik / Sergey Pyatakov

Model of the Helios spacecraft: Valentin Glushko conceived the idea already in 1929.

The provisional end

Valentin Glushko was later commissioned by NPO Energia (formerly the Experimental Design Bureau OKB-1) to resume the N1-L3 programme. The Energia carrier rocket was launched successfully on 15 May 1987, which encouraged Glushko to pursue his vision of the Helios spacecraft with the Energia and Vulkan rockets as launch vehicles. His plan was to send a crew in the Helios spacecraft into lunar orbit, where they would circle around the moon before returning back to Earth. However, the cosmonauts he wanted as his crew were not available, as they were selected to serve as commanders of the Buran spacecraft. These cosmonauts completed a special training programme at the Yuri Gagarin Cosmonaut Training Centre in Moscow between November 1985 and May 1987. However, a delay in the Buran programme's implementation made the cosmonauts available for the Helios, which allowed them to eventually make it into outer space. Glushko would have without a doubt taken good care of his protégés in his Helios spacecraft, but he unfortunately ran out of time. He died in 1989.

Alexander Krassotkin

UR-700 carrier rocket (1964)

The Soviet space shuttle Buran only flew once, without a crew, in November 1988. The programme floundered after Valentin Glushko's death in 1989 and was finally discontinued in 1993.

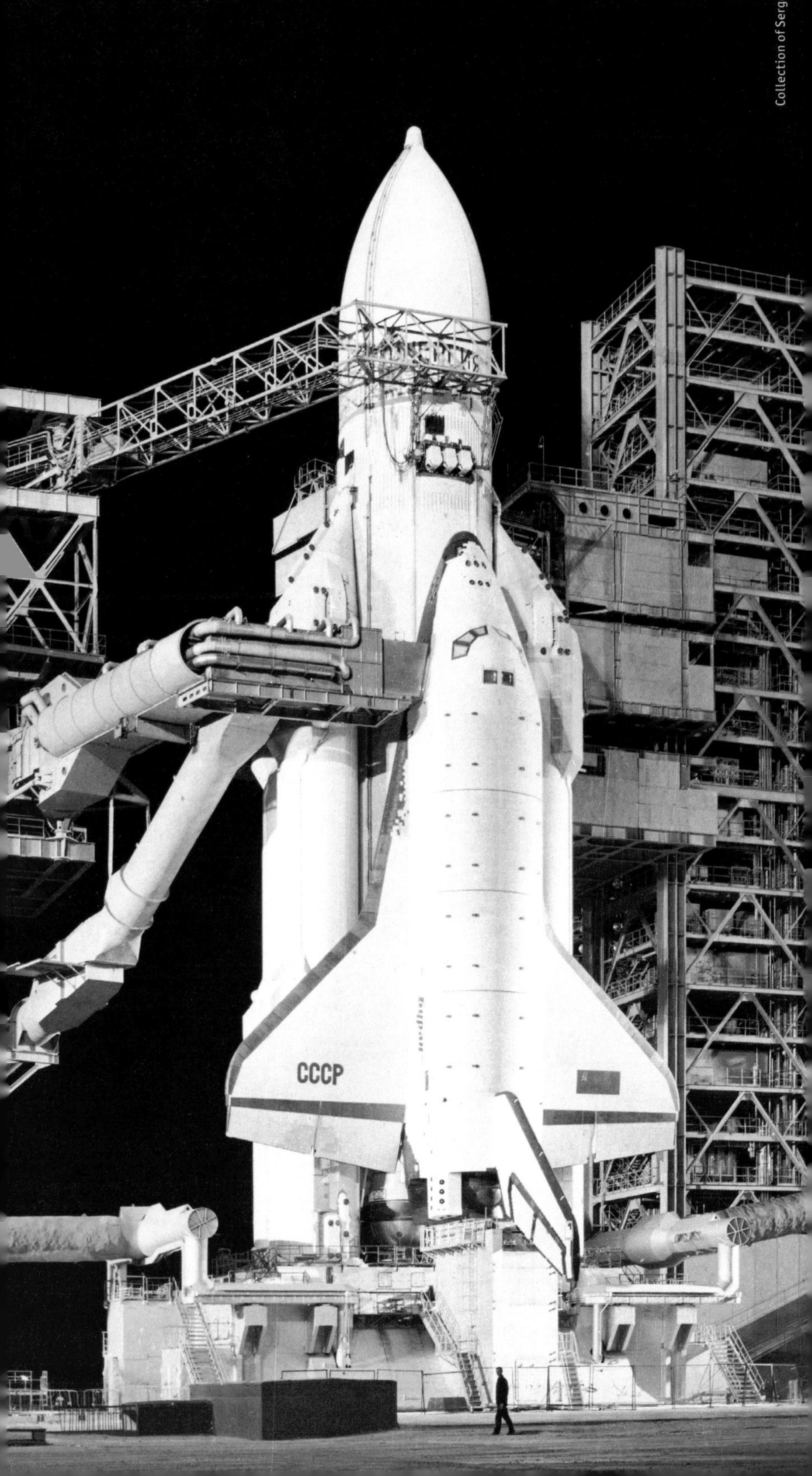

Collection of Sergey Abramo

Roscosmos

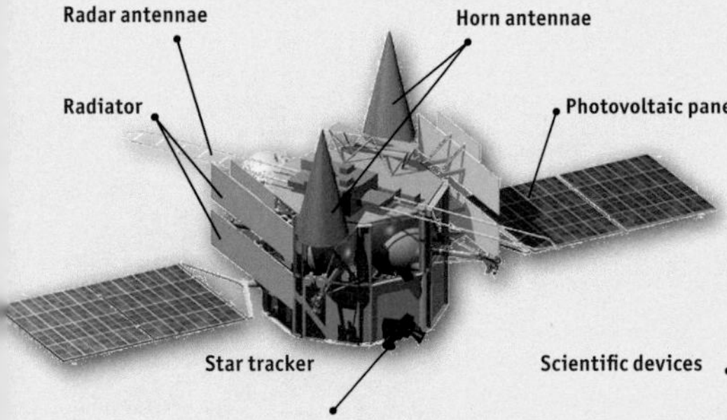

Luna 26

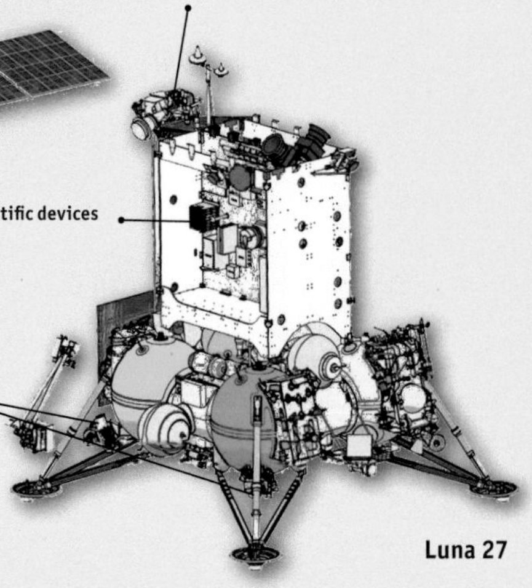

Luna 27

The future

In 2007, Roscosmos, the Russian Federal Space Agency, announced that it would send the first cosmonauts to the moon by 2025. They would then establish a permanent moonbase there a few years later. In 2014, the design of the three-stage Russian lunar programme was made public:

1. The automated interplanetary stations Luna 25 (see pp. 332–335), Luna 26, Luna 27, and Luna 28 will be sent to the moon by 2025. They will study the chemical and physical properties of the regolith on the lunar poles. Moreover, they will search for the most appropriate location in which a test site and moonbase can be established in the future.

2. Between 2028 and 2030, three crewed spacecraft will enter and stay in a lunar orbit without landing on the moon.

3. Between 2030 and 2040, the cosmonauts will land on the potential lunar testing site and install the first infrastructural components, which will be manufactured using lunar substances, and build a lunar observatory for monitoring the Earth.

The plan also entailed the construction of a habitable base and equipment by 2050. However, there was a delay in the plan's implementation in 2015 due to financial difficulties.

The next steps:

a) First, an observation satellite will be sent into lunar orbit. It will photograph the Soviet moon rover (Lunokhod) as well as the landing sites of the Apollo modules.

b) Afterwards, a moon rover will drill holes and bring back lunar soil samples to Earth. The mission will be carried out as part of the Russian lunar programme.

Pline

Lander of the planned Luna 27 mission, model

Pline

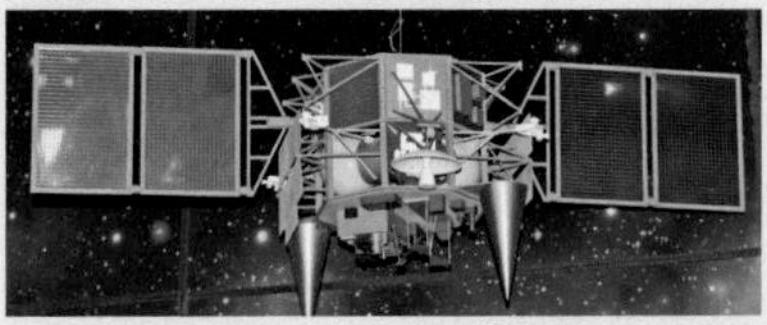

Orbiter of the planned Luna 26 mission, model

Lunar Orbiter LOK and Lunar Lander LK

RKK Energiya

1960–1965
Experimental Design Bureau (OKB-1)
Landing weight: 5,560 kg
Ascent weight: 3,800 kg
Height: 5.2 m

The Soviet lunar orbiter LOK was an enhanced version of the Soyuz space capsule for crewed missions to the moon. It featured a spherical orbiter section, a bell-shaped return capsule, and an equipment module. The LK lander had two stages: the descent stage and ascent stage. The descent stage comprised a steel base with a radius of 2.27 metres and a landing mechanism. The Block E propulsion unit was placed inside the base and ensured the return launch from the lunar surface. The landing module encompassed an equipment module with landing radar and an integrated research system with a drilling device. The ascent module was located at the top of the landing module and comprised the equipment module of the pressure cabin, the steering motor, and the propulsion unit block. The cosmonaut section, measuring 2.3 × 3 metres, offered space for one cosmonaut in the Krechet space suit, in standing position, in front of the dashboard and control panel. The breathing air's composition and pressure enabled the opening of the pressurised helmet for eating. The flight navigation system made it possible to complete flights with descent, ascent, and docking manoeuvres in automatic mode. In the upper part of the compartment, there was a docking device with a viewing area to aid the docking with the lunar orbiter. On the left side wall, there was a hatch through which the crew could step onto the lunar surface or into space.

Lunniy Korabl (LK-3) during the approach

LK-3 on the lunar surface

Eberhard Marx

LK-3 during the launch to lunar orbiter LOK

Vladimir Nekrasov

Port

Starboard

Top view

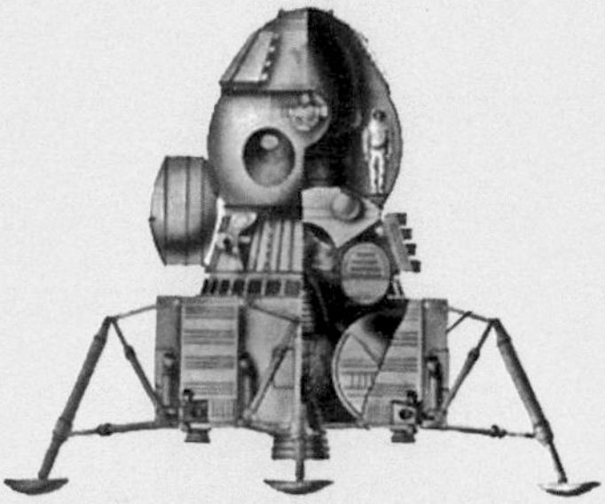

Elevation of a model

Eberhard Marx

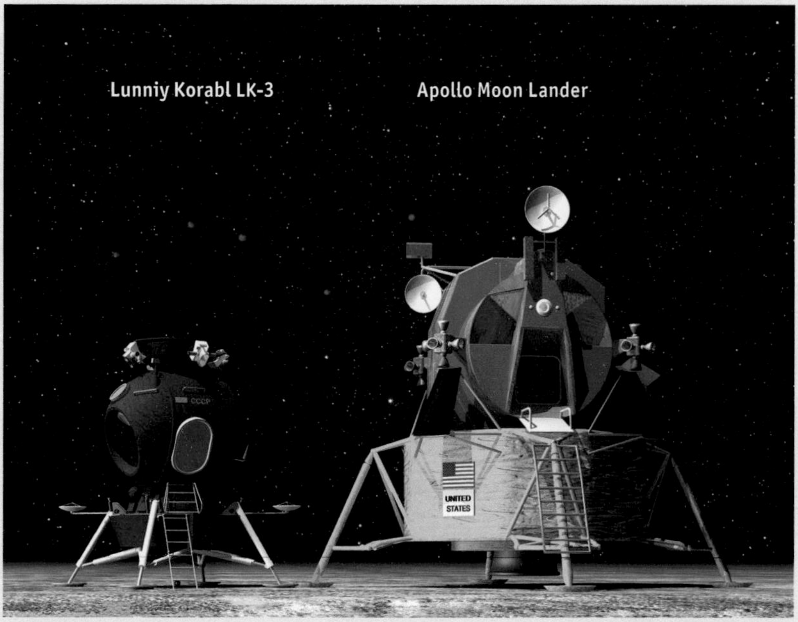

Temporary Missions on the Moon: The Highs and Lows from Apollo 11 to Luna 24

1969 to 1976

The 1970s were a particularly successful time for space exploration. This is when the launching of Earth satellites had become routine, and flights to the moon and missions to explore interstellar space were successfully completed, one after the other, as part of NASA's Voyager programme, for example.
Although the early lunar probes completely missed their target, reaching the moon increasingly became a matter of course, until soon a human was stepping onto the lunar surface for the first time. The journey to the moon was not a short sprint, but rather an exhausting marathon that required an enormous strength of will and myriad technical accomplishments. A large number of missions were required to lay the technical groundwork for the first astronauts to arrive on the moon.
When the Americans achieved the first crewed landing on the moon in July 1969, the Soviet Union had to admit defeat, in spite of the tireless work it had put towards the same goal. Whenever a rocket would fall from the sky over Baikonur and burst into a fiery inferno, the next rocket would be promptly placed on the launch pad a short time later. But technical defects in the N1 rocket type continuously impeded the Soviet Union's progress during the most decisive stage of the race to the moon. The Soviets covered up their repeated failures in a manner that had become habitual during the Cold War: they explained that remote-controlled probes were far more capable of performing scientific operations on the moon. Once it became clear that they had no prospects of landing a human on the moon, there was nothing left to do than fully commit to their alternative programme. And so, while the Americans sent one crew after the other on lunar missions, the Soviets sent return probes and the first (remote-controlled) rovers to the grey celestial body. It was a neck-and-neck race, not only between the US and the Soviet Union, but also between humans and machines. When this race finally came to an end, this also marked the end of human presence on the moon, and Earth's grey satellite found its peace for the first time in many years. The easing of Cold War tensions also took an effect, albeit with a delay of a few years, on space flights to the moon.
The landing of Apollo 11 introduced the first work of architecture, in the strictest sense, on the moon. After all, a true work of architecture needs both human presence and a form of housing. The first primitive hut on the moon was at once a vehicle and a shelter, inhabited by people. This is why this book, which explores this period of the space race, is presented as an architectural guide.

The historic Apollo 11 landing being celebrated with a patriotic bounty of blue, white, and red – and Cuban cigars.

NASA

The first-ever footprint on the moon:
the image is one of the most famous photographs in human history.

Apollo 11
US

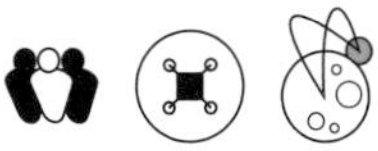

Launch: 16 July 1969, 1:32 pm
Landing: 20 July 1969, 8:18 pm
Launch from moon: 21 July 1969, 5:54 pm
Launch site: Cape Canaveral
Landing site: 0.67408°N 23.47297°O

NASA

Who hasn't seen the image of the American flag in the grey lunar landscape against the backdrop of vast, dark, empty space? It has become a symbol of that historic event when a human stepped onto the moon for the first time. This was the achievement of the Apollo 11 mission, which took place over the course of five days in late July 1969. The mission's objective was to land a crewed spacecraft on the moon and to return it safely to Earth. On the moon, the astronauts would leave the lunar capsule for a short period of time. The mission fulfilled the challenge posed to the American nation by President John F. Kennedy in 1961. At the same time, it marked the end of the race to the moon between the two competing political systems of the Cold War.

On the outside, the multi-part spacecraft resembled its forerunner, which had successfully, though with a few difficulties, achieved a moon landing. Even before Apollo 10 entered lunar orbit, the next Saturn V rocket, carrying Apollo 11, was rolled onto the launch pad so that the next mission could be prepared over the subsequent two months. The crew comprised Neil Armstrong as commander, Michael Collins as command module pilot, and Edwin 'Buzz' Aldrin as lunar module pilot. The latter two would become the first humans to walk on the moon.

The crew journeyed through space in a small capsule measuring 6.6 cubic metres. After orbiting the Earth one and a half times and docking the command module to the lunar module, they finally arrived on the moon. It wasn't until the approach that difficulties arose. It turned out that the onboard computer was navigating the capsule away from

NASA

Cinematic image: almost all conspiracy theories claiming the moon landing was a staged hoax begin with this photograph. The flag appears to billow, even in the absence of wind.

the planned landing site and towards a boulder field. And so, Armstrong manually piloted the landing module, Eagle, steering it towards a site not far from the intended destination. However, this manoeuvre used up the fuel supply faster than planned, and the astronauts needed to quickly decide whether they could land the capsule within twenty seconds, or if the landing needed to be aborted immediately. They opted for the landing, which took place at 8:17 pm on 20 July 1969.

After the landing, a low-risk but somewhat improvised Apollo mission took place. Armstrong and Aldrin stayed on the moon for twenty hours, spending two and a half hours outside the lunar module. During this time, they used several scientific tools to carry out measurements: aluminium foils to analyse the solar wind and a laser reflector to measure the distance between the Earth and moon, for example. They also had a phone conversation with President Richard Nixon and hoisted the American flag, which would become a routine procedure in the subsequent Apollo missions. They had to make sure the escape hatch didn't close during their time on the moon, since the pressure build-up would have made it impossible to open from outside. They departed the moon in the ascent stage and successfully completed the docking manoeuvre to join the ascent stage to the command module, Columbia.

Some of the scientific tools were left behind on the moon: the ascent stage's gold-coated, spider-like landing module, for example. However, the flag could not withstand the strong wind caused by the engines during the launch of the ascent stage. As such, the symbol of the American nation can no longer be seen on the moon. The space capsule splashed down in the Pacific Ocean five days after the mission's launch, signalling to the public the mission's successful conclusion. But the mission was not yet over for the crew: they needed to stay in quarantine for 17 days until it was certain that they had not brought back dangerous and unknown micro-organisms with them from their excursion.

NASA

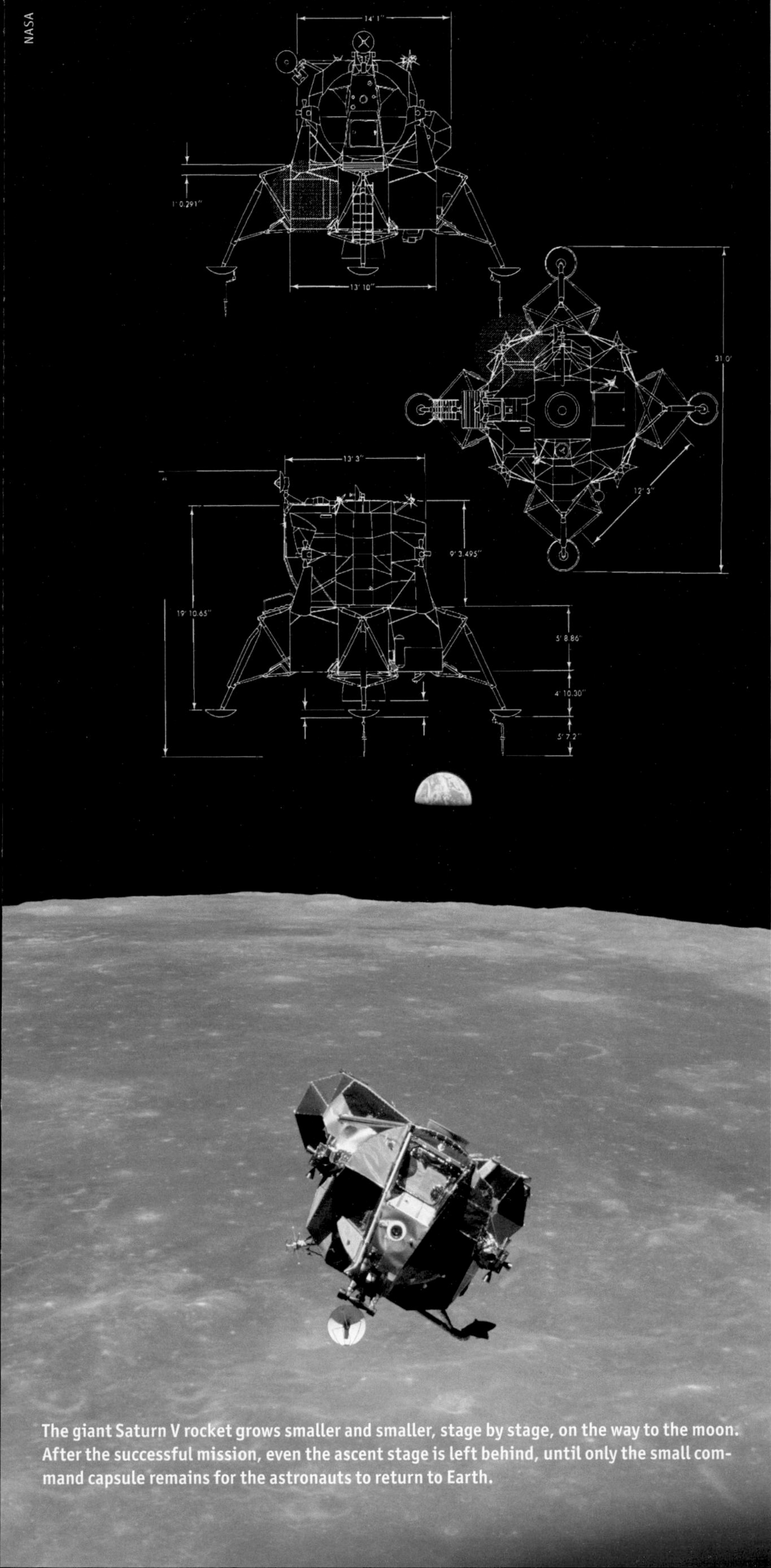

The giant Saturn V rocket grows smaller and smaller, stage by stage, on the way to the moon. After the successful mission, even the ascent stage is left behind, until only the small command capsule remains for the astronauts to return to Earth.

LUNAR ORBIT REST PERIOD ATTITUDE

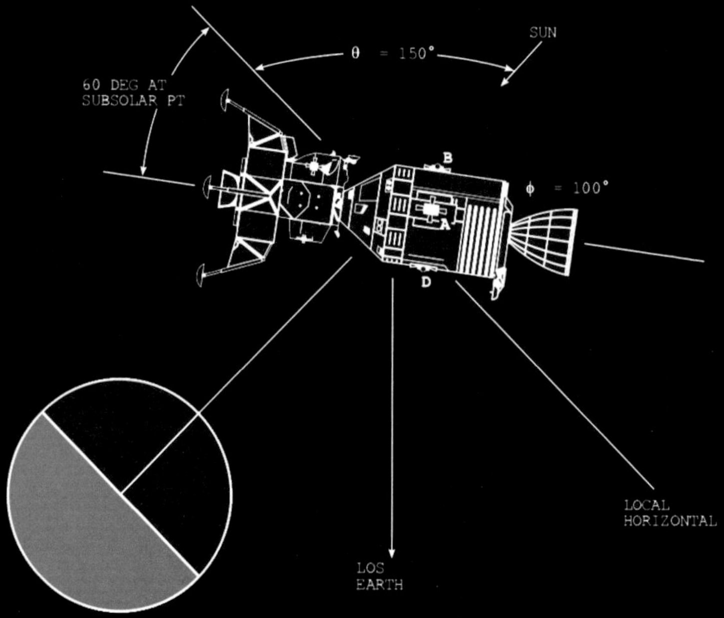

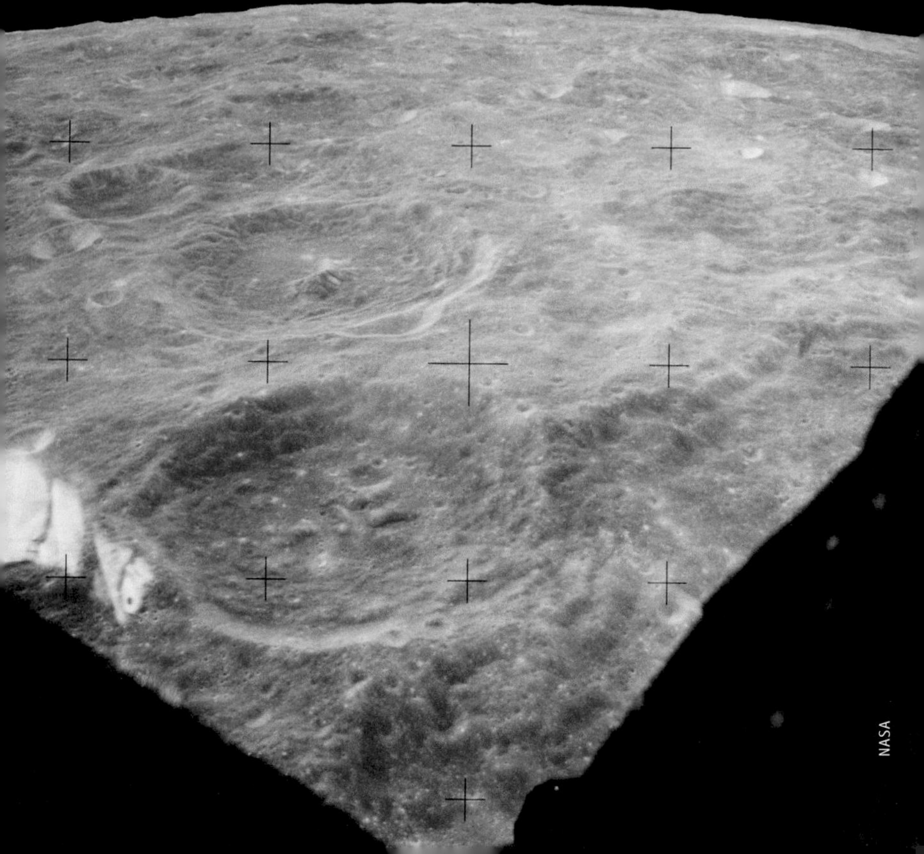

NASA

NASA

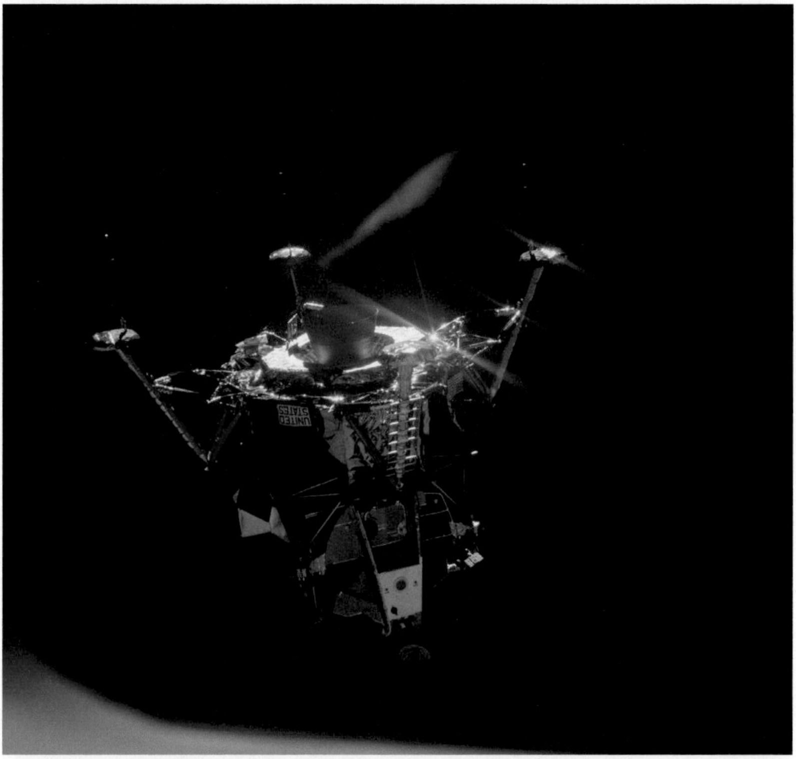

The orbiter is separated and circles around the moon, while the landing module (top) heads towards the lunar surface.

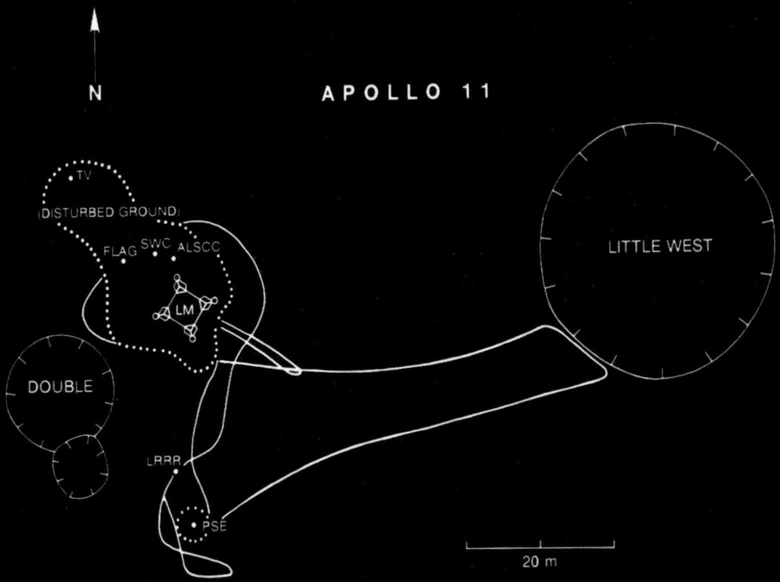
N
APOLLO 11
TV
(DISTURBED GROUND)
FLAG
SWC
ALSCC
LM
LITTLE WEST
DOUBLE
LRRR
PSE
20 m

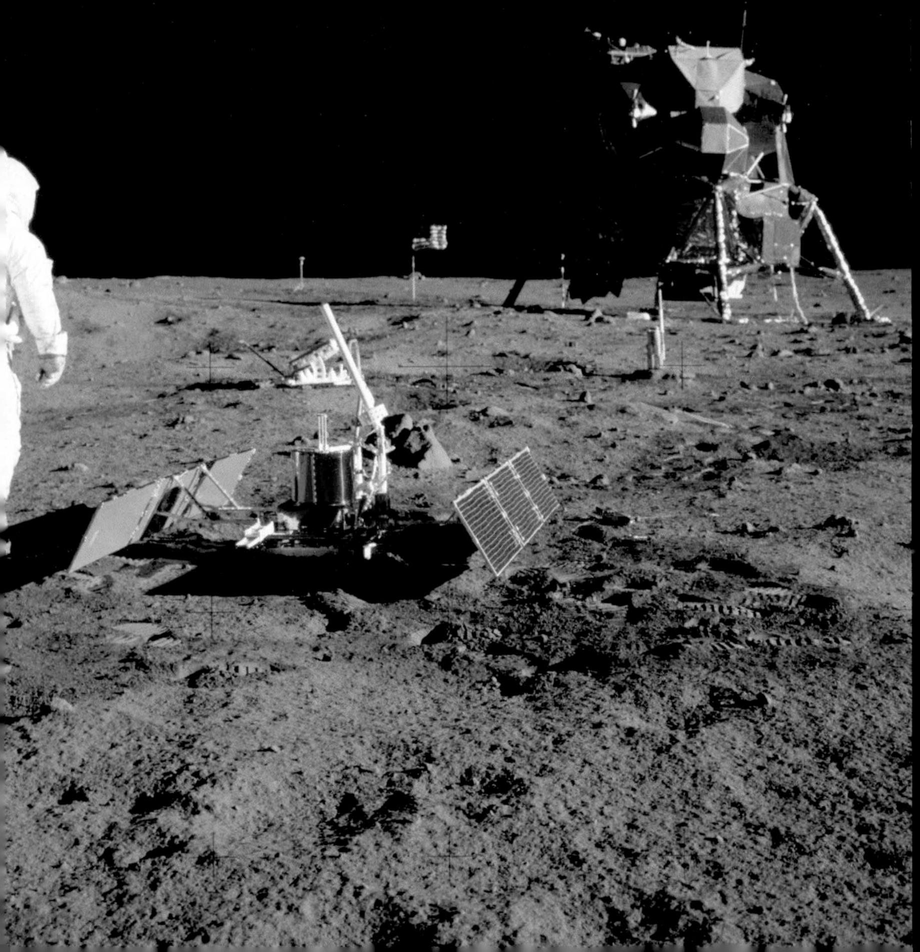

Just as famous as the footprint depicted earlier: the image of Buzz Aldrin, *A Man on the Moon*

Certification authorities would never approve such a narrow door.
astronaut Alan Bean squeezes through the tiny hatch of the landing module.

Apollo 12
US

Launch: 14 November 1969, 4:22 pm
Landing: 19 November 1969, 6:54 am
Last contact: 20 Nov. 1969, 2:25 pm
Launch site: Cape Canaveral
Landing site: 3.01°S 23.42°W

Five months after the first humans stepped onto the moon, the same, Apollo spacecraft was deployed once more. The aim of the mission was once again to achieve a crewed moon landing. The spacecraft's outer appearance did not change: the American space engineers saw no reason to change a proven design. During the successful mission, the astronauts remained on the moon for 11.5 hours longer than their predecessors. Eight of these hours were spent outside the space capsule. Apollo 12 touched down, according to plan, just 163 metres away from where Surveyor 3 had landed two and a half years earlier. As such, the astronauts were able to dismantle parts of the old probe and return them back to Earth as souvenirs.
This mission involved a number of (artistic) artefacts. Pete Conrad and Alan Bean created high-quality self-portraits with elaborate reflections and dramatic lighting during their trip to the moon. Moreover, American avant-garde artists of the 1960s had collaborated on an artwork leading up to the mission: the *Moon Museum*, by Forrest Myers, Robert Rauschenberg, David Novros, John Chamberlain, Claes Oldenburg, and Andy Warhol. Myers produced small ceramic tiles, each measuring 1.9 × 1.3 centimetres, on which miniature drawings by the above-mentioned artists were collected. Their plan was to send this small 2D museum to the moon. It is said that a copy of the tile was covertly attached to one of the legs of Apollo 12. This would make it the first work of art for architecture on the moon, and Apollo 12 itself the first exhibition building.

MoMA

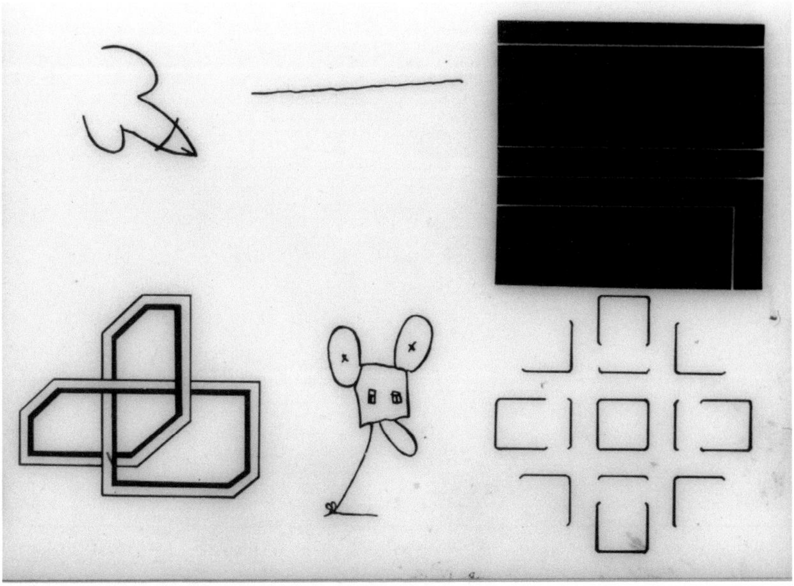

This copy of *Moon Museum* can be found not on the moon but in the Museum of Modern Art in New York.

APOLLO 12

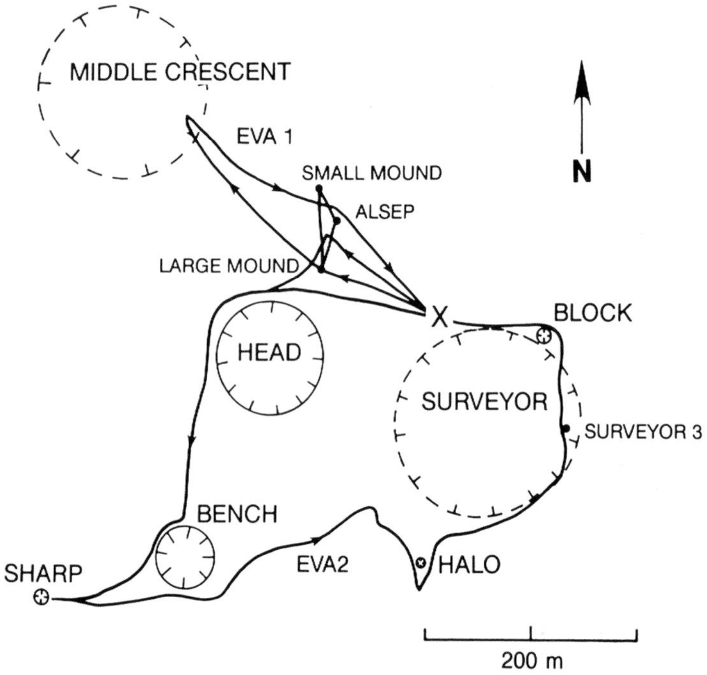
MIDDLE CRESCENT
EVA 1
SMALL MOUND
ALSEP
LARGE MOUND
N
X
BLOCK
HEAD
SURVEYOR
SURVEYOR 3
BENCH
SHARP
EVA2
HALO
200 m

NASA

Within eyeshot of the landing site: Apollo 12 (back) and Surveyor 3 (front)

NASA

Even after the historic moon landing, representing years of technological achievement ...

... the astronauts need to unload and set up the experiments by hand.

As many instruments as in an orchestra:
the astronauts’ tools and equipment for the extravehicular activity (right)

NASA

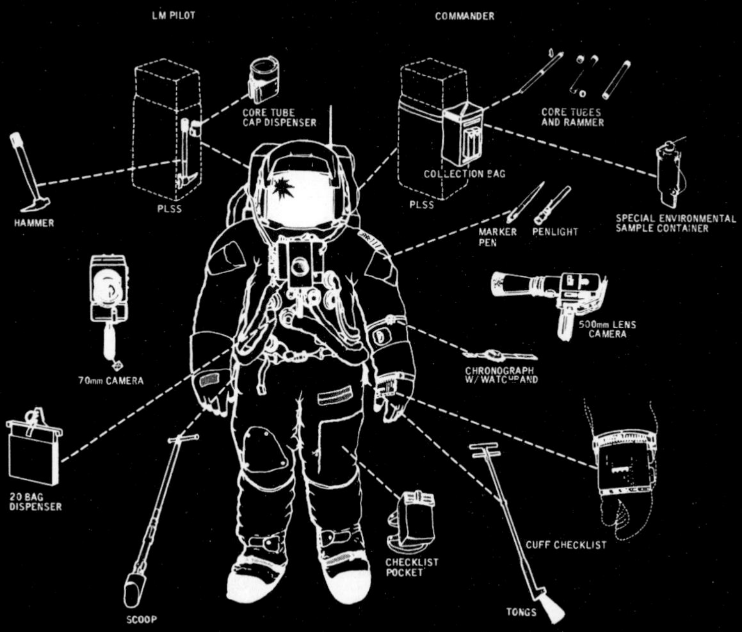
LM PILOT
COMMANDER
CORE TUBE
CAP DISPENSER
CORE TUBES
AND RAMMER
COLLECTION BAG
HAMMER
PLSS
PLSS
SPECIAL ENVIRONMENTAL
SAMPLE CONTAINER
MARKER
PEN
PENLIGHT
500mm LENS
CAMERA
70mm CAMERA
CHRONOGRAPH
W/ WATCHBAND
20 BAG
DISPENSER
CUFF CHECKLIST
CHECKLIST
POCKET
SCOOP
TONGS

Re-purposing the high-gain antennae. James T. Burns slipped this humorous drawing into a presentation for NASA. Burns had accompanied the Apollo programme over the years as a graphic designer for the Radio Cooperation of America (RCA).

James T. Burns

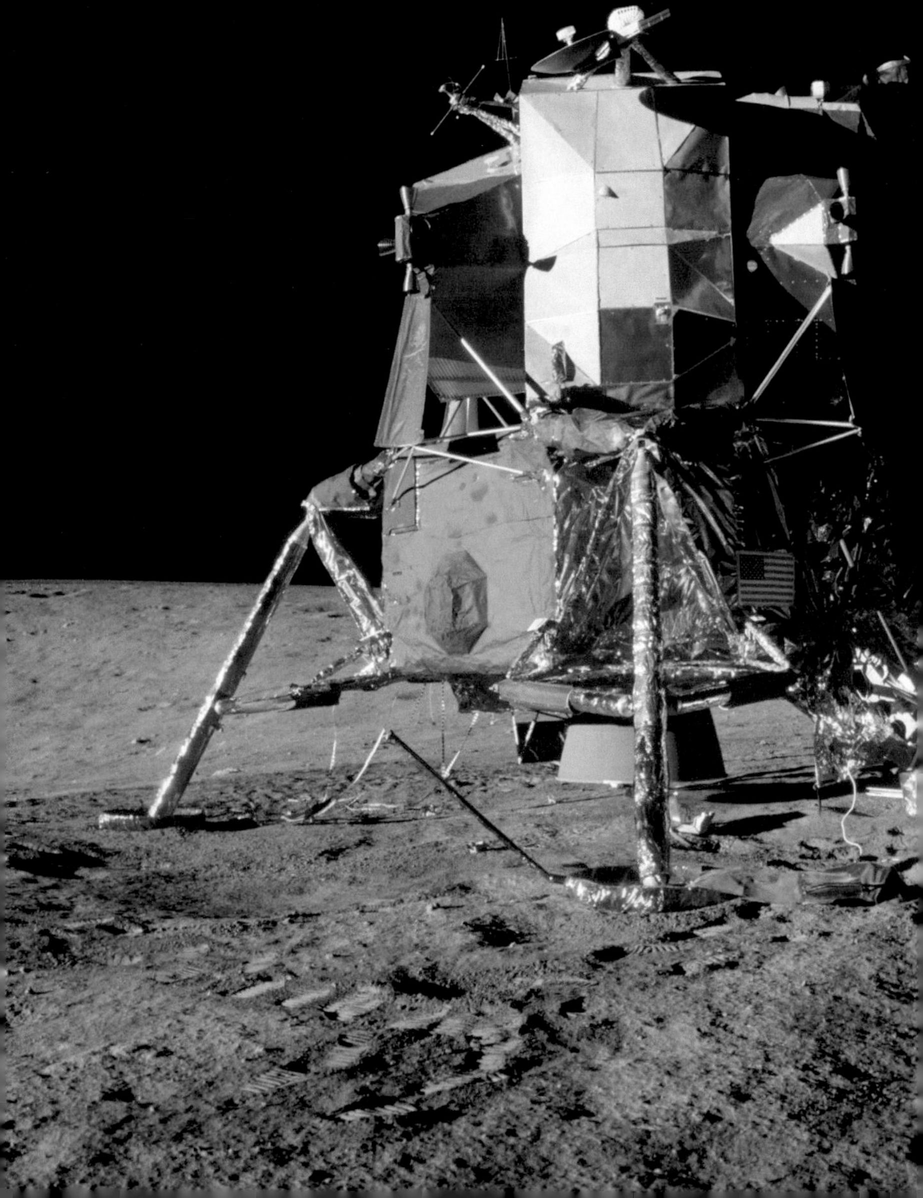

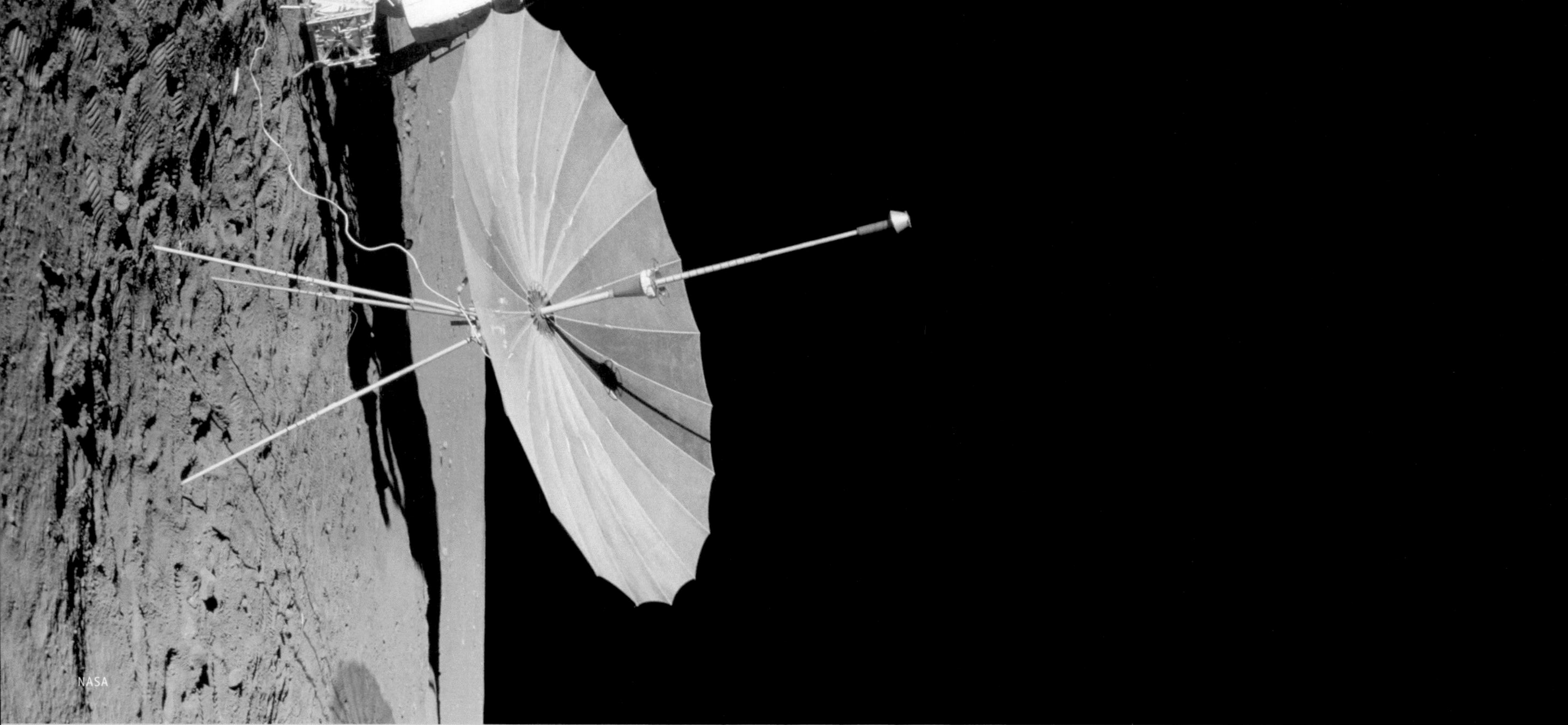
NASA

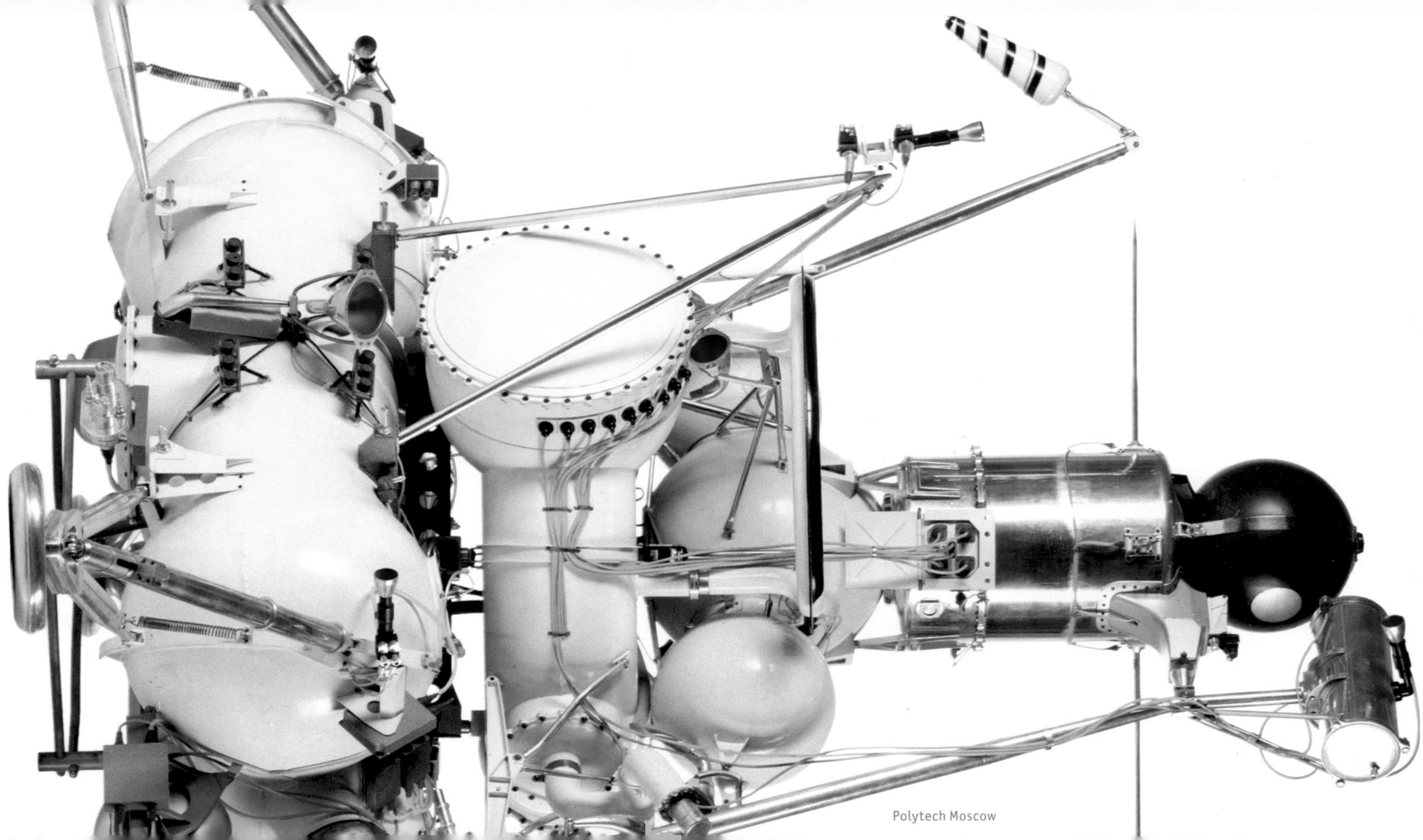

Polytech Moscow

Luna 16
Soviet Union

Launch: 12 September 1970, 1:25 pm
Landing: 24 September 1970, 5:25 am
Last contact: 14 Sept. 1979, 10:02 pm
Launch site: Baikonur
Landing site: 0.6833°S 56.3°O

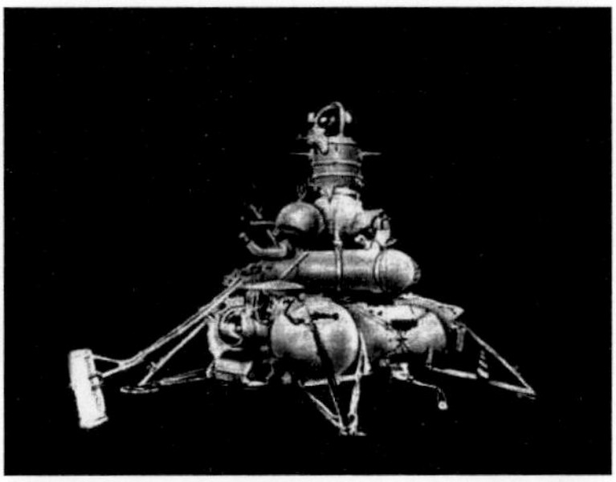

Engel 1988

When Luna 16 touched down in the Sea of Fertility (Mare Fecunditatis), it became the first probe from the Soviet space programme to achieve a successful lunar landing. The outer appearance of the spacecraft was identical to that of Luna 15, aside from a small number of minor details. Although Apollo 11 and 12 had already brought back lunar soil samples to the West, the uncrewed Luna 16 probe was the first to bring a sample back to the Soviet Union. To this end, the unclad space probe lowered its extendable arm from the top of its tower-like body onto the lunar surface after the landing. Its drilling mechanism pierced through the lunar surface and extracted a number of soil samples. Once it had collected enough material, the arm automatically returned to the top of the probe and stowed the vial with the samples inside the spherical capsule. The tower-shaped upper stage took off from the moon the next day to transport the samples back to Earth. The round metal capsule was jettisoned from the rest of the spacecraft during the return flight and landed, undamaged, in the Kazakh Steppe. The probe's landing module, like those of the Apollo missions, still lies on the moon. It continued to transmit data on the temperatures and radiations on the moon to the space agency's research centre on Earth, even after the return flight of the upper stage. Contact with the landing module was lost nine years later. Although the success of Luna 16 has been overshadowed by the Apollo missions, it still has its place in the history of space flight to the moon. It was the first robotic probe to land on the moon and bring back lunar soil samples to Earth.

Creamy colours in space: Luna 16 reflecting the fashion trends of its time

Moscow Polytech

Assembly shortly before the launch: probe and landing module

Simulation on Earth: soil samples in the samples capsule

The beetle-like robot roams ebulliently across the lunar surface in this illustration, filled with dramatic scenery all around. The rover does not look back, its black eyes always pointing forwards.

Moscow Polytech

Luna 17 *Lunokhod 1*
Soviet Union

Launch: 10 November 1970, 2:40 pm
Landing: 17 November 1970, 5:25 am
Last contact: 14 Sept. 1971, 1:05 pm
Launch site: Baikonur
Landing site: 38.23° N 35.00° W

The Soviet space programme followed up on the success of Luna 16 two months later: on 17 November 1970, the moon rover Lunokhod 1 became the first moving vehicle to land on the moon. Luna 17, which featured the same spherical and cylindrical containers found in its predecessors, flew to the grey moon with the rover on its back. The vehicle, measuring just over 1.5 metres in height and 1.35 metres in width, was more compact than the Apollo modules, which measured several metres in height and width. It looked like a hybrid of a tank and an insect. It was composed of an elliptical tub, resting on eight wheels, which were in turn attached to four axles. Various antennae were arranged around the body. Two camera eyes marked the front side of the zoomorphic structure, and a convex plate of armour rounded off the top of the body. The plate was kept stable by a ring of struts that remained visible and formed a kind of cornice. This 'roof' was attached to the body by a hinge and could be opened to reveal solar panels inside, which supplied power to the vehicle. Once it had landed on the moon, the rover transmitted photographs of the surroundings to the control centre on Earth for an hour before rolling down a ramp onto the lunar surface. Five people steered the vehicle via remote control, based on television images transmitted to Earth every twenty seconds. Although the rover's mission was initially planned to last just 30 days, it ended up roaming the lunar surface for almost a year, covering a distance of 10.5 kilometres. It transmitted over 20,000 images and over 200 panoramas to Earth. It also took a wide range of measurements of the conditions on the lunar surface. Lunokhod 1 used its polonium reserves to generate heat and withstand the moon's cold nights, each lasting 14 Earth days. The robot was decommissioned once the polonium reserves had been depleted. It was auctioned off to a private person in 1993, as was its carrier, Luna 17. However, both vehicles continued to remain on the moon, especially since the location of Lunokhod 1 could not be determined until 2010. Since then, its laser reflector, a contribution from French scientists, has been used again to measure the distance between the Earth and Moon.

Lunokhod 1 in a stamp of the German Democratic Republic (1970)

Moscow Polytech

For 11 months, the five-person crew guided their rover across the lunar surface on their computers, via remote control, with a three-second delay.

Avduyevski 1979

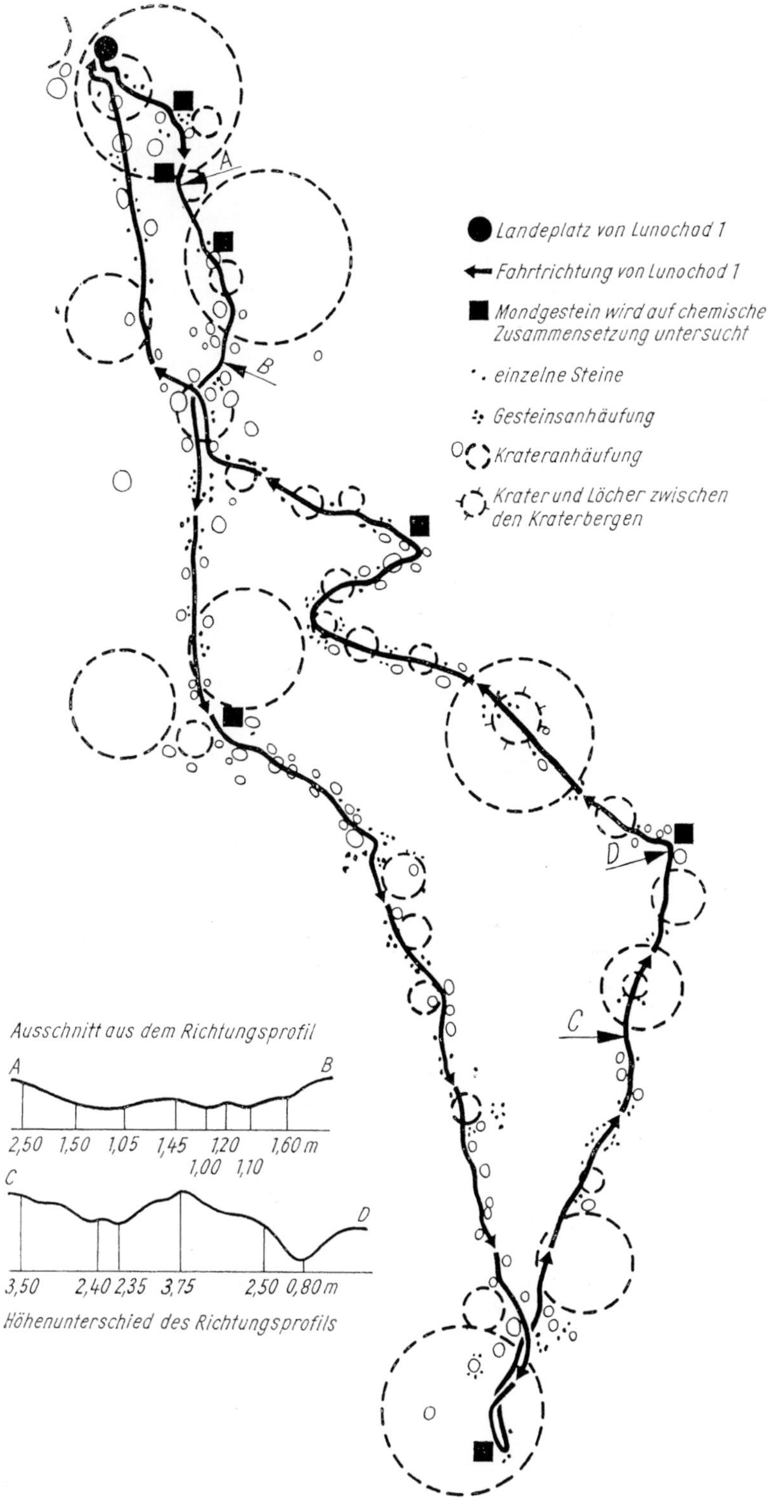

As the sole participant in the race, Lunokhod 1 set the record for the ten-kilometre distance with a finishing time of eleven months.

Saturn I was the first heavy-lift space launcher of the US that could transport payloads into Earth orbit.

NASA

The Saturn V rocket flew into Earth orbit twice, with Apollo 9 and Skylab, the first American space station. Otherwise, it flew exclusively to the moon.

NASA

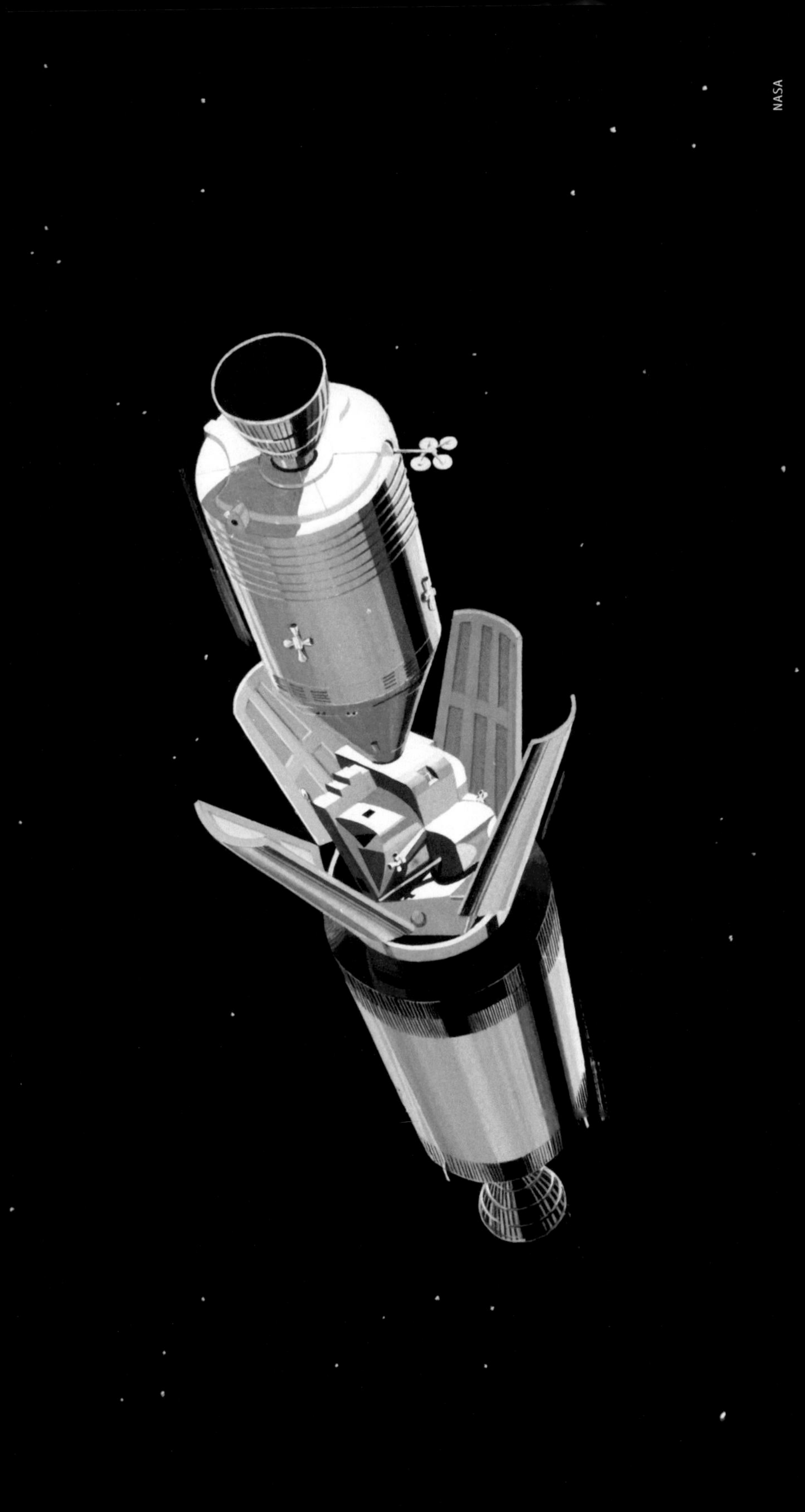

NASA

The command module and lander (ascent stage) dock together in space.

Apollo 13 S-IVB
US

CPM

Launch: 11 April 1970, 7:13 pm
Landing: 14 April 1970
Launch site: Cape Canaveral
Crash location: 21.64°S 165,36°W

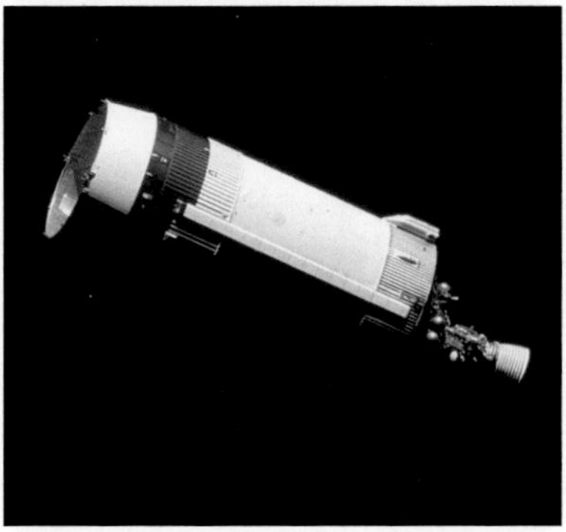
NASA

The Saturn V rocket, which carried all Apollo missions to the moon, was actually composed of three constituent rockets, called 'stages'. Each stage was jettisoned during the flight once it had used up its propellant and the spacecraft had reached the required altitude. Only the third, uppermost stage, which contained the Apollo modules, flew beyond Earth orbit. The landing module was housed in the belly of this stage, and the command module in the head. During the docking manoeuvre, when the two Apollo modules would be coupled together for the rest of the journey, the third rocket stage, S-IVB, would spring open its hatch to release the landing module. At this point, the command module would have already separated from the S-IVB and rotated by 180 degrees, so that its tip could connect with the fixing mechanism on the roof of the landing module. Afterwards, the last stage of the Saturn V rocket would usually fly off into a different lunar orbit in order to avoid colliding with Apollo. From Apollo 13 onwards, the third rocket stage was assigned an additional function: it was intentionally crashed into the moon, rather than being parked in a lunar orbit, while the Apollo attempted a soft landing. Earlier missions had brought seismological tools to the moon to measure the effects of meteorite impacts on the lunar surface and to study the conditions of the moon and its core. But these instruments still needed to be calibrated with a control variable. The S-IVB was used for this purpose from Apollo 13 onwards. NASA scientists determined the precise times and locations of the crash, and the seismological measurements became increasingly accurate.

NASA
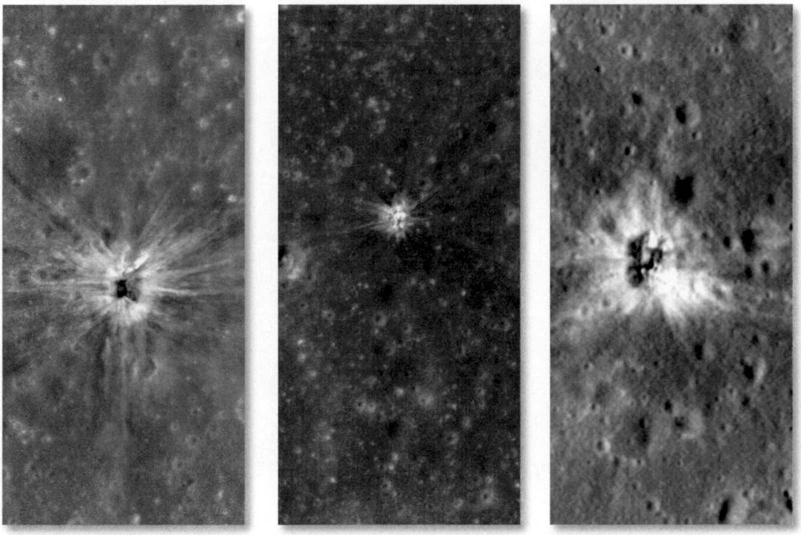

Aerial shots of the impact craters created by the S-IVB stage of Apollo 13, 14, and 16

Apollo 9 landing module in the S-IVB stage

NASA

NASA

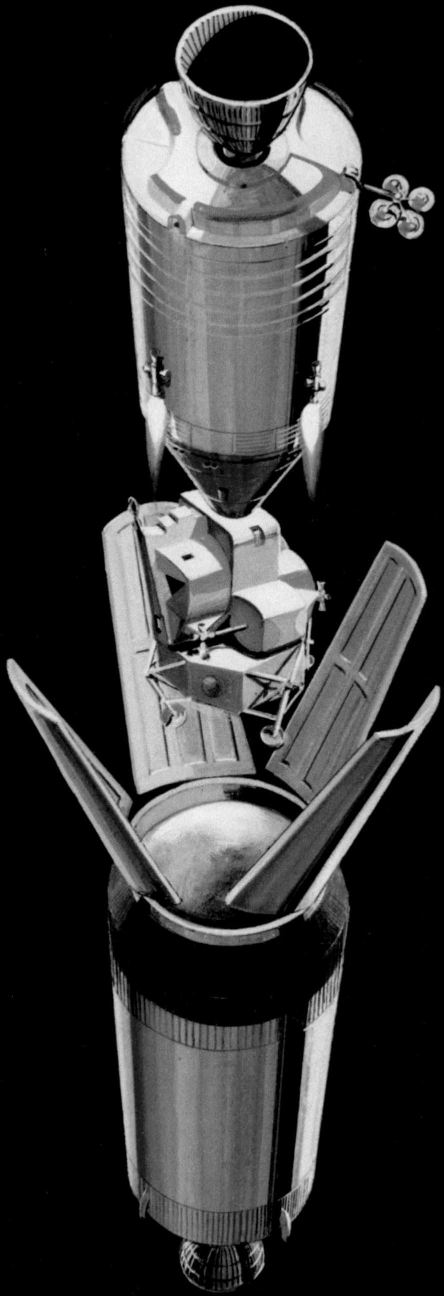

Command module and lander docked together

The return probe's ascent: illustration with the Earth in the background

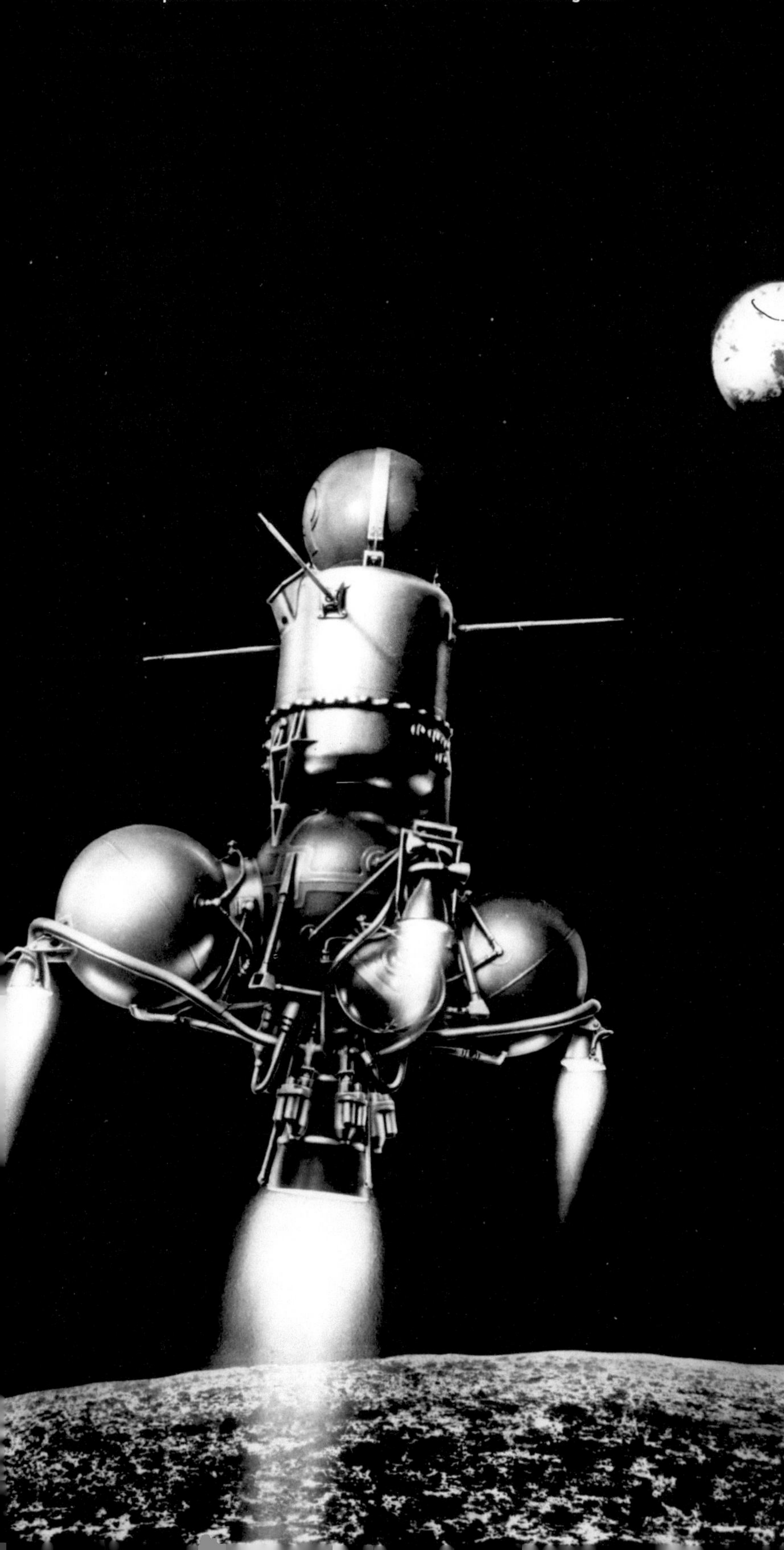

Moscow Polytech

Luna 18
Soviet Union

Launch: 2 September 1971, 1:40 pm
Landing: 11 September 1971, 7:48 am
Last contact: 11 Sept. 1971, 7:48 am
Launch site: Baikonur
Landing site: 3.56°N 56,5°O

Luna 20
Soviet Union

Launch: 14 February 1972, 3:27 am
Landing: 21 February 1972, 7:19 pm
Launch site: Baikonur
Landing site: 3.53°N 56.55°O

Luna 18 orbited the moon 54 times and completed 85 communication sessions with the command centre on Earth 85 times. After nine days in lunar orbit, the probe began its descent towards the moon, aiming to collect more soil samples from the lunar surface. But the probe, like Luna 15, crashed into a previously undetected mountain. The space engineers thereafter concluded that complex mountainous regions are unsuitable as landing sites. However, half a year later, Luna 20 landed just two kilometres away from the scene of the crash and carried out the mission its forerunner had failed to complete. The probe was initially parked in a medium Earth orbit, from which it was then sent to the moon. Luna 20, the eighth probe launched by the Soviet Union, brought back a second lunar soil sample to Earth. Although only 30 grams of material was delivered – far less than the amount from the first delivery – it had been collected from a different site and enabled new scientific findings. The first sample primarily consisted of basalt rocks, while the more recent sample contained a larger share of feldspar. Parts of the sample were shared with French and American scientists.

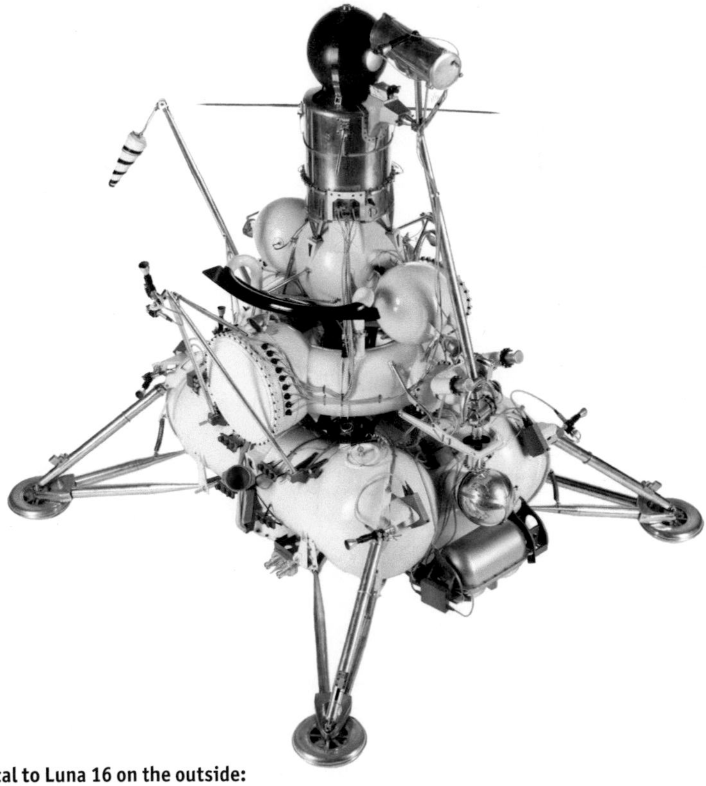

Identical to Luna 16 on the outside: Luna 18 and 20

Luna 19
Soviet Union

Launch: 28 September 1971, 10:00 am
Landing: crashed at unknown time
Last contact: 20 October 1972
Launch site: Baikonur

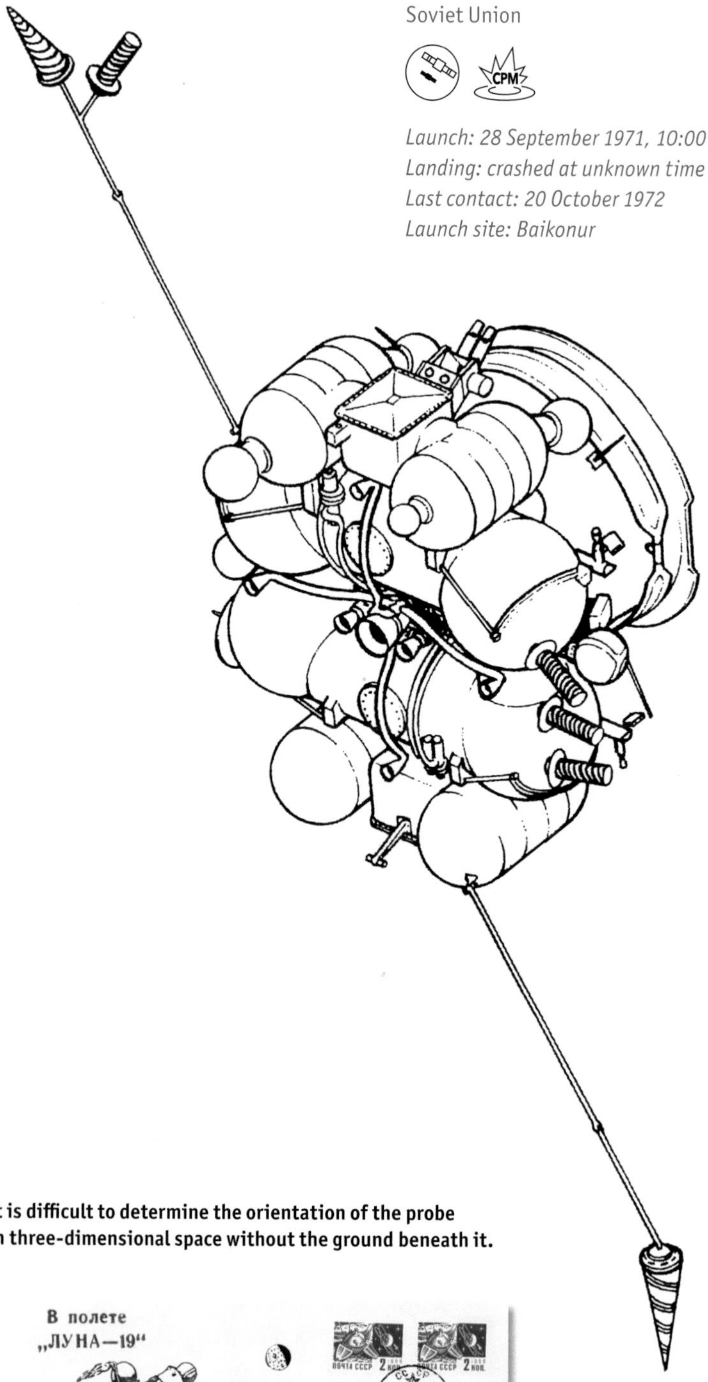

It is difficult to determine the orientation of the probe in three-dimensional space without the ground beneath it.

Postal cover commemorating the launch, stamped on 28 September 1971

Luna 22
Soviet Union

Launch: 29 May 1974, 8:57 am
Landing: crashed at unknown time
Launch site: Baikonur

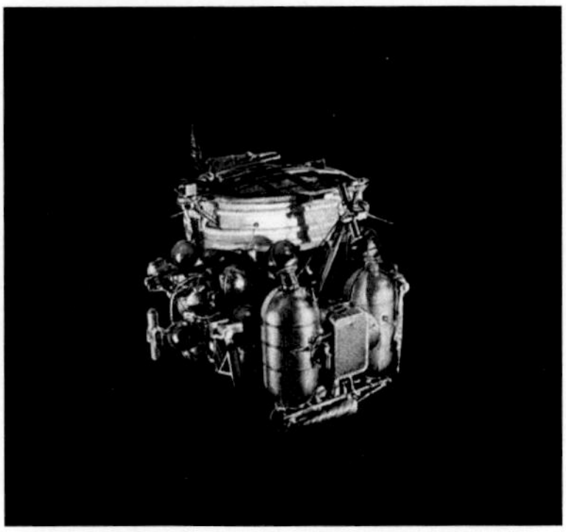

Luna 19 represented a new orbiter series (E-8-LS) of the Soviet uncrewed lunar programme. At first glance, the probe, together with the bus, looks like a complex constellation of roots. The open design recalls the architecture of the Centre Pompidou in Paris, whose construction began almost at the same time as the launch of the probe. A flat cylindrical body in the centre recalls the tub found in the Lunokhod. But below the body, instead of wheels, there are spheres of different sizes, much like the ones found in Luna 15. As such, the new Luna model can be seen as a hybrid of the tub-shaped lunar vehicle and the spherical sample-return probes. Most of the instruments and antennae are mounted on the roof or on the wide ring, which is accentuated by a kind of cornice. Luna 19, like the sample-return probes and lunar rover, was launched into space on the E-8 bus. It orbited the moon and proceeded to photograph the lunar surface – a task that had become a standard procedure for all lunar missions. Moreover, it investigated the gravitational field and mass of the moon and took radiation measurements. The probe was deactivated once the mission was completed. It thereafter crashed into the moon at an unknown time. The structurally identical Luna 22 suffered the same fate. Like its forerunner, Luna 22 took photographs of the moon, but it also investigated the chemical composition of the lunar surface. Moreover, it searched for a lunar magnetic field and observed meteoroid activity in the vicinity. The probe orbited the moon almost 4,000 times over a period of 1.5 years. It began its orbit at an altitude of over 220 kilometres before drawing ever closer to the moon. It was deactivated once it had used up all of its fuel. Powerless against the gravity of the moon, it crashed into an unknown location in the grey hilly landscape.

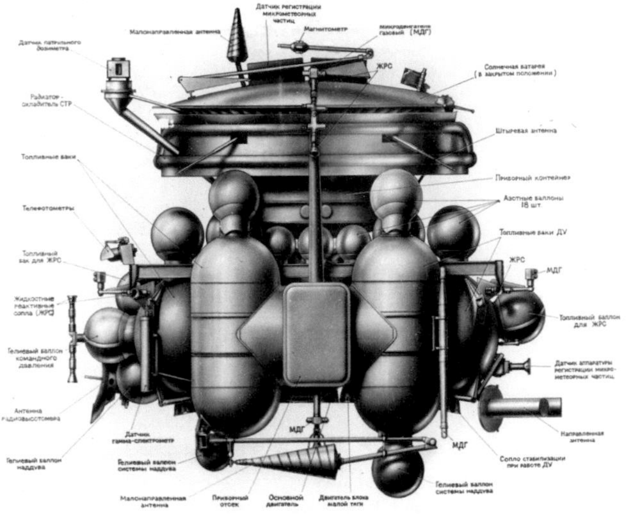

Moscow Polytech

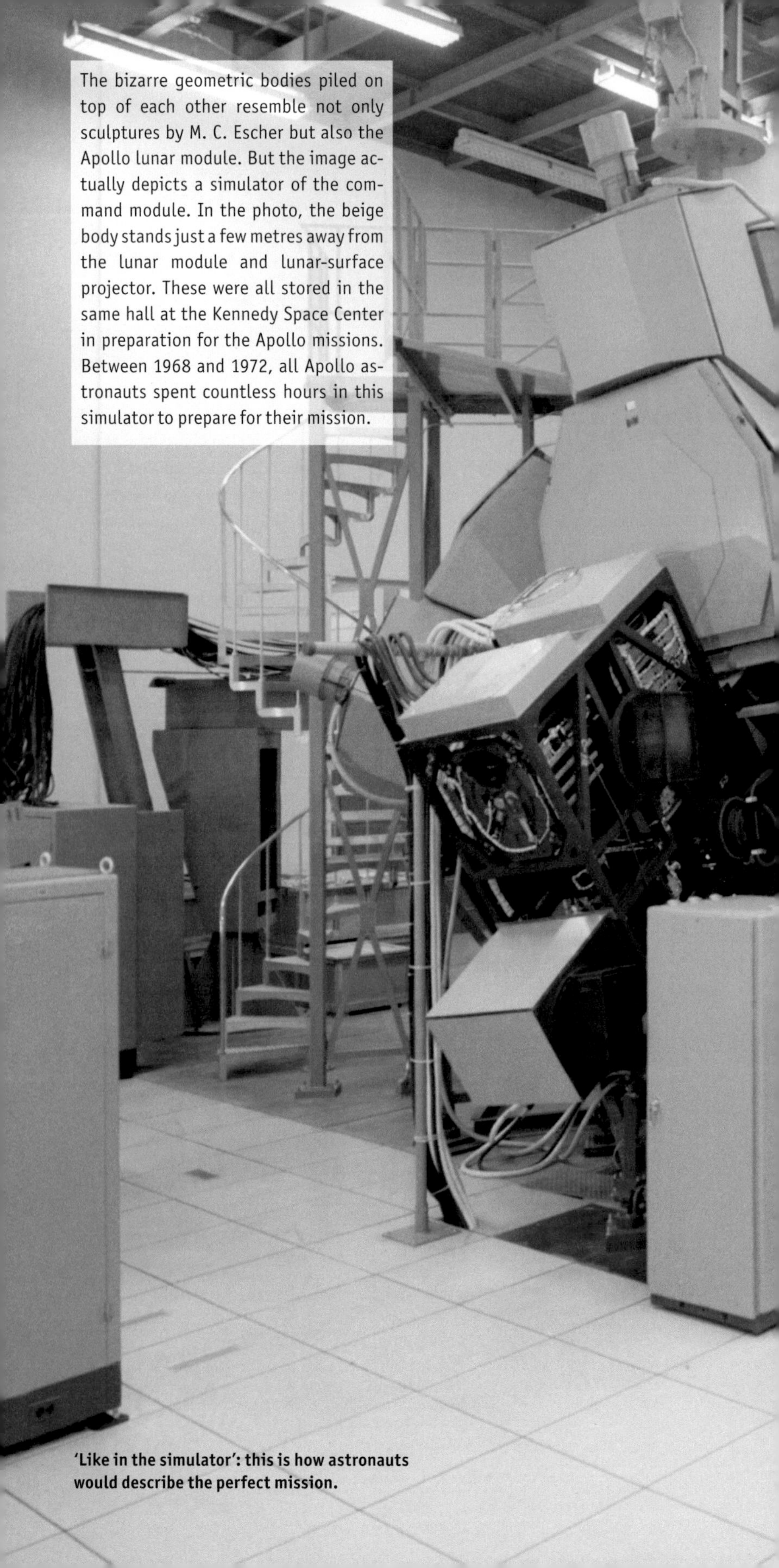

The bizarre geometric bodies piled on top of each other resemble not only sculptures by M. C. Escher but also the Apollo lunar module. But the image actually depicts a simulator of the command module. In the photo, the beige body stands just a few metres away from the lunar module and lunar-surface projector. These were all stored in the same hall at the Kennedy Space Center in preparation for the Apollo missions. Between 1968 and 1972, all Apollo astronauts spent countless hours in this simulator to prepare for their mission.

'Like in the simulator': this is how astronauts would describe the perfect mission.

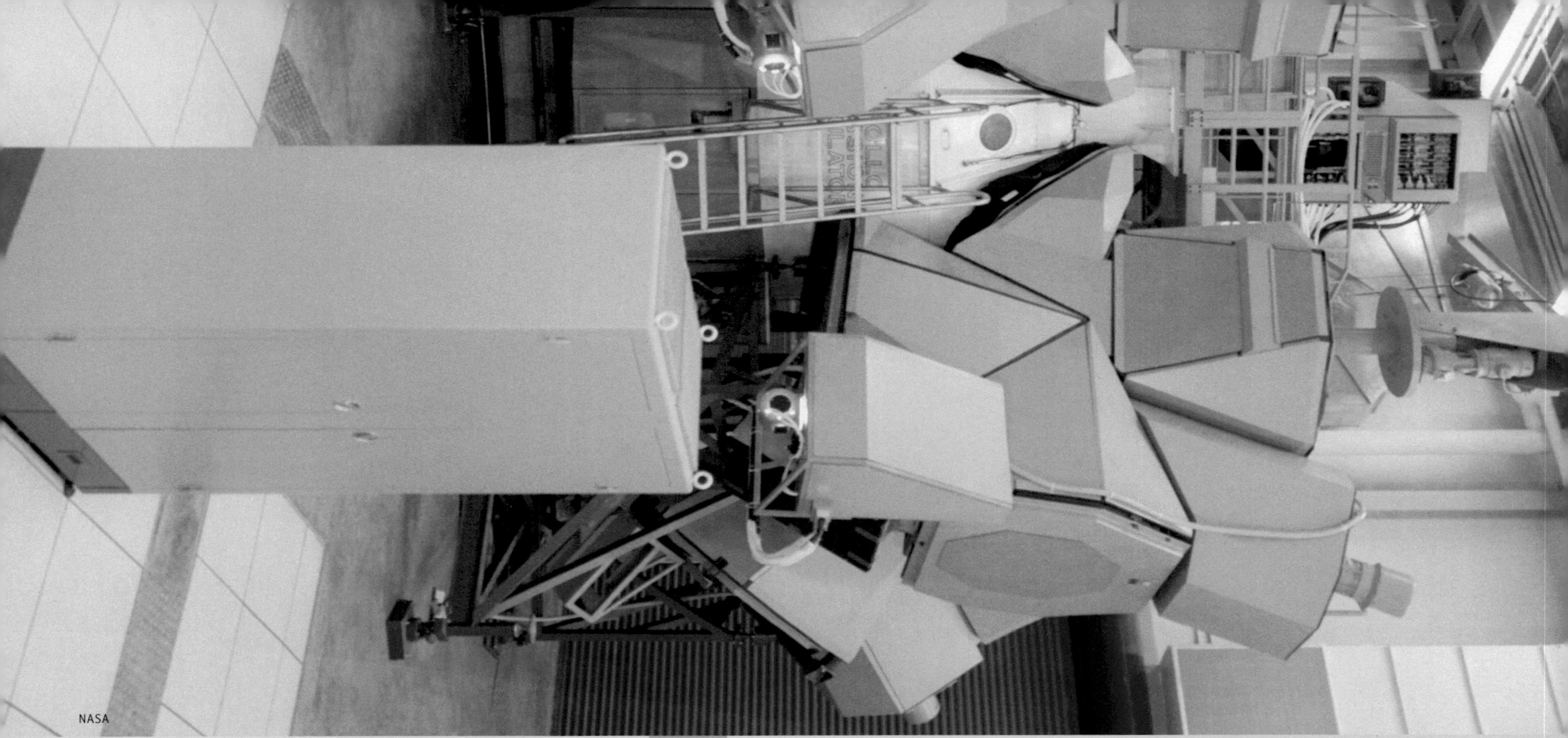

NASA

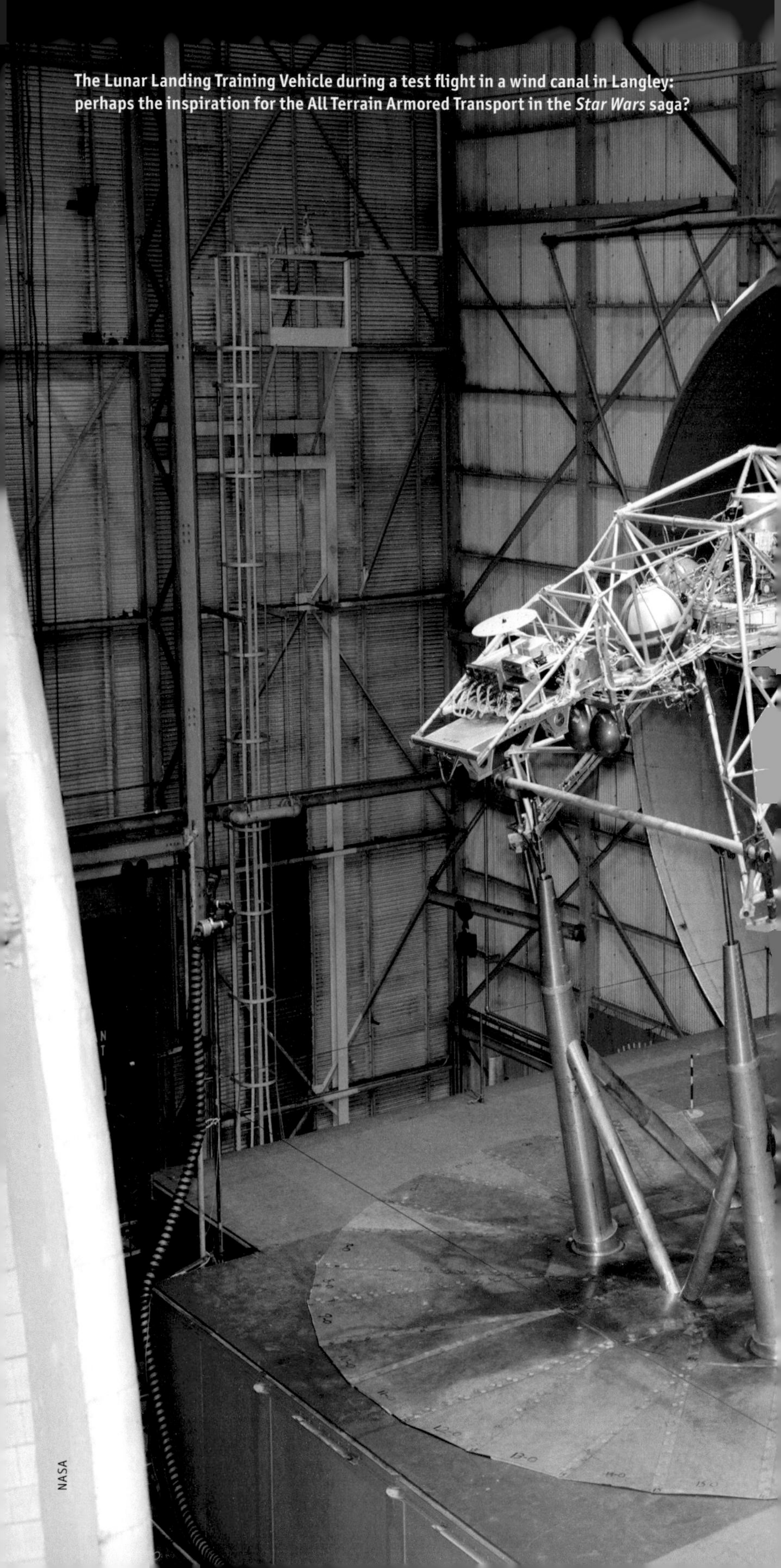

The Lunar Landing Training Vehicle during a test flight in a wind canal in Langley: perhaps the inspiration for the All Terrain Armored Transport in the *Star Wars* saga?

NASA

NASA

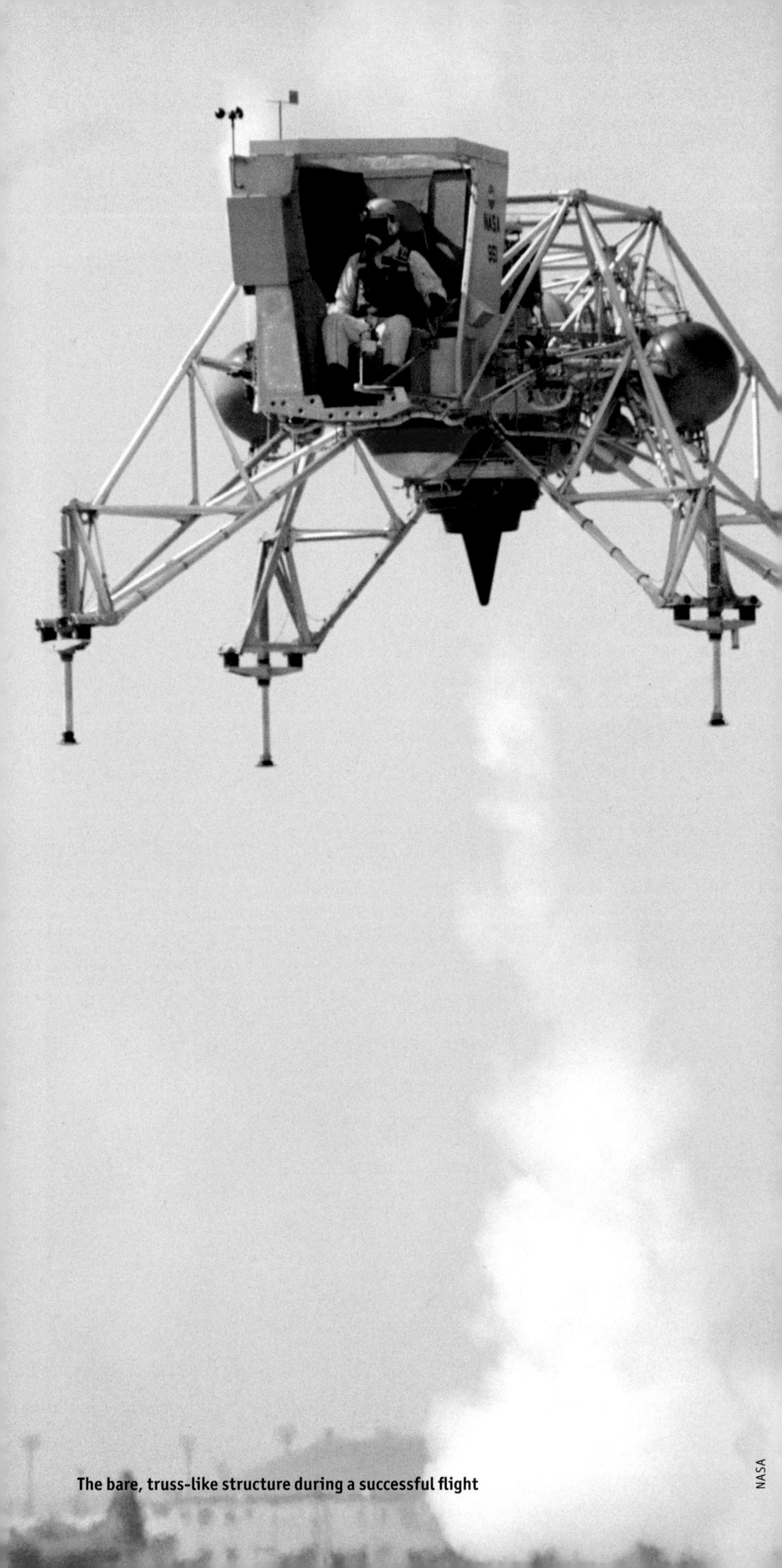

The bare, truss-like structure during a successful flight

NASA

Lunar Landing Training Vehicle

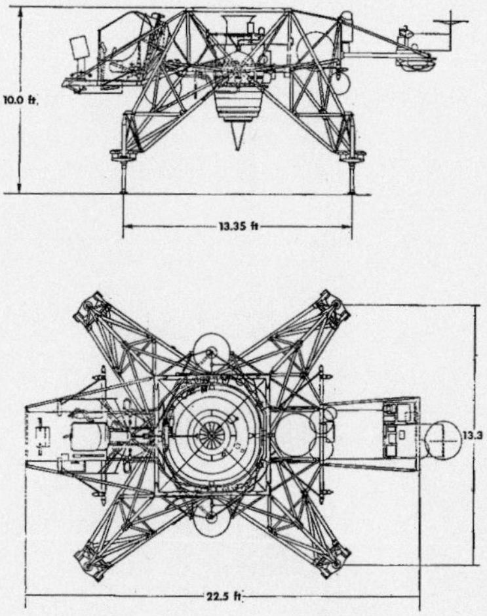

The 'flying bedstead', as it was called by astronauts, is one of the forgotten relics of the Apollo programme. The open structure, whose floor plan resembles that of the Eiffel Tower, simulated the behaviour of the Apollo landing module. Several engines applied a counter thrust to cancel 5/6 of the vehicle's weight. The pilot sat enthroned in a loggia-like booth to operate the bare structure, composed of aluminium trusses and fuel tanks. The vehicle enabled soon-to-be astronauts to practise the landing procedure under artificial conditions simulating the ones on the moon. The Lunar Landing Training Vehicle (LLTV) was virtually identical to the Lunar Landing Research Vehicle (LLRV). Neil Armstrong himself performed over 50 test landings in the LLTV, and he later attributed his successful moon landing to the practice flights he had completed in the simulator. Two units of the flying devices survive today. These can be visited either at the Johnson Space Center in Houston or the Air Force Flight Test Museum in Edwards, California.

NASA

Failed test flight: Neil Armstrong parachutes to the ground on 6 May 1968.

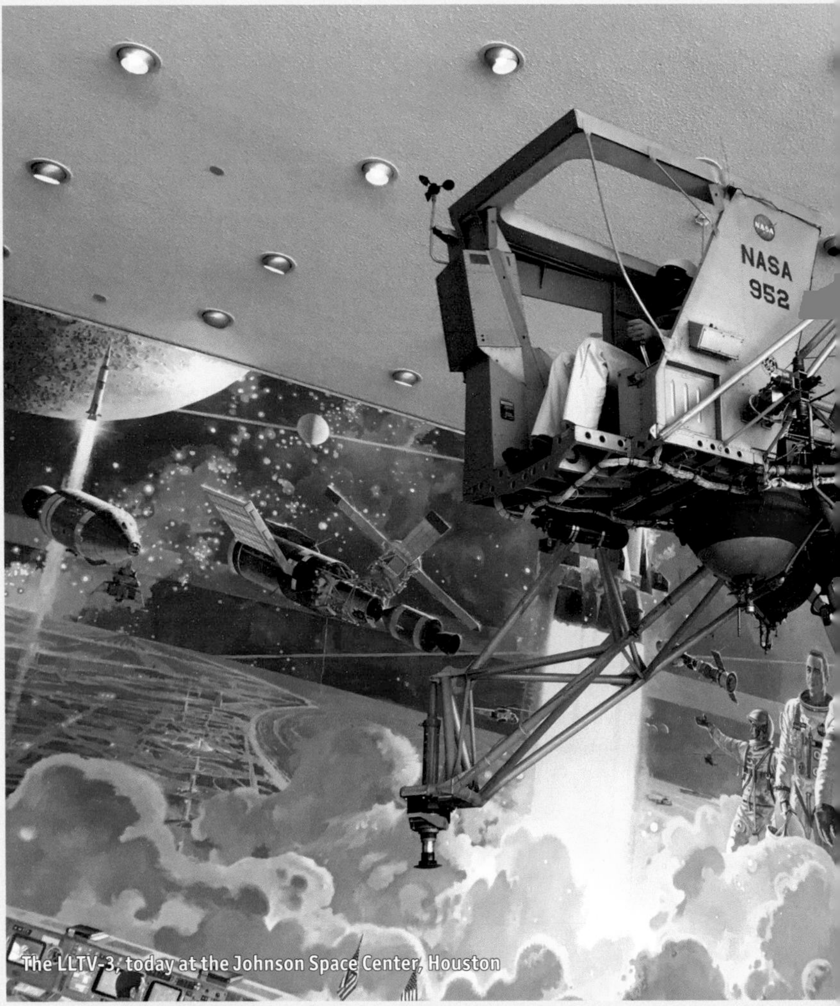

The LLTV-3, today at the Johnson Space Center, Houston

NASA

Could also serve as a roof truss of a pavilion: an earlier model of the LLTV

NASA

NASA

Neil Armstrong before the launch in LLTV-2 (1969)

Assembly of the Apollo lunar module (1968)

A naked frame of an Apollo lunar module can be found at the Cradle of Aviation Museum in Long Island.

Bianca Weeko Martin

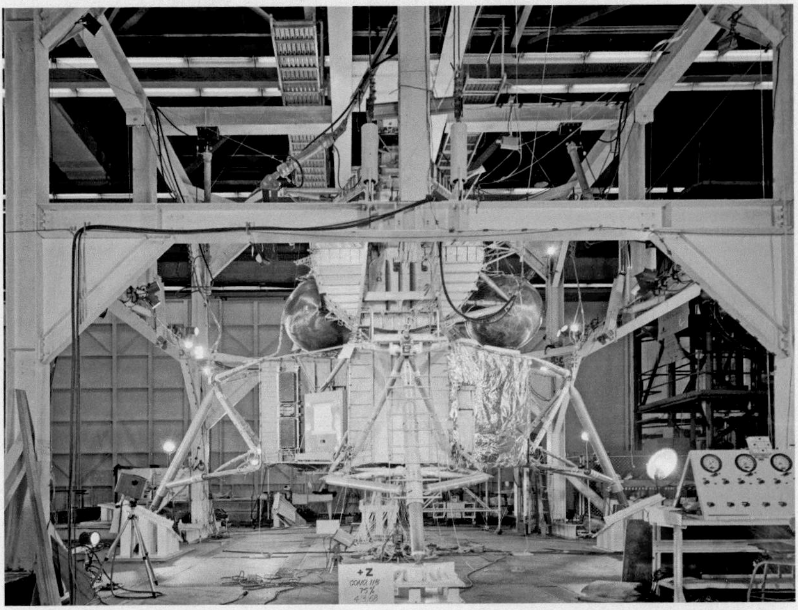

The Grumann LM, painted white, resembles the sci-fi spaceships of Hollywood movies.

Grumman LM

On 25 May 1961, US President John F. Kennedy proclaimed that the first person to walk on the moon should be an American. One year later, Grumman Aircraft Engineering Corporation, headquartered in Bethpage, New York, was commissioned to realise this vision. And it was the landing module Eagle, also known as LM 5 (Landing Module 5), that allowed Neil Armstrong and Buzz Aldrin to become the first humans to step onto the moon. There had been four forerunners to this model, including LM-4, also known as Snoopy, which became the first Grumman landing module to touch down on the moon, as part of the uncrewed Apollo 10 mission. The Grumman landing modules were the first spacecraft designed specifically for use in a vacuum and therefore had several original properties. Its idiosyncratic form is particularly striking. It comprises a hexagonal base and a thin barrel laid sideways on top, though the latter is hardly visible due to the large number of instruments mounted on it. Experiments for the extravehicular activities and the lunar rover were housed in the lower section of the landing module, and the ascent module was stored above. The outer shell is made of wafer-thin metal panels, embedded with struts that improve the structural stability and create a rhythmic appearance. Though the landing module itself weighed little, it was designed to carry a payload of up to five tonnes to the moon. The insulation cladding over the raw framework shimmers gold, silver, and black. Both parts of the landing module were left behind on the moon after the successful mission. The landing vehicle did not ascend at all after the landing, while the landing stage was jettisoned after the successful docking manoeuvre with the 'mothership'.

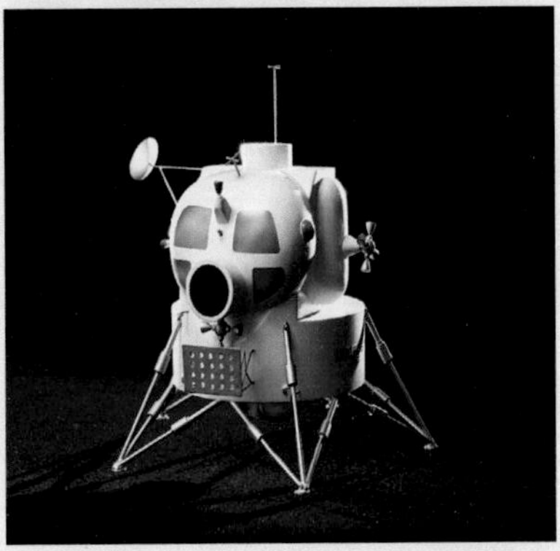

First model of the lunar module (1962)

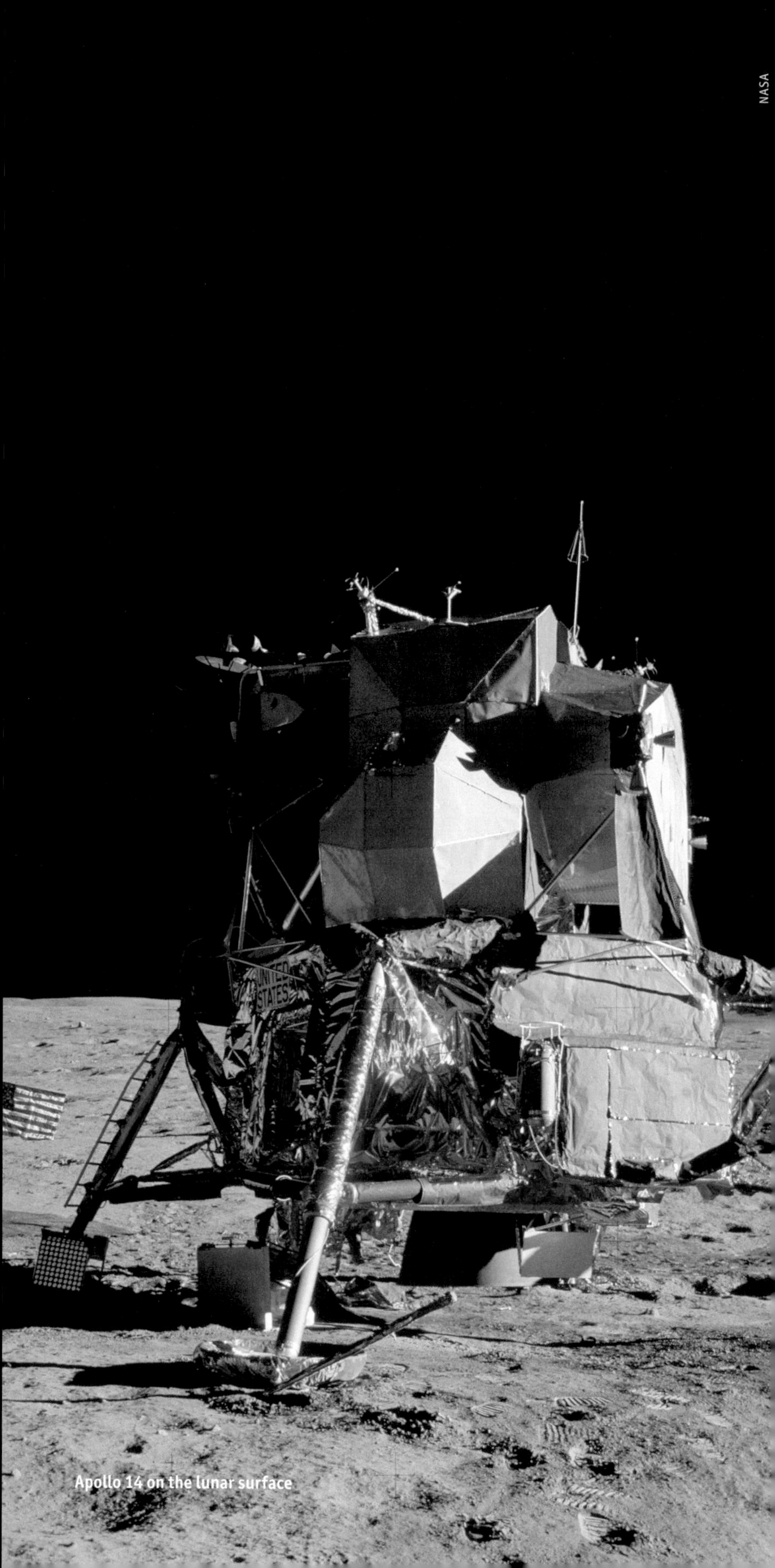

NASA

Apollo 14 on the lunar surface

Apollo 14
US

Launch: 31 January 1971, 9:03 pm
Landing: 5 February 1971, 9:18 am
Launch from moon: 6 Feb. 1971, 6:48 pm
Launch site: Cape Canaveral
Landing site: 3.65°S 7.47°W

NASA

The Apollo 14 mission ran into a complication during its journey: the docking manoeuvre, which was initiated after the carrier rocket's upper stage had been jettisoned, did not succeed right away. After this point, however, the three astronauts, Stuart Roosa, Alan Shepard, and Edgar Mitchell, were able to land without further problems. They carried out their extravehicular operations with a new kind of vehicle, the Modular Equipment Transporter (MET), a small trolley for transporting instruments and devices. The MET was the forerunner to the 'moon buggy' that would be used in the following Apollo missions. It enabled the astronauts to more quickly cover larger distances with their equipment. The equipment included a seismometer and antenna, for example, which the astronauts would use to set up an automated scientific laboratory, which would be controlled from Earth. The mission also entailed several experiments involving explosives. Mitchell used a device called the 'thumper', which detonated small explosions to create shock waves on the lunar surface. Moreover, the ascent module itself served an experimental tool: after the astronauts had transferred back to the command module, they jettisoned the no longer required ascent module. As such, it remains on the moon to this day, along with the rocket stages and landing module, in ruins.

NASA

Training in Earth's gravitational field: setting up the instruments of Apollo 14

The lunar module's foot, coated in a golden insulation foil

NASA

NASA

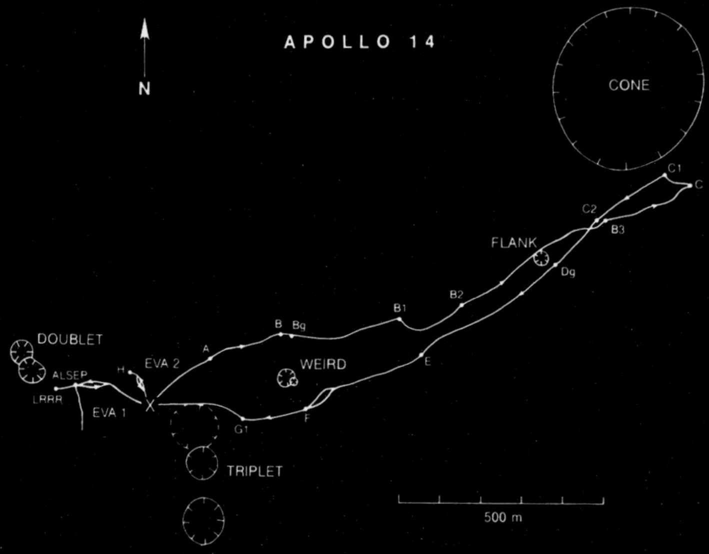

Foils, tubes, cables: the immediate vicinity of the Apollo landing site

NASA

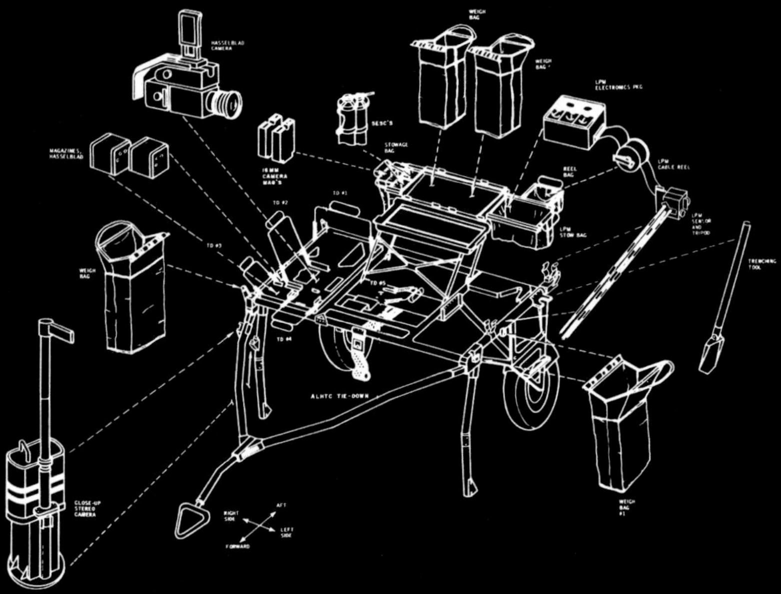

Technical drawing of the Modular Equipment Transporter (MET)

NASA

Lost in the endless grey landscape of the moon: Alan Shepard with the MET

NASA
SUN

Apollo 15
US

Launch: 26 July 1971, 1:34 pm
Landing: 30 July 1971, 5:11 pm
Launch from moon: 2 August 1971, 5:11 pm
Launch site: Cape Canaveral
Landing site: 26.13°N 3.63°O

NASA

Apollo 15 took place earlier than initially planned, since the total number of Apollo missions was cut down due to financial difficulties. The mission therefore entailed a much wider range of scientific instruments, and the astronauts – David Scott, Alfred Worden, and James Irwin – also received geological training before the launch. Apollo 15 saw the first use of the Lunar Roving Vehicle (LRV). This was a sparsely fitted vehicle with a simple frame, whose form recalled the very early days of the automobile. It was folded up and stored on the exterior of the landing module during the flight and was unfolded after the arrival on the moon. Its most striking features included the protective copper guards over the wheels and the umbrella-shaped directional antenna, which transmitted to Earth the first tracking shot ever captured on the moon. After a four-day flight, the landing module touched down near Hadley Rille, which was a challenging procedure due to the landing site's mountainous terrain. Scott and Irwin set up a wide range of experiments during three extravehicular operations lasting a total of eight and a half days. Meanwhile, Worden remained in lunar orbit, capturing images of the moon. Scott and Irwin travelled 1.5 km away from the landing site in the LRV, or moon buggy, to the St. George crater, where they set up an additional remote-controlled station for experiments. This made it possible to triangulate all the stations set up by Apollo astronauts and synchronise their data. Scott also carried out a simple experiment to verify a theory posited by Galileo: that the mass of an object does not influence its rate of fall in a vacuum. He dropped a hammer and feather at the same time, and confirmed that they hit the ground simultaneously. The solar wind collector served as an apposite backdrop to this experiment, standing at a slight angle like the Leaning Tower of Pisa, where Galileo had carried out his astronomical studies over 300 years earlier. Finally, the astronauts installed a commemorative plaque by the Belgian artist Paul Van Hoeydonck to honour the astronauts who had died in pursuit of space exploration. The work comprised a list of names of all astronauts who had lost their life during a mission as well as the *Fallen Astronaut*, a miniature figure measuring 8.5 cm (see p. 200). A subsatellite from the mission was left behind in lunar orbit, and it remained operational until a year after the crew returned back to Earth. The mission delivered 77 kilograms of lunar rock to Earth. But it also became embroiled in a scandal: the Apollo 15 postal covers incident. Without informing NASA, the astronauts had taken 600 postcards, postmarked with stamps before the launch, to the moon. After their return to Earth, the postcards were sold for several thousand dollars each. NASA investigated the affair, and one year later, it was clear that the three astronauts' careers were over. They had taken unauthorised objects with them on the mission and profited from their increase in value conferred by their sojourn on the moon.

Like the Leaning Tower of Pisa: solar-wind collector of the Apollo 15 mission

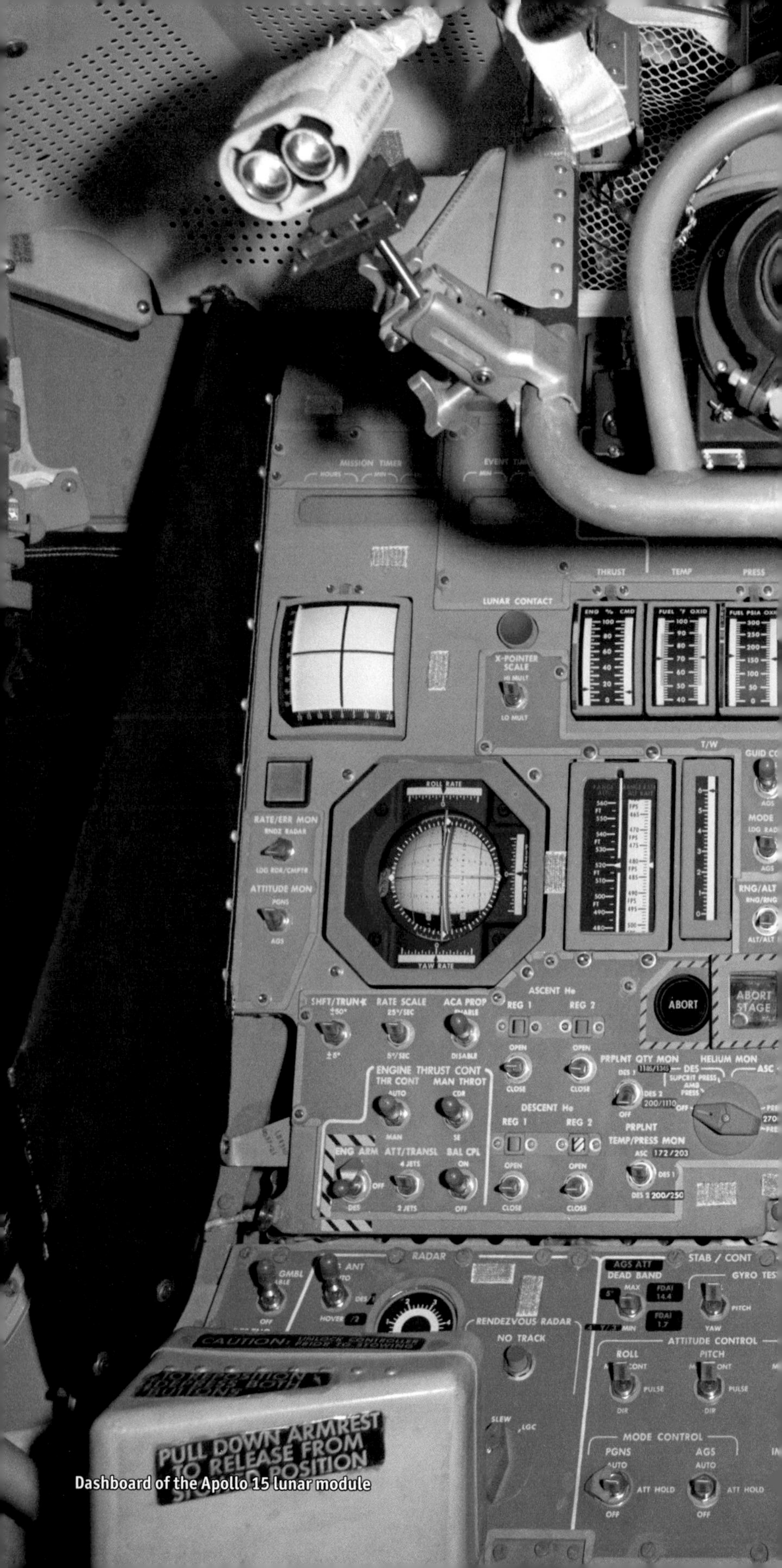

Dashboard of the Apollo 15 lunar module

PULL DOWN ARMREST
TO RELEASE FROM
STOWED POSITION
SYSTEM A
SYSTEM B
ASC FUEL
ASC OXID
LGC THRUSTER PAIR CMDS
QUAD 1
QUAD 4
QUAD 2
QUAD 3
MAIN SOV
SYS A
SYS B
ACA PROP
GLYCOL
QUANTITY
SUIT FAN
ECS
PRESS
PART PRESS CO2
RATE/ERR MON
ATTITUDE MON
EVENT TIMER
LIGHTING
SIDE PANELS
LUNAR CONTACT
TEMP MONITOR
RCS SYS A/B-2
HTR CONT
3
NASA

NASA

Hammer, feather, and the *Fallen Astronaut*: legacies of the Apollo 15 mission

Apollo 15 subsatellite and the astronaut crew in the lunar rover: photoshooting on Earth versus flight sequence in space

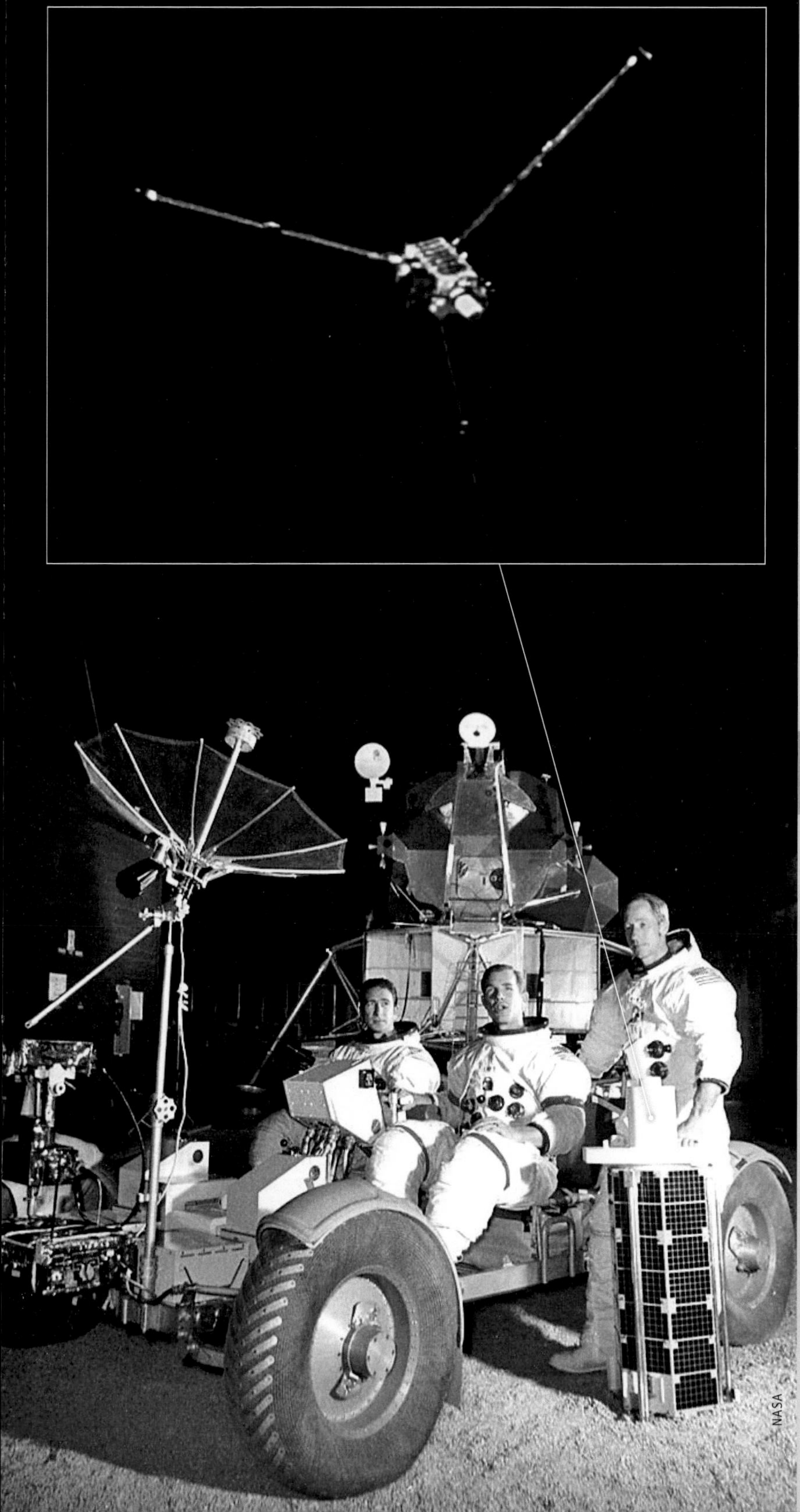

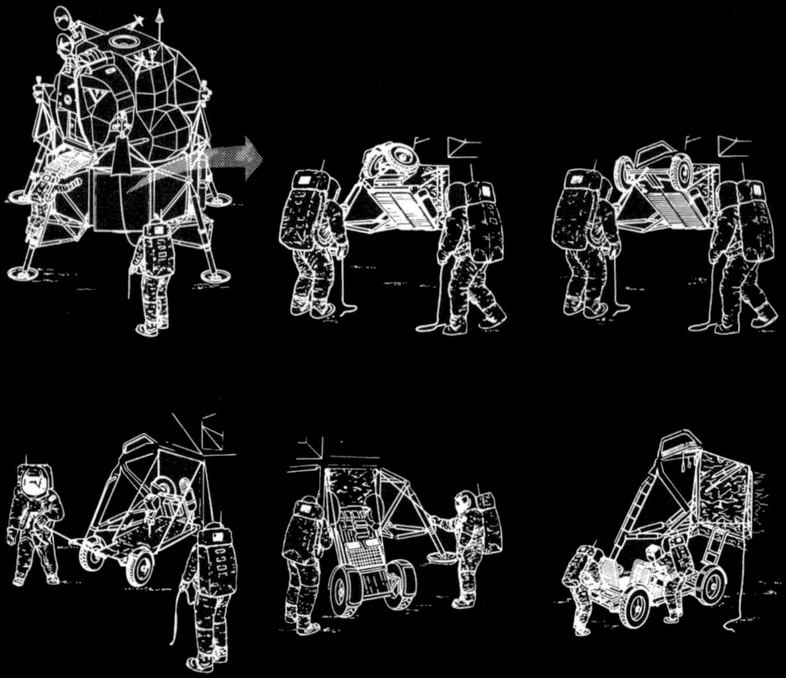

DIY instructions for unfolding a lunar vehicle

NASA

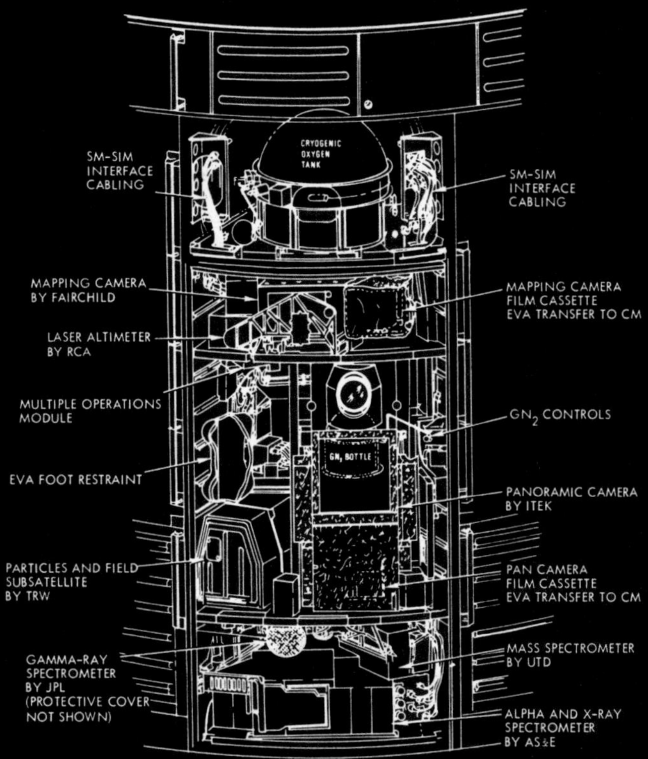
CRYOGENIC OXYGEN TANK
SM-SIM INTERFACE CABLING
SM-SIM INTERFACE CABLING
MAPPING CAMERA BY FAIRCHILD
MAPPING CAMERA FILM CASSETTE EVA TRANSFER TO CM
LASER ALTIMETER BY RCA
MULTIPLE OPERATIONS MODULE
GN2 CONTROLS
GN2 BOTTLE
EVA FOOT RESTRAINT
PANORAMIC CAMERA BY ITEK
PARTICLES AND FIELD SUBSATELLITE BY TRW
PAN CAMERA FILM CASSETTE EVA TRANSFER TO CM
GAMMA-RAY SPECTROMETER BY JPL (PROTECTIVE COVER NOT SHOWN)
MASS SPECTROMETER BY UTD
ALPHA AND X-RAY SPECTROMETER BY AS&E

Inside the Apollo 15 command module

Lunar Roving Vehicle

NASA

The Lunar Roving Vehicle (LRV) was deployed on three of the six lunar-landing missions carried out as part of the Apollo programme. The American astronauts were able to cover ever-greater distances in this electrically powered vehicle during Apollo 15, 16, and 17. The rover, which performed virtually without any malfunctions, had only taken 17 months to develop. During Apollo 15, the LRV covered a distance of almost 28 kilometres in just over three hours. Two missions later, it covered 39.9 kilometres in almost four and a half hours. The astronauts covered this distance over a period of three mission days, carrying out experiments and taking samples at locations over 7.5 kilometres away from the landing site. The rover, measuring 3.1 metres by 2.3 metres, was primarily composed of welded aluminium alloy tubes and weighed 35 kg on the moon. To use the vehicle, the astronauts merely had to unfold the rover, mounted inside the landing module with tension belts and cables, and unpack the two seats, made of nylon straps. The two-seater on four wheels was controlled with a joystick, mounted on a T-shaped centre console. The vehicle had 0.25 horsepower per wheel, allowing Eugene Cernan to ramp up the speed to a breakneck 18 km/hr and thereby set the unofficial world record for the highest speed in a vehicle on the moon.

NASA

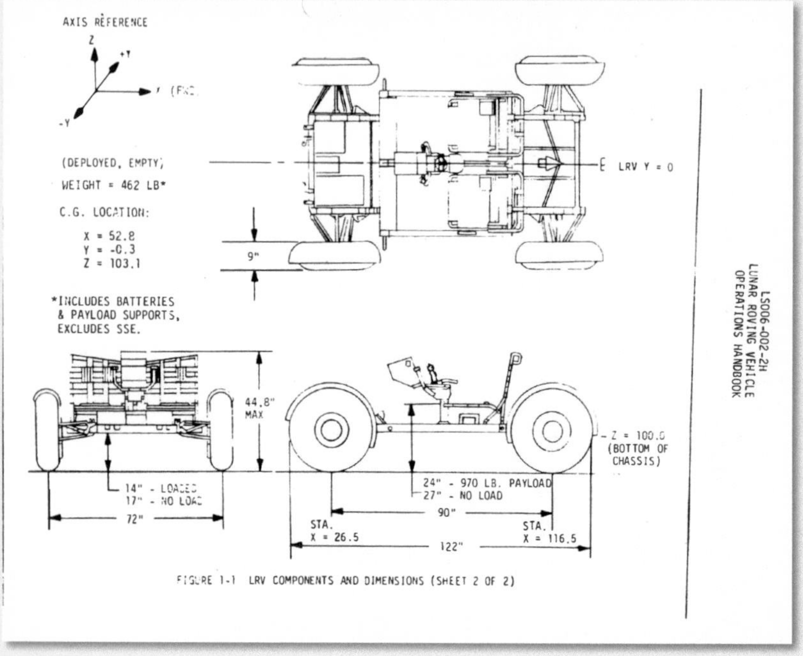

Top: Apollo 15 crew with Corvettes, produced by General Motors (GM), behind the Test Moon Rover, also developed by GM. The wondrous vehicle bears no resemblance to the streamlined sports cars. And why should it? There is no headwind on the moon!

NASA

NASA

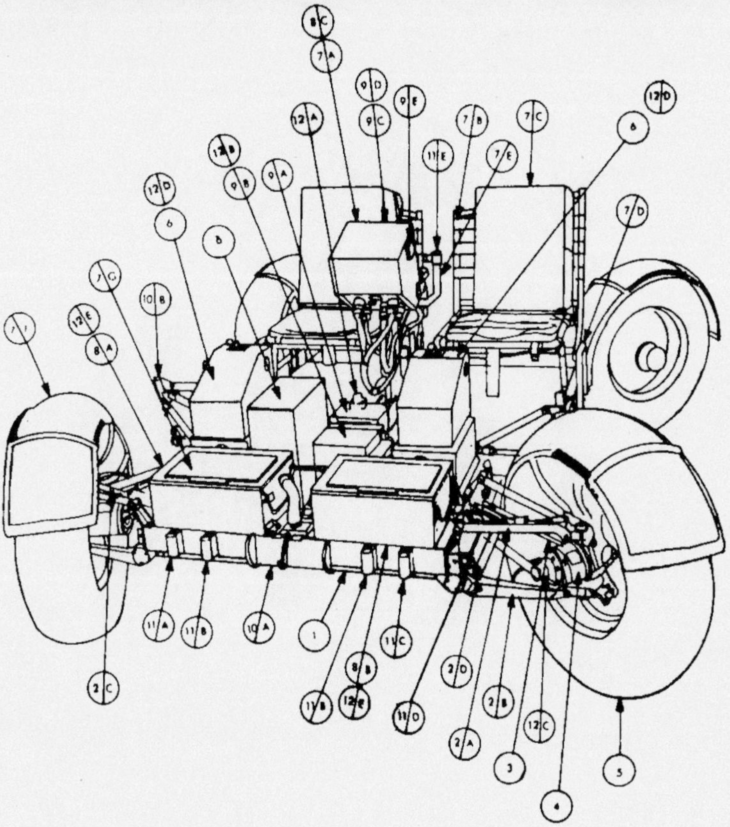

Plain appearance, technical marvel: the Lunar Roving Vehicle (LRV)

The final remaining part of Apollo 16 – the reetry capsule – shortly before the splashdown

Apollo 16
US

NASA

Launch: 16 April 1972, 5:54 pm
Landing: 21 April 1972, 2:23 am
Launch from moon: 24 April 1972, 1:25 am
Launch site: Cape Canaveral
Landing site: 8.9730°S 15.5002°O

Apollo 16 was NASA's sixth crewed mission to the moon. The early stages of the mission were marked by several difficulties. Not only was the launch of the Saturn V rocket delayed due to maintenance works, the navigation system also failed after three days in space, forcing the crew – Ken Mattingly, John Young, and Charles Duke – to navigate the space capsule using a sextant. Moreover, after the separation of the two Apollo modules, the main engine's swivel drive failed. Nonetheless, the astronauts successfully achieved a moon landing and were able to sleep in their hammocks, which had been fitted into the landing modules after the Apollo 11 mission. The mission produced the first ultraviolet images ever captured in space, which were analysed back on Earth. After the landing, the astronauts covered a distance of 25 kilometres on the moon buggy and collected 95 kilograms of lunar rocks, including a rock weighing 11 kilograms, the heaviest collected as part of the Apollo programme. These samples disproved the hypothesis that the lunar surface around the landing site had been formed by volcanic eruptions. The ascent module was launched successfully from the moon, in spite of a minor incident whereby Young accidentally snapped a cable from its frame. The rover's television camera filmed the return launch, which proceeded more or less seamlessly, although the astronauts, after transferring to the CSM, failed to initiate the ascent module's intentional de-orbit. The ascent module remained in lunar orbit for another year before crashing into the moon. The final stages of the mission included the release of a small satellite during the return flight. Moreover, Mattingly briefly left the command module to carry out an additional extravehicular activity: he retrieved films from a compartment in the spaceship, which was about to be jettisoned before the splashdown.

NASA

The astronaut jumps, almost clumsily, into the air to salute the viewer.

NASA

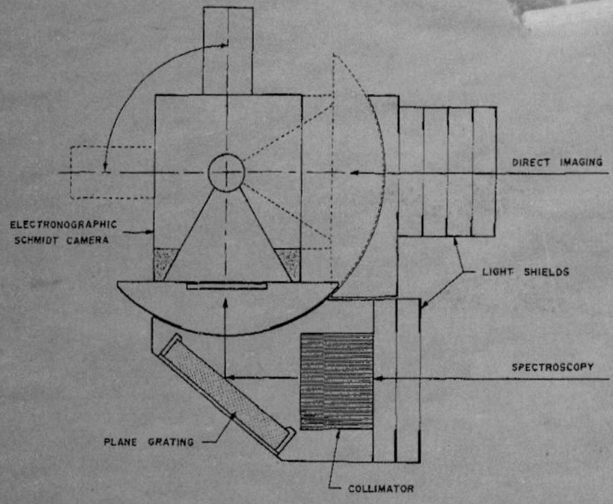

A golden camera, invented by George Carruthers (right), was taken on the Apollo 16 mission. It was an ultraviolet camera with a spectrograph and became the world's first lunar observatorium. To this day, it remains standing in the shadows of the Apollo 16 lunar module.

NASA

The Apollo 16 ascent stage's cladding crumpled like paper during the launch from the lunar surface (bottom). The walls did not have a uniform thickness. Usually, walls in a vacuum can be as thin as a drinks can, and lunar missions generally tried to minimise weight as far as possible. Evidently, the Apollo 16 engineers had pushed the principle of reduction to the very limit.

NASA

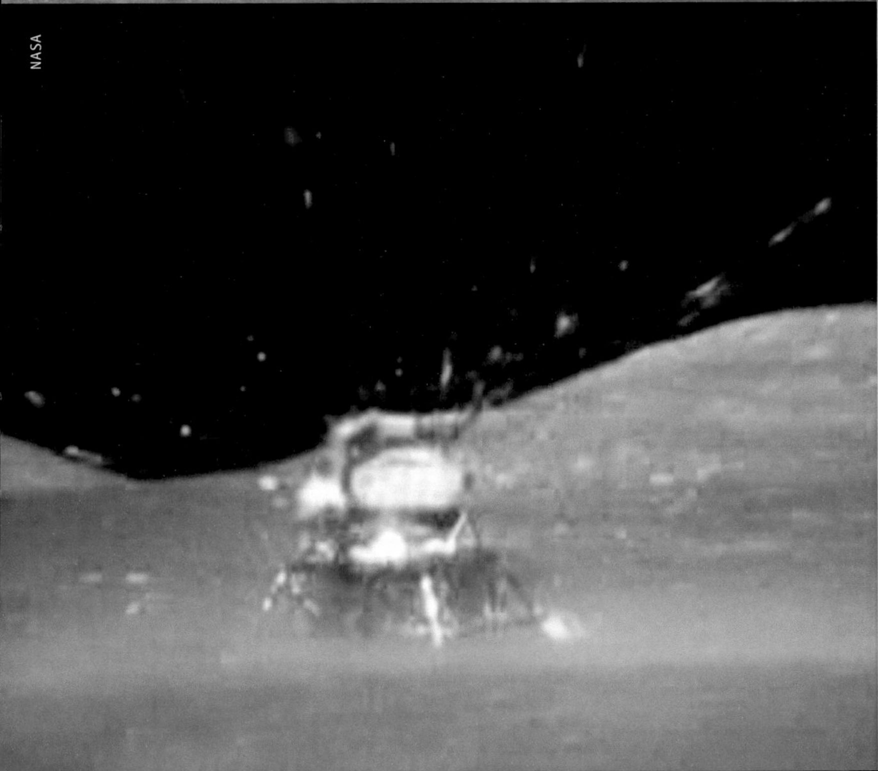

NASA

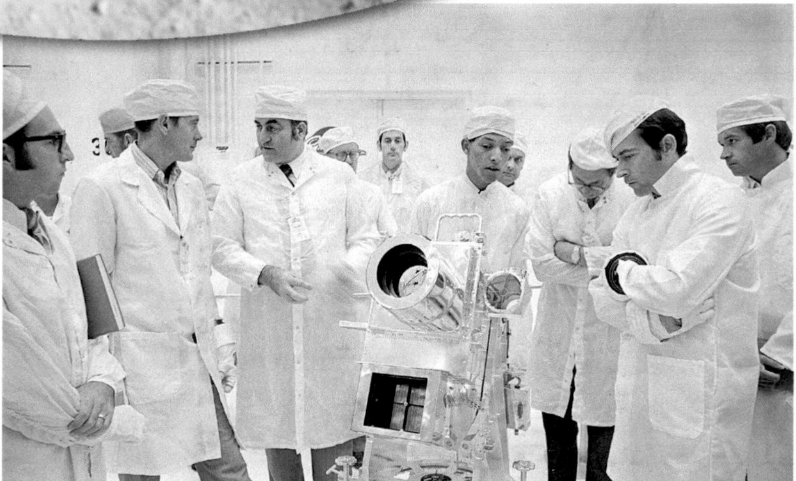

George Carruthers (centre) explaining his camera to the Apollo 16 crew (1971)

NASA

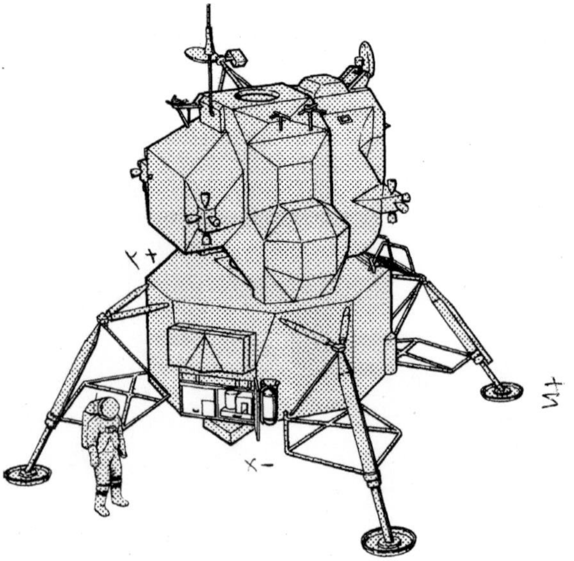

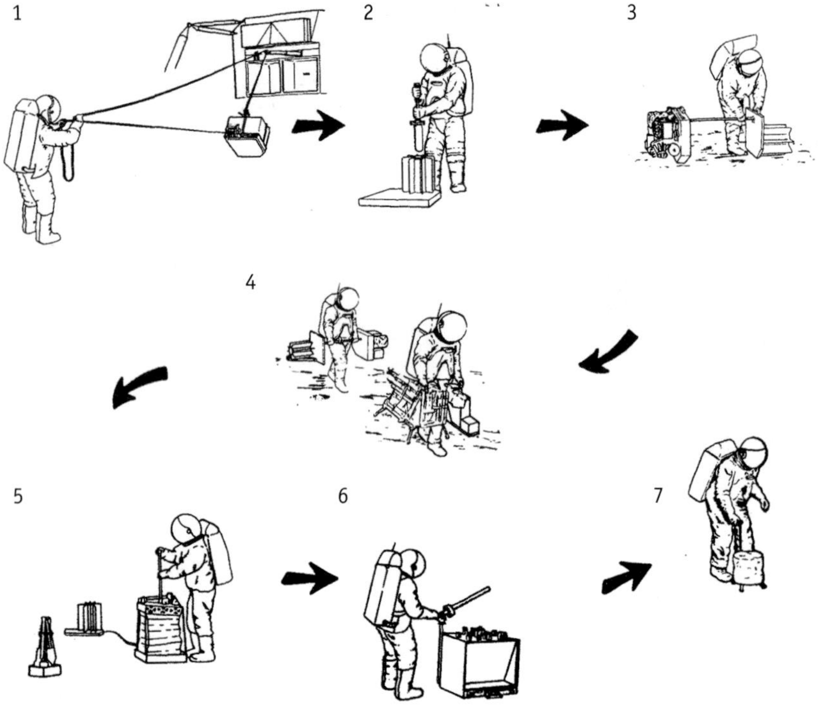

The ALSEP (Apollo Lunar Surface Experiments Package) was a set of instruments for long-term scientific experiments on the moon. The ALSEP package was assembled from a catalogue of experiments, with a different combination assembled for each mission. The collected data was transmitted to Earth. The package was stored in a compact storage area in the landing module. After each moon landing, the astronauts unpacked the experiments according to the following procedure: (1) First, they unloaded the heavy package from the landing module. (2) Then they charged the batteries with plutonium and (3) mounted most of the experiments on a carrying bar. (4) They carried the package to a location around 100 metres away from the landing site in order to (5) set up the first Central Station and connect it to the batteries. (6) Then they manually oriented the antenna towards the Earth and connected the rest of the experiments to the power circuit.

NASA

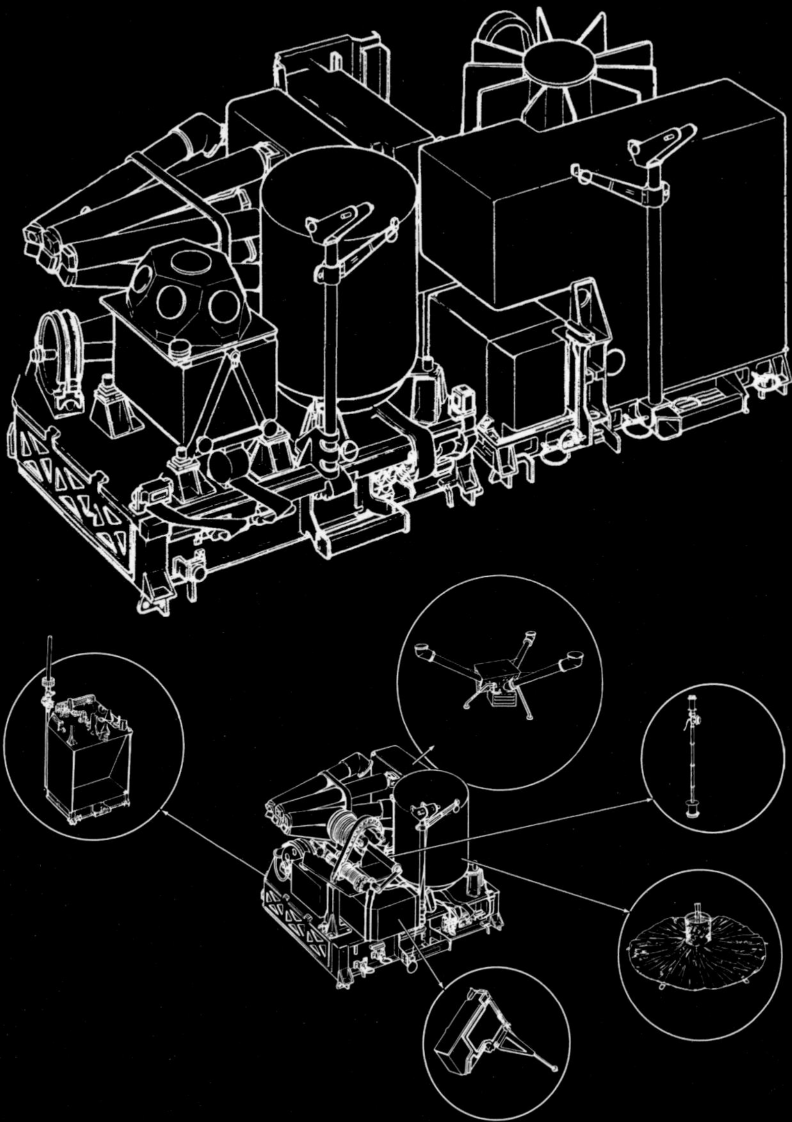

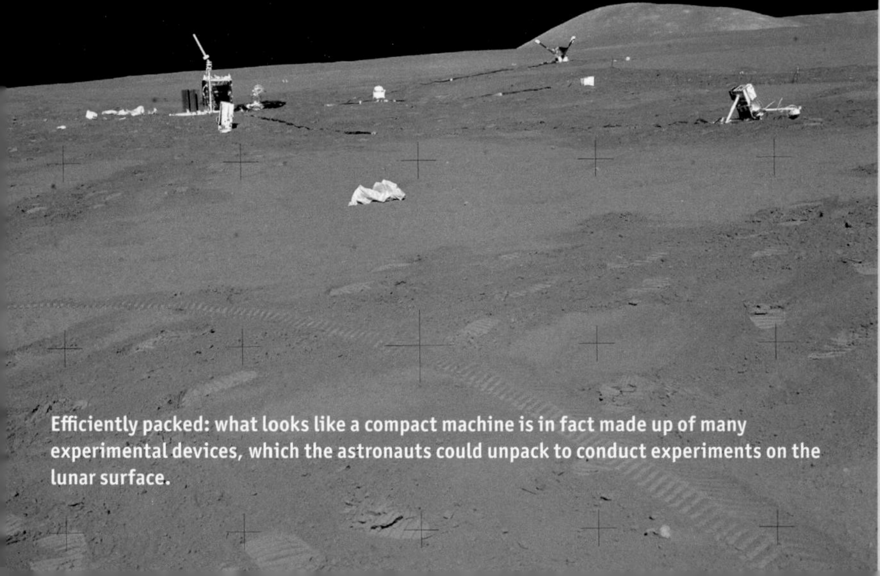

Efficiently packed: what looks like a compact machine is in fact made up of many experimental devices, which the astronauts could unpack to conduct experiments on the lunar surface.

1

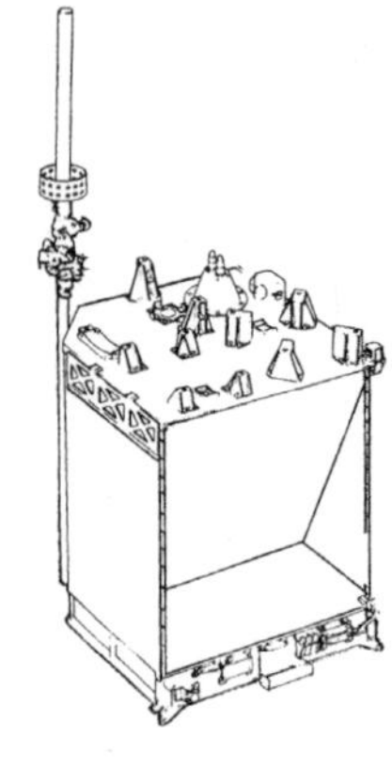

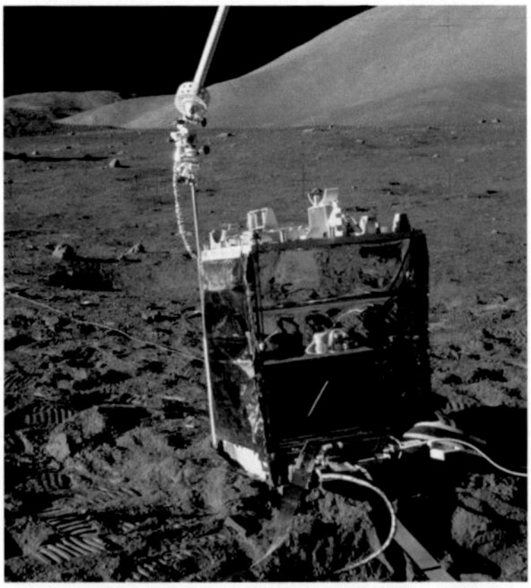

NASA

2

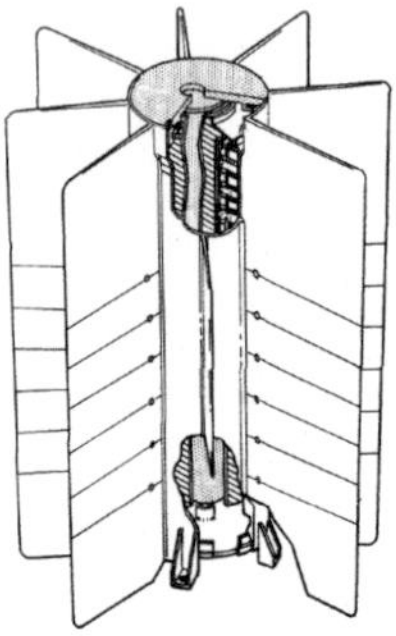

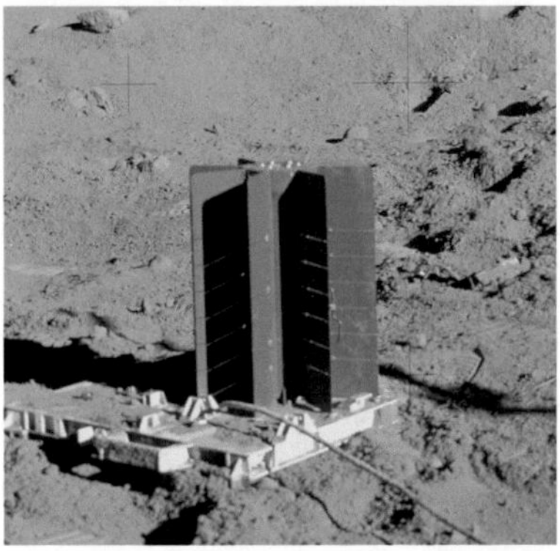

3

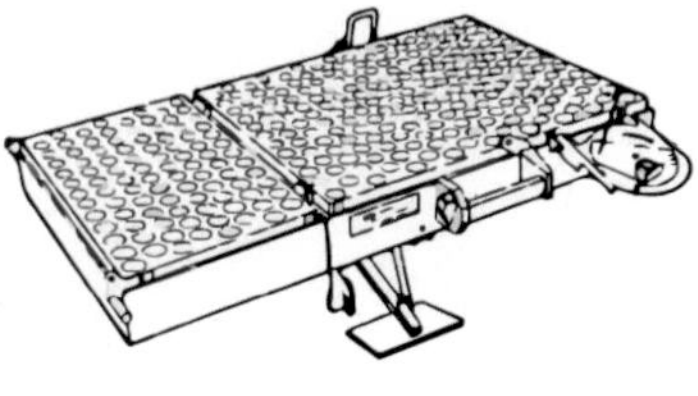

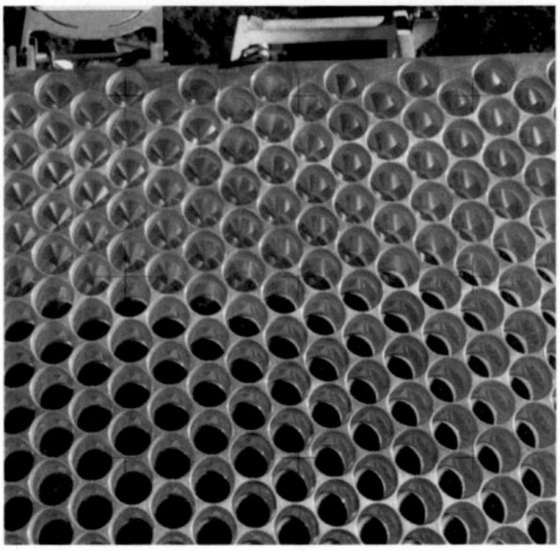

The ALSEP was part of every mission from Apollo 12 onwards. The EASEP (Early Apollo Scientific Experiments Package), a prototype with a smaller range of experiments, had been taken on the Apollo 11 mission. The basic package entailed the same instruments. (1) The cabinet-like Central Station was the centrepiece, supplying power to the other instruments and establishing a connection to the control centre on Earth. (2) The radioisotope thermoelectric generator (RTG) converted the heat released by the plutonium's decay into electricity. For safety reasons, the RTG was not activated until after the arrival on the moon. It had the form of a water wheel and generated 70 W of power for the experiments. (3) The laser retroreflector, transported to a pre-determined location by the rover, enabled a very precise measurement of the distance between the Earth and moon. The reflector shown above, which was deployed during Apollo 15,

NASA

4

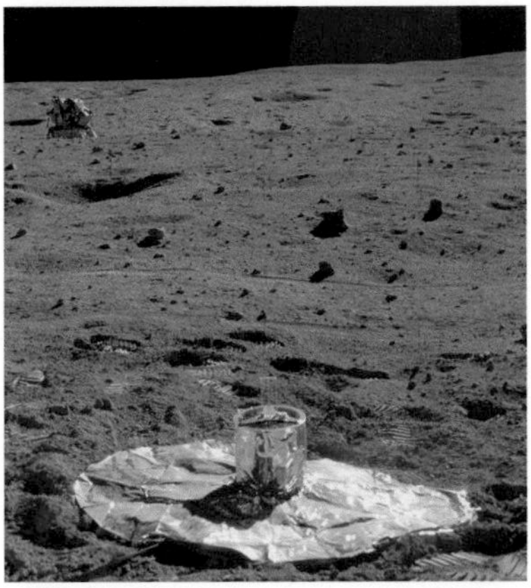

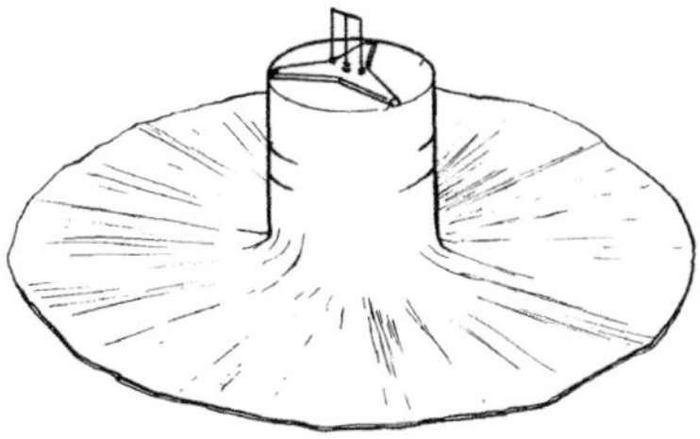

5

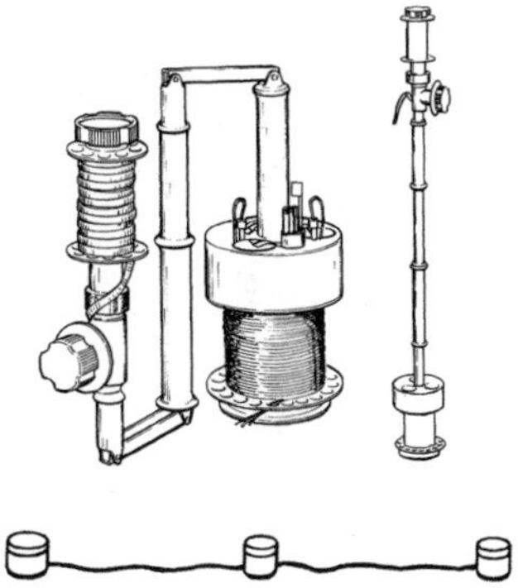

6

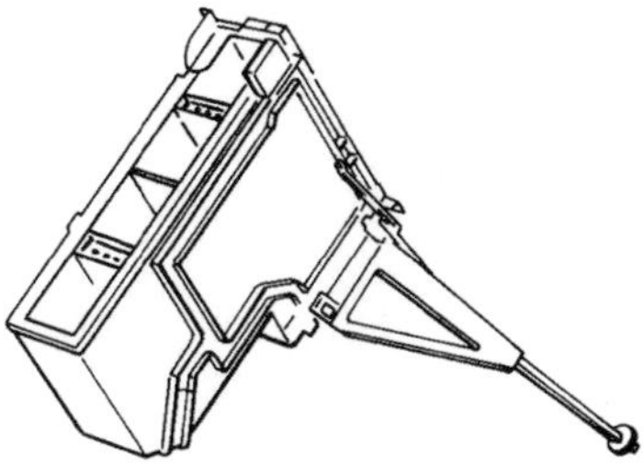

had 300 prisms, though it ultimately produced the same results as its smaller forerunners. All crewed Apollo missions entailed the Passive Seismic Experiment (PSE). (4) The device, resembling a hat with a wide brim made of silver foil, contained a seismic sensor under it. Several such sensors were used to measure natural and artificially caused vibrations on the lunar surface and to obtain knowledge about the conditions deep underground. During Apollo 13, the artificial shocks were created by the launch of the third rocket stage, S-IVB, while during Apollo 14 and 16, they were created by (5) the 'thumper', which detonated mortar shells to cause small shocks. Three geophones measured and recorded the resulting pressure waves. (6) Heavier artillery was used during Apollo 16 as part of a seismic experiment. Three mortar shells were used to hurl explosives and geophones by distances of up to 900 m once the crew had left the moon.

7

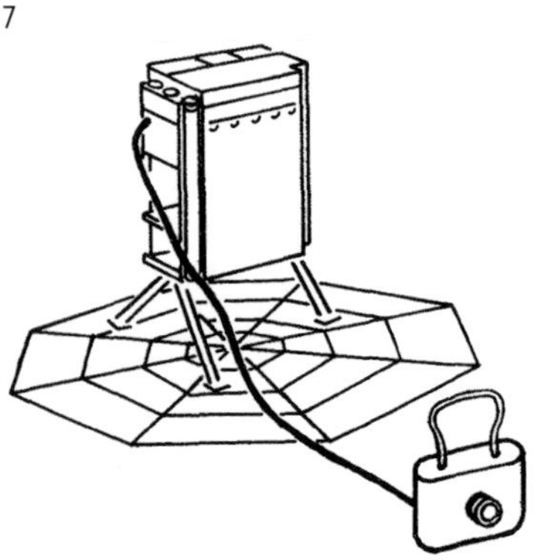

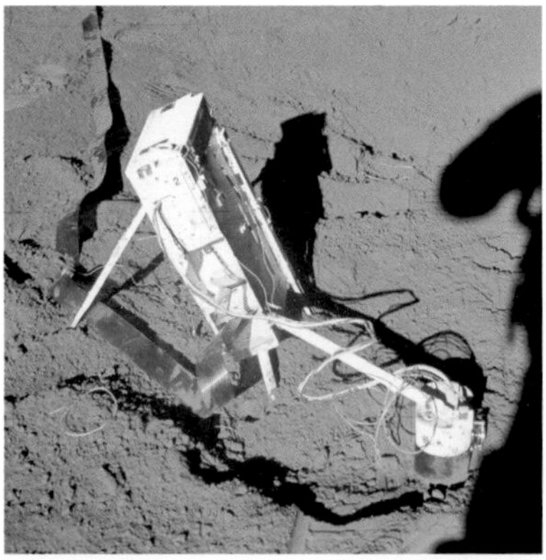

NASA

8

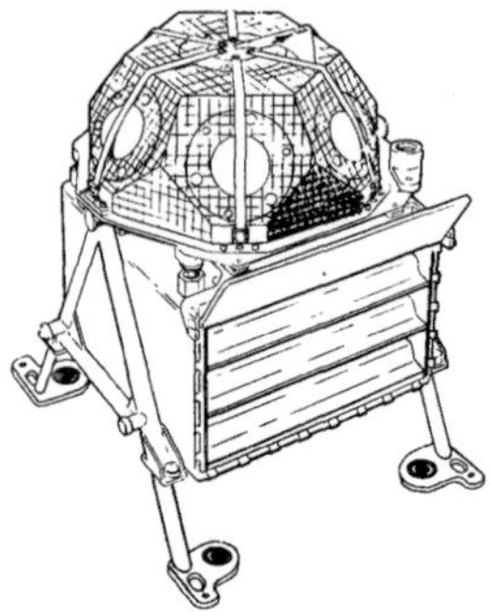

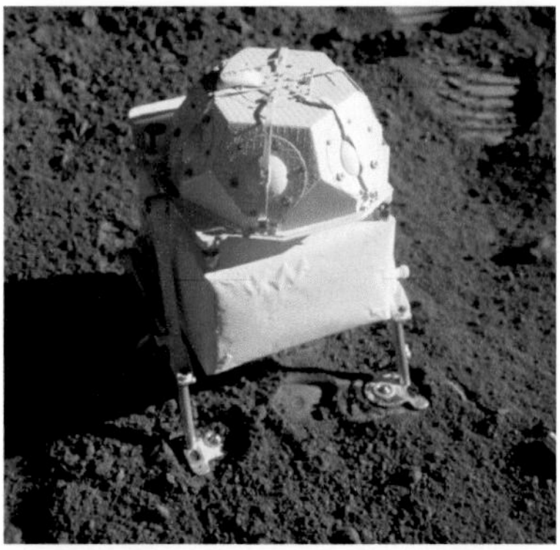

9

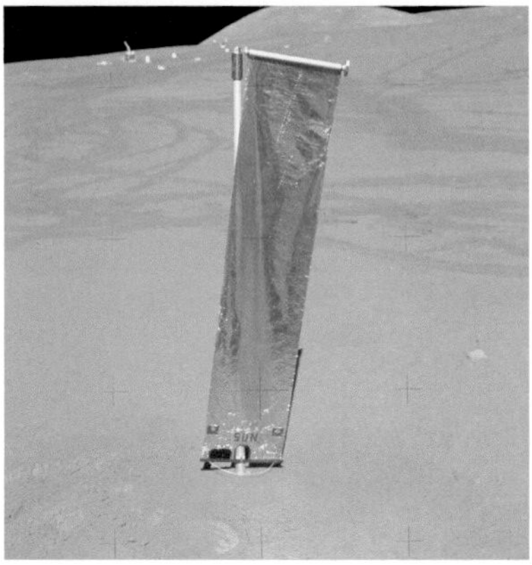

(7) The suprathermal ion detector accompanied Apollo 12, 14, and 15. This device detected positively charged ions and solar wind on the moon in order to measure possible interactions and to test the electric properties of the lunar surface. The sensors were so sensitive that they could even measure the fuel emissions of the rocket. The cabinet-like structure was mounted on a grounding system that resembled a spider web. (8) The cubic form with a polygonal top section rather resembled the famous robot R2-D2 from the *Star Wars* saga. This was a solar wind spectrometer (SWS). It was used to determine the conditions of the solar wind and their influence on the lunar environment. The SWS was only taken on Apollo 12 and 15. (9) All other Apollo missions, aside from Apollo 17, used aluminium foils to measure solar winds. The long, rectangular sails were unfurled and oriented towards the sun for different lengths of time in order to analyse

NASA

10

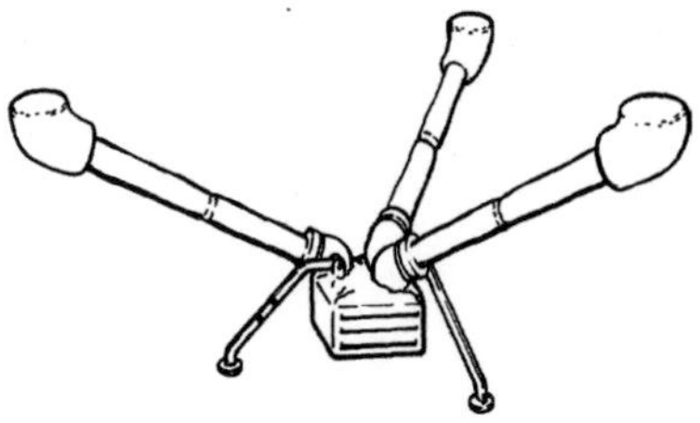

11

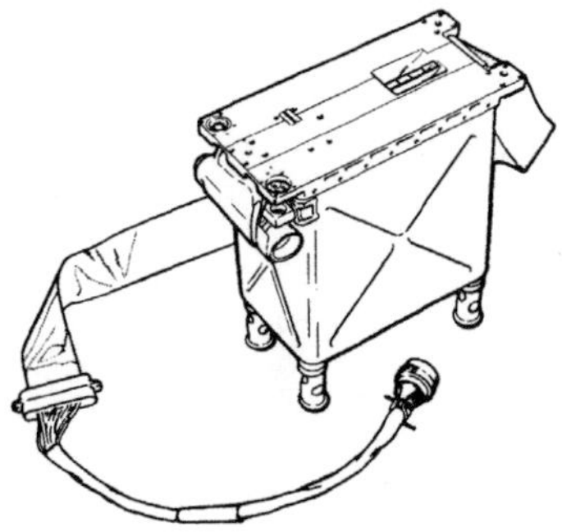

12

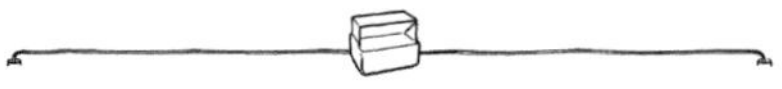

its radiance. (10) The surface magnetometer almost looked like an Apollo module after a crash landing. Three thick feelers protruded from a small box like an upended tripod. They measured the magnetic properties of the moon. The investigation of the magnetic field in turn led to new hypotheses regarding the electrical properties of the moon's interior. (11) The box, which looked like a hybrid of a radiator and a petrol can, measured charged particles in the lunar environment, particularly electrons and ions in the solar wind. It was part of the ALSEP package taken on Apollo 14 and 15. (12) The heat-flow measuring device analysed the heat radiating from the moon's interior to determine whether the moon's interior is fluid and whether the heat arises due to radioactive decay. To use this device, the astronauts drilled to a depth of between 2 and 2.5 metres beneath the lunar surface to bury a temperature sensor in each hole.

The astronaut Ronald Evans (Apollo 17) ventures into outer space in order to retrieve the film cassettes from the command module and bring them into the safety of the re-entry capsule. Although the task seems simple, Evans spent over an hour outside the capsule.

NASA

NASA

What looks like a lock screen on a smartphone is in fact one of the most famous photos in the history of space flight: the Earth, as depicted in the *Blue Marble*.

Apollo 17
US

NASA

Launch: 7 December 1972, 5:33 am
Landing: 11 December 1972, 7:54 pm
Launch from moon: 14 Dec. 1972, 10:54 pm
Launch site: Cape Canaveral
Landing site: 20.1908°N 30.7717°O

The final Apollo mission took place sooner than initially planned. Financial pressure brought NASA's crewed lunar programme to an early end, with only 17 of the 20 planned missions completed. The public's interest in the crewed lunar programme had begun to wane after the success of Apollo 11, which had marked the victory of the US over the Soviet Union in the race to the Moon. NASA's interest also dwindled, and funds were reallocated to the budding space-station programme. The mission emblem for Apollo 17 featured the head of Apollo Belvedere and an outline of the American eagle filled with stars and stripes – a fitting design to mark the close of the programme. Apollo 17 was the longest lunar mission, lasting 13 days, and the first to include a geologist, Harrison Schmitt, among the crew. Its primary objective was to investigate the older lunar rocks in a highland not far from the Sea of Serenity (Mare Serenitatis). It was also the first crewed space mission to be launched at night. During the flight, the astronauts took an image of the Earth from a distance of 29,000 kilometres. This would become a famous and oft-reproduced image, bearing the title *Blue Marble*. After the successful landing, the astronauts hoisted the American flag – now a standard procedure – and set up the ALSEP, with its cubic base and wide range of instruments and experiments. The extravehicular operations, covering a distance of 34 kilometres and spread out over three days, proceeded without any major problems. Though a fender broke off the Lunar Roving Vehicle, the astronauts found a makeshift solution to repair it. Harrison Schmitt and Eugene Cernan also collected 111 kilograms of lunar rock from nine pre-selected geological stations and bought it back to Earth. They left behind a plaque, commemorating the success of the Apollo missions and containing a message of peace to humanity. After this mission, no human has travelled beyond Earth orbit, let alone stepped on the moon.

Command module of the Apollo 17 mission

NASA

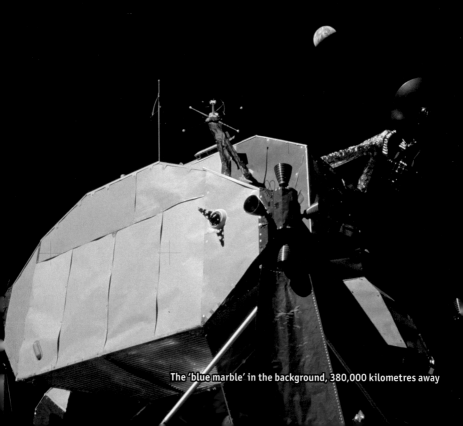

The 'blue marble' in the background, 380,000 kilometres away

NASA

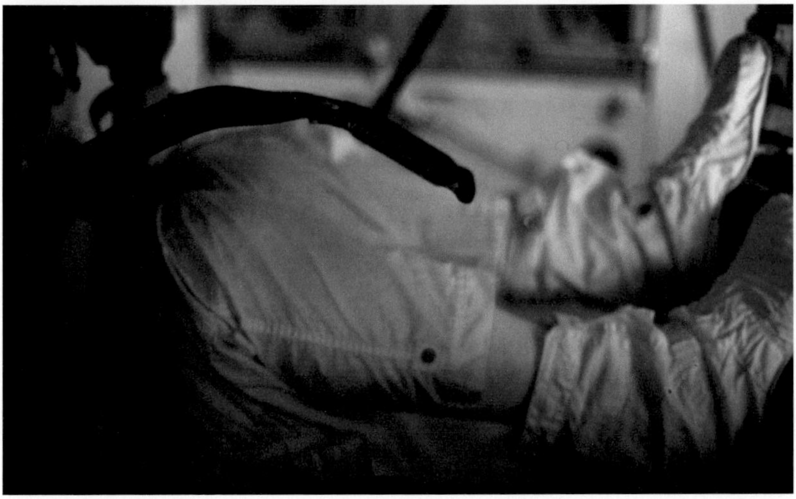

Beware when flying on the Apollo! Although the ticket costs millions, the limited amount of legroom recalls journeys with budget airlines. Nonetheless: drinks are for free!

NASA

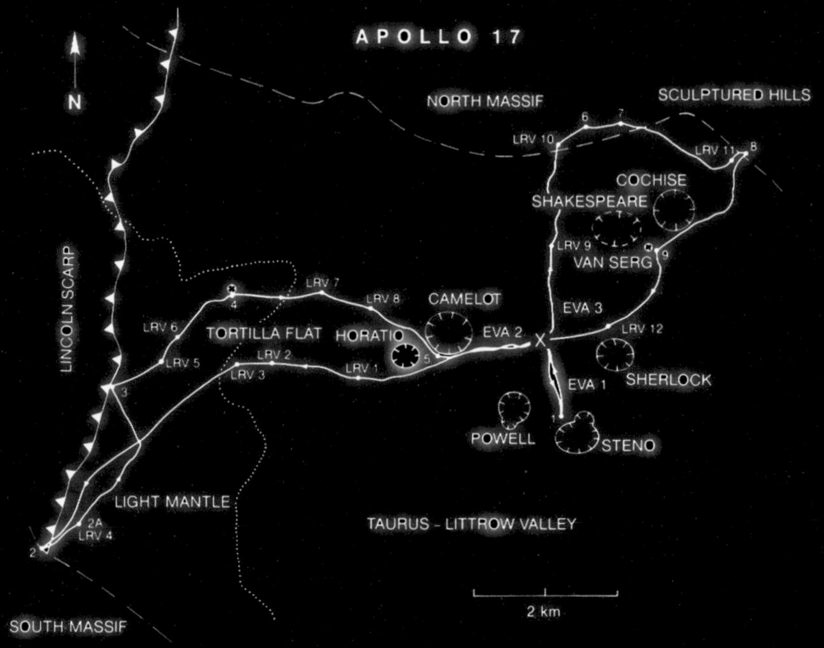

NASA

An improvised contraption based on tape and lunar maps: today the makeshift fender can be seen at the National Air and Space Museum in Washington.

NASA

Lunokhod 2 in the Tsiolkovsky State Museum of Cosmonautics, Kaluga

Paul Meuser

Paul Meuser

Luna 21 *Lunokhod 2*
Soviet Union

Launch: 8 January 1973, 6:55 am
Landing: 15 January 1973, 10:35 pm
Last contact: 3 June 1973
Launch site: Baikonur
Landing site: 25.85°N 30.45°O

Lunokhod 2 successfully landed on the moon around two years after the Soviet Union had landed its first uncrewed lunar vehicle. The newer lunar rover had a less zoomorphic appearance due to its camera eyes, embedded in rectangular frames. At this point, the Soviet space programme, with its mechanical probes, seemed virtually unbeatable when it came to uncrewed flights to the moon. The second lunar rover was assigned a challenging mission. It was to explore a difficult terrain, between the Sea of Serenity (Mare Serenitatis) and the Taurus Mountains, near a rift valley in the lunar surface. To this end, it was equipped with a wide range of new technical devices. In particular, a new television camera system – which entailed an additional camera on the roof – together with floor sensors, a gyroscope system, and an inclinometer helped to navigate the vehicle. However, one day, in spite of the improved vision, the five-person remote-control team back on Earth overlooked a crater. A defect in the communications mechanism led to several failed radio signals, and the rover eventually got stuck. The controllers hastily attempted to free it from this unfortunate situation, but during this process, they caused a pile-up of dust on the vehicle that covered the solar panel and instruments inside. The robot overheated as a result, and several systems needed to be shut down. In the end, communication could not be recovered. Although the accident shortened the duration of the mission, Lunokhod 2 covered a distance of 39 kilometres and thus held the record for longest distance covered by a rover on a celestial body until 2014.

Lenin's face decorating Lunokhod

Karina Diemer

František Hajdekr

Moon connoisseurs can delight in replicas of the second Soviet lunar rover, both in Kaluga (previous double spread) and in Moscow (left). Model builders across the world are still fascinated by the eight-wheel vehicle to this day (above).

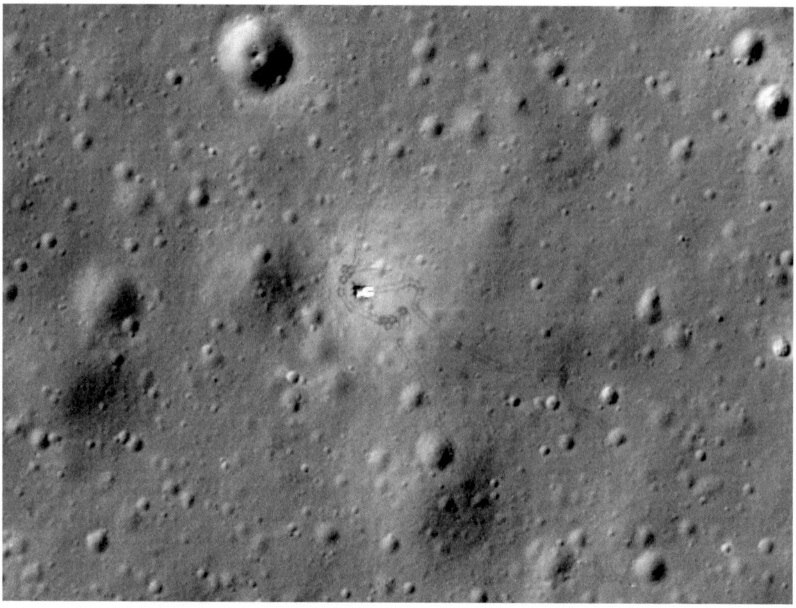

Moscow Polytech

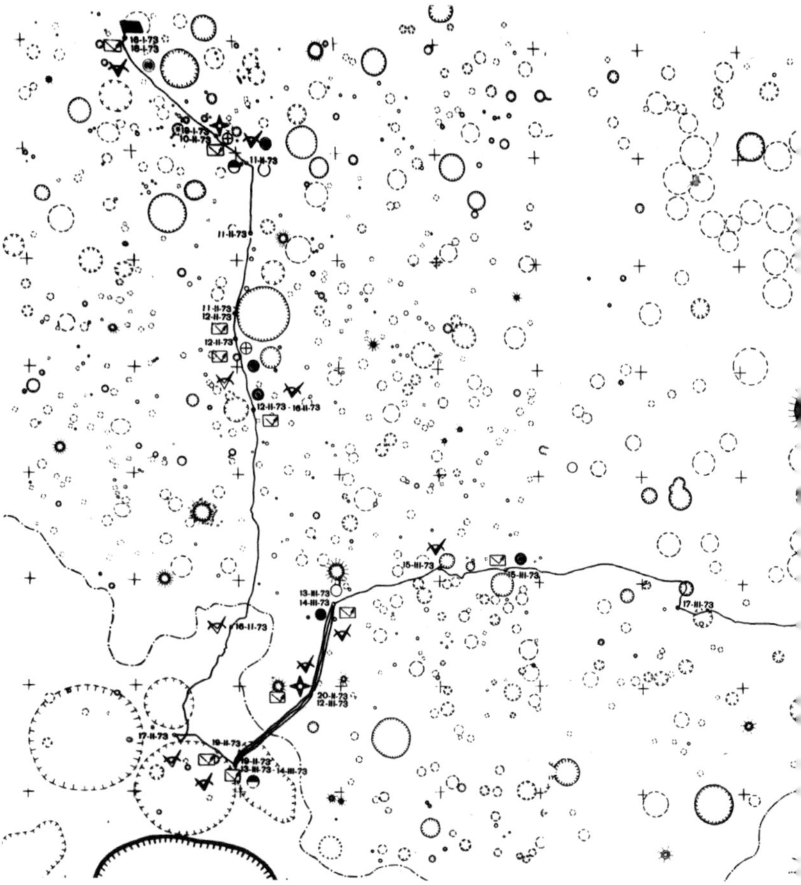

The Luna 17 lander and Lunokhod 2 look tiny between the craters. The LRO photographed both at their most recently documented locations (see pp. 296–301). In spite of the highest possible resolution, the pixels only leave the outlines of the two vehicles visible.

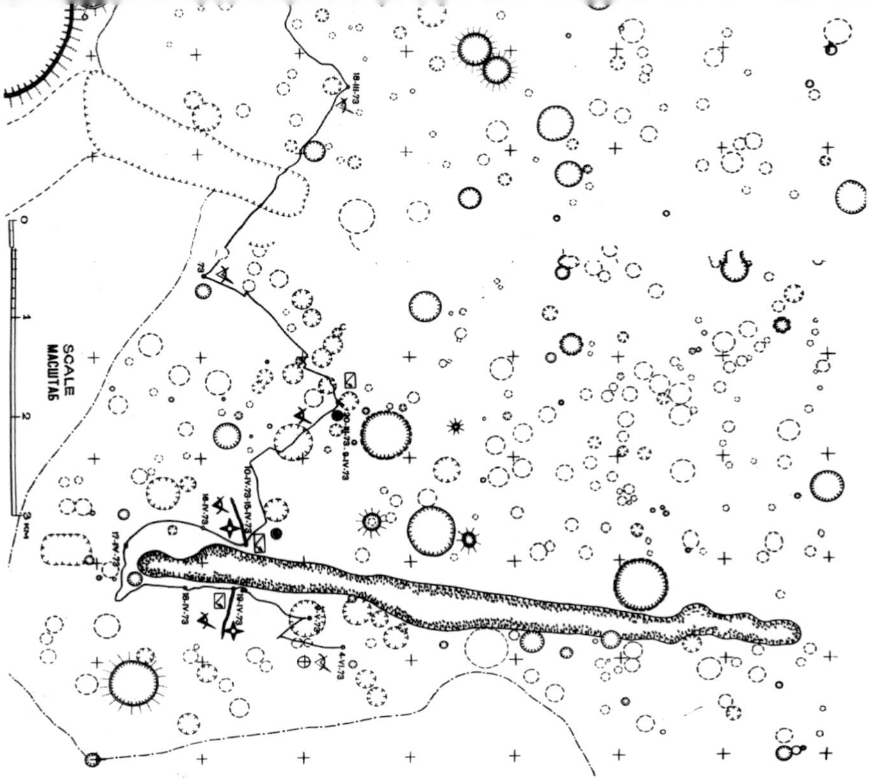

Moscow Polytech

Moscow Polytech

NASA

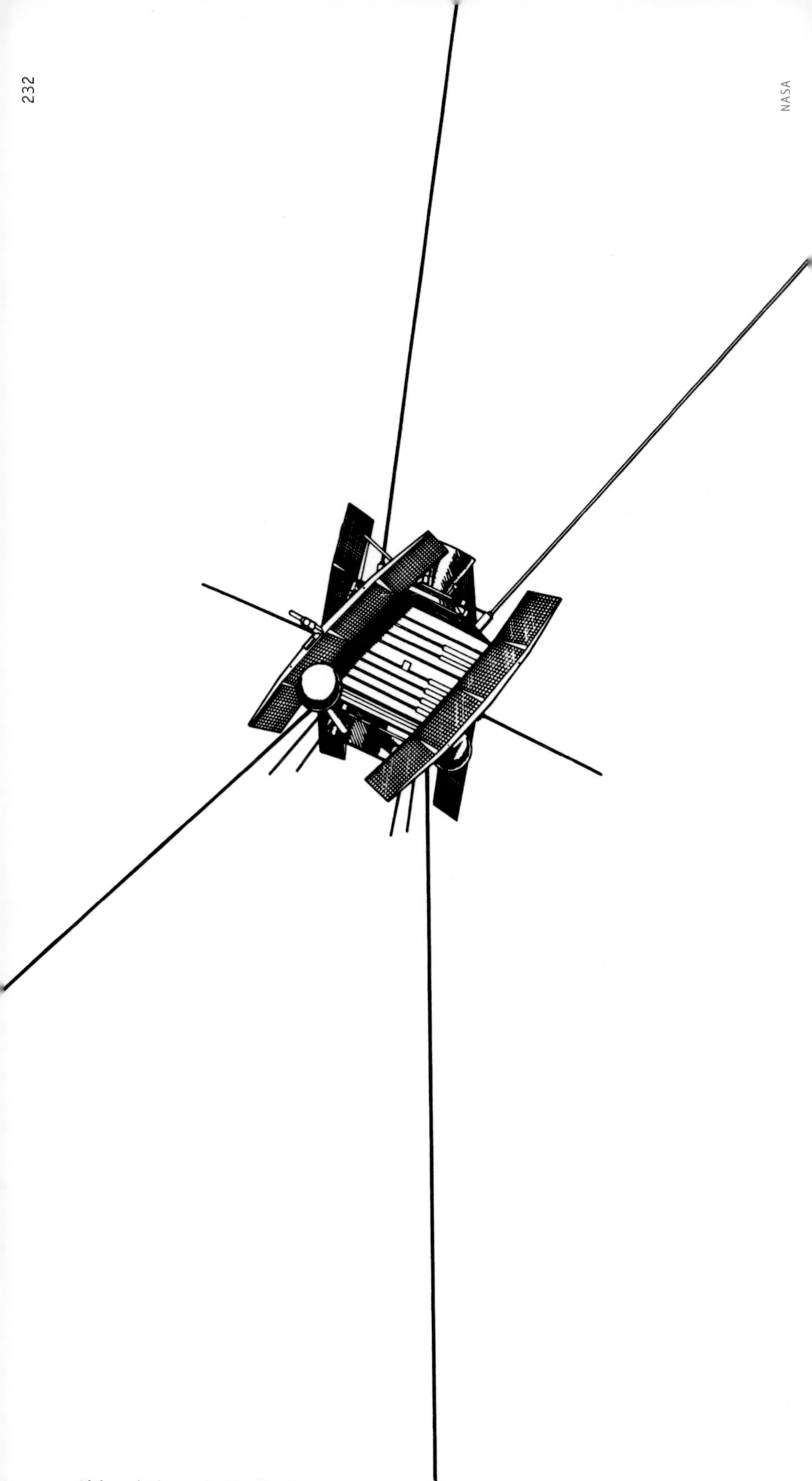

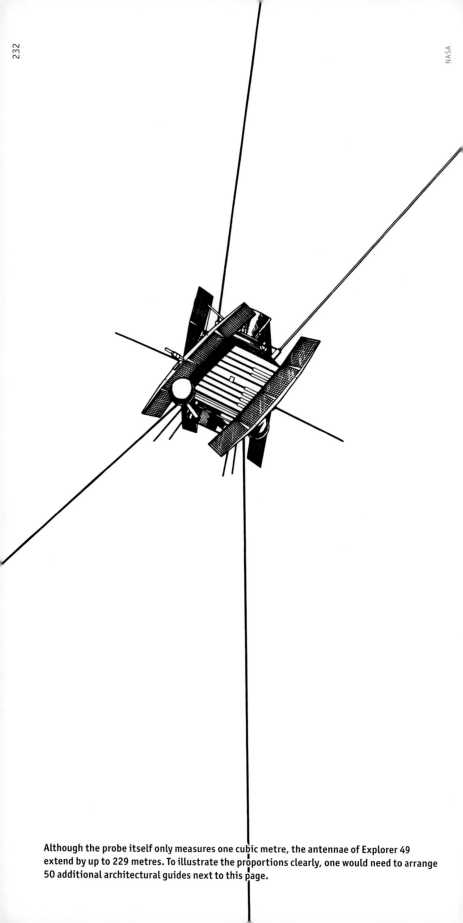

Although the probe itself only measures one cubic metre, the antennae of Explorer 49 extend by up to 229 metres. To illustrate the proportions clearly, one would need to arrange 50 additional architectural guides next to this page.

Explorer 49
US

Launch: 10 June 1973, 2:13 pm
Landing: crashed at unknown time
Last contact: August 1977
Launch site: Cape Canaveral

NASA

By 1973, the Apollo programme had already been terminated, and a new space race was already in full swing, the objective being to establish the first crewed research station in outer space. Nonetheless, Explorer 49 was launched on a modified Delta 1000 rocket towards the moon on 10 June. Like Explorer 35, it was designed to carry out studies of space from a lunar orbit. But unlike its predecessor, Explorer 49 was deployed to research radio waves, which is why it was also called the Radio Astronomy Explorer. It was the second satellite of this type, hence the technical abbreviation RAE-B. The satellite's small and dainty appearance was deceptive: it weighed a hefty 328 kilograms. The cylindrical body had a volume of less than one cubic metre, but its extensions had a far reach: four long and curving solar panels were placed around the cylinder, attached such that they could stay oriented towards the sun. A vibration damper and four V antennae, all of which were 229 metres long, protruded from the central body. They were arranged perpendicular to each other, the vibration damper pointing directly at the moon and the antennae reaching out into space. The antennae were extended gradually; they didn't reach their full length until 1.5 years after the launch of the mission. The spidery satellite collected data for four years until last contact was made in August 1977. It subsequently crashed into the moon at an unknown time.

NASA

Explorer 49

Luna 23
Soviet Union

Launch: 28 October 1974, 2:30 pm
Landing: 6 November 1974
Last contact: 9 November 1974
Launch site: Baikonur
Landing site: 12.6669°N 62.1511°O

The Soviet space programme launched another Luna sample-return probe in October 1974. Unfortunately, it suffered the same unhappy fate as most other Luna probes, though it didn't fail until it was very close to its destination. The landing was almost successful: the braking rockets ignited as planned, and the probe landed softly. However, it proceeded to tip over for unknown reasons. It was therefore unable to gather soil samples as planned, nor did it launch to return back to Earth. As such, the probe still lies in a site that is appropriately named Sea of Crises (Mare Crisium). It became the second probe, after Luna 15, to crash at this location. In spite of the crash, the Soviet

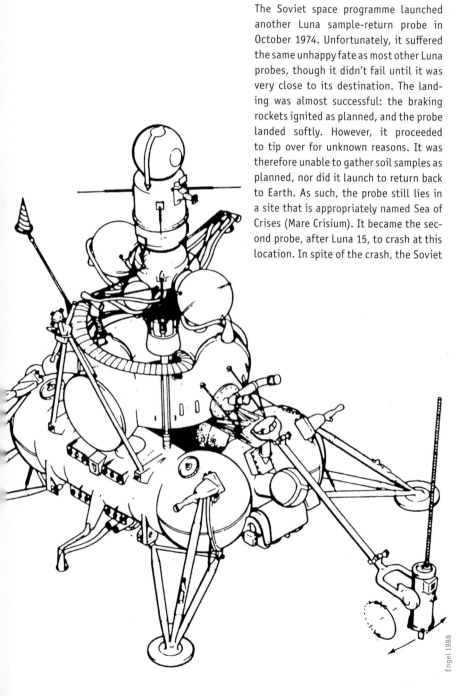

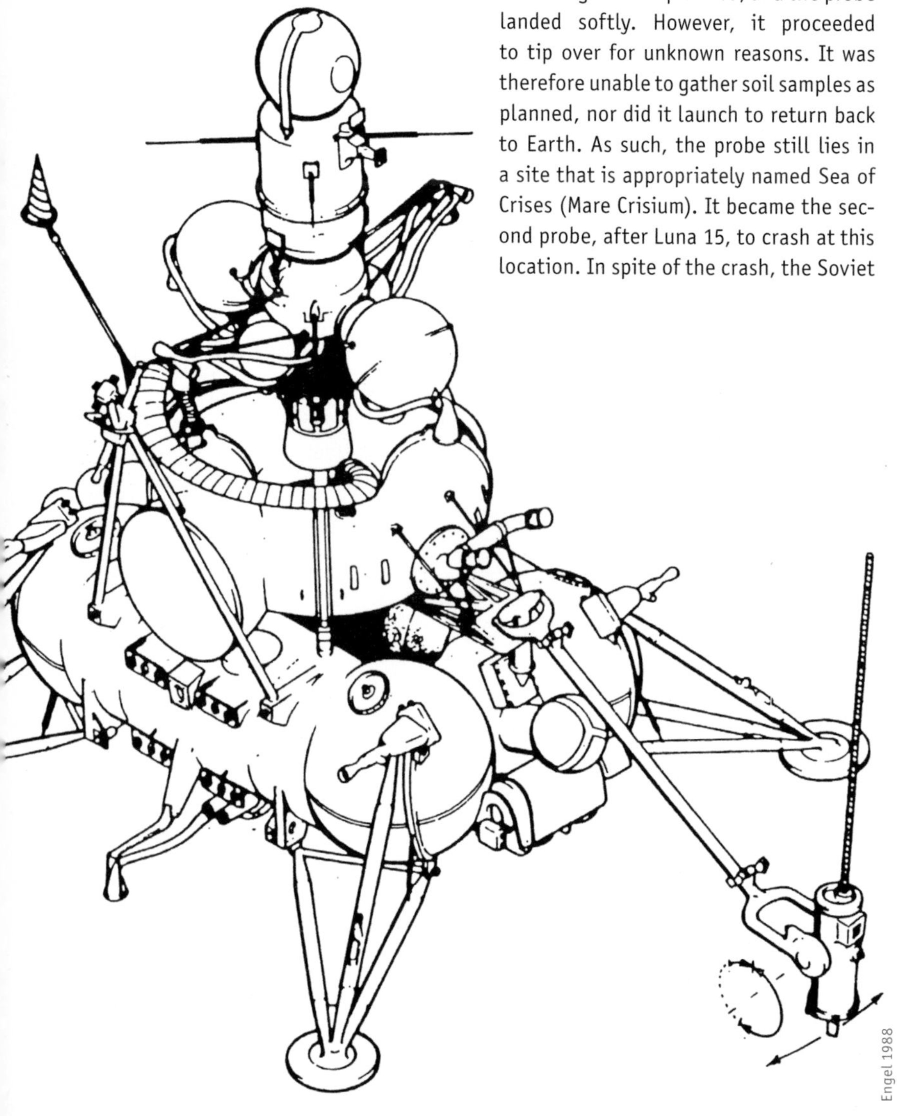

Engel 1988

Luna 24
Soviet Union

Launch: 9 August 1976, 3:04 pm
Landing: 18 August 1976, 6:36 am
Launch from moon: 19 August 1976, 5:25 am
Launch site: Baikonur
Landing site: 12.7145°N 62.2129°O

scientists took advantage of the remaining – though restricted – possibilities to carry out secondary experiments using the instruments on board.

Luna 24, structurally identical to its predecessor, marked the successful conclusion of the Soviet space programme after all its efforts, setbacks, and tireless persistence. After this mission, the space programme shifted its focus to new challenges, in particular, the establishment of a research station in space. Luna 24 remained operational on the moon for 13 days. It drilled deeper into the lunar surface than all its forerunners had done and delivered 170 grams of lunar soil and rock to Earth. This amount exceeded what had been obtained during the two previous successful missions combined.

21 Soviet relics remain on the moon to this day, recalling both the failures and successes of over 60 launch attempts. Luna 24 was also the final lunar mission that took place as part of the Cold War. This would also mark the end of the duopoly of the US and Soviet Union over missions to the moon. Several other nations would later initiate space travel programmes of their own and shoot for the moon. But first, lunar exploration went through a dormant period of almost 15 years. It wouldn't be until the early 1990s that the next works of moon architecture would land once again.

Svobodat

Luna 24 model at an exhibition in Prague (1980)

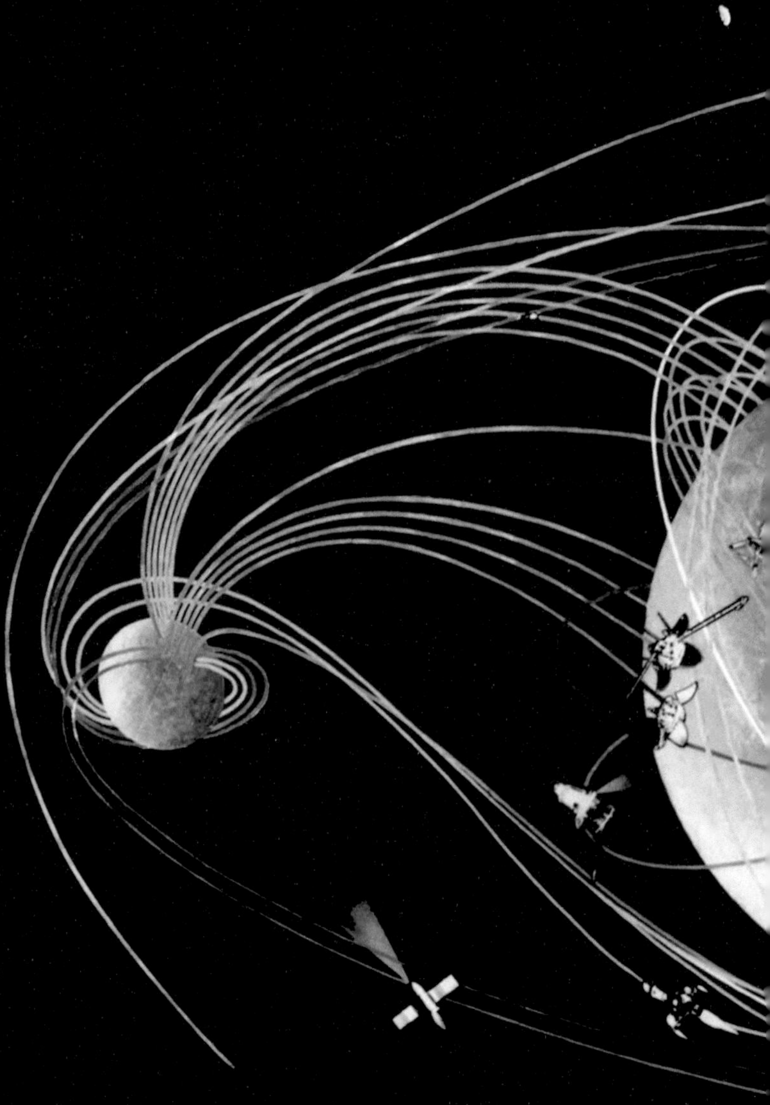

17 years of the Soviet lunar programme at a glance

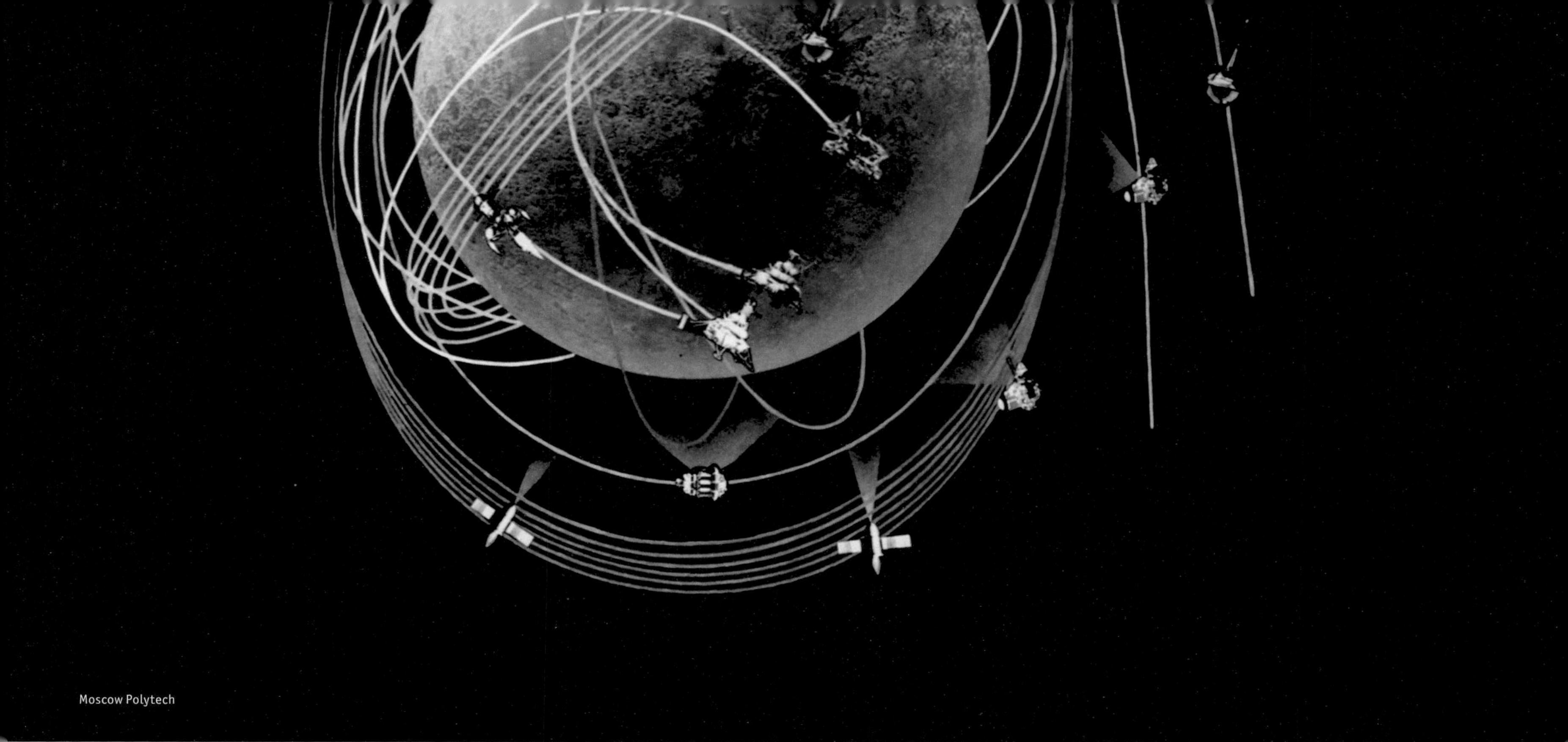

Moscow Polytech

Internationalisation of the Lunar Programme
New Players on the Moon: Japan, Europe, China, India, and Israel
1993 to the Present

The US and the Soviet Union shared a duopoly of the moon for 40 years. Exploration of space is a costly undertaking. Launching into space not only carries a certain technological prestige, it also represents an achievement of the national economy. NASA began to feel the symbolic resonance of space flight after the successes of the Apollo missions. With the desired goal – a crewed landing on the moon – already accomplished, funding for further missions to the moon was ended early so that the money could be invested in other endeavours.

Japan was the first country to follow the footsteps of the Cold War-era pioneers. Today, the Land of the Rising Sun is at the forefront of unconventional technologies and innovations in robotics. However, as much as Japan embodies the future of technology, there are two other nations participating in today's race to the moon: China and India. The two nations demonstrate that they are now playing in the top league of space flight, where the European Space Agency (ESA) has, in comparison, had major difficulties. Although ESA completed a successful project in the new millennium, it has not yet produced any follow ups. In the year of this book's publication, Israel became the seventh nation to go down the path to the moon.

These international participants are engaged in a competition to establish a lunar base and make use of untouched resources on the moon. And new competitors are joining the fray from one year to the next. The story of space flights to the moon is far from over. New developments arise continuously, which has made it a challenge to keep this book as up to date as possible. This is why the book does not end with the following chapter, describing the latest lunar missions, whose soft or hard landings still lie ahead, but with a look into the present and future of flights to the moon.

Launch of a Japanese Epsilon rocket on the launch pad of the Uchinoura Space Center

JAXA

JAXA

Hiten
Japan

Launch: 24 January 1990, 11:46 am
Landing: 10 April 1993, 6:03 pm
Launch site: Uchinoura Space Center
Crash location: 34.3°S 55.6°E

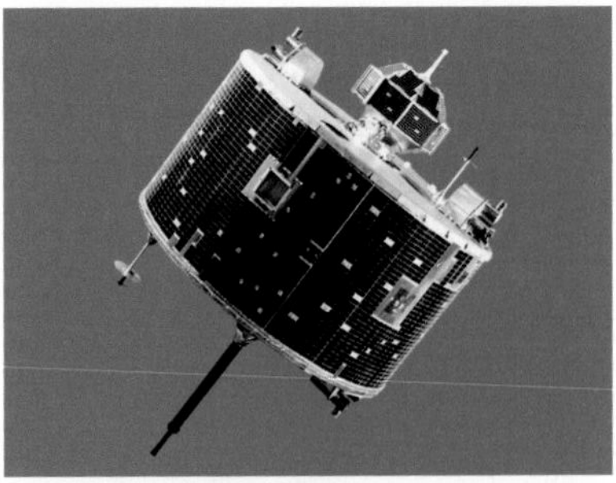

Hagoromo *(Subsatellite)*
Japan

Release: 18 March 1990, 7:37 pm
Last contact: 20 March 1991

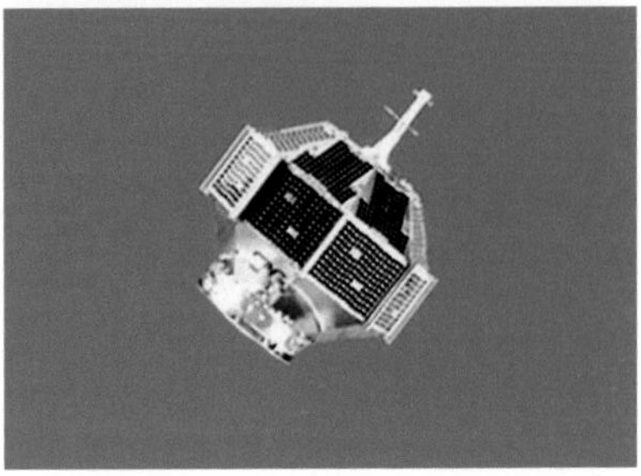

In 1990, Japan became the third nation to launch a probe to the moon, half a generation after Luna 2 had impacted the lunar surface in 1959. The mother satellite Hiten and its subsatellite Hagoromo were launched on the Mu-3 AII carrier rocket from Uchinoura Space Centre on the Japanese island Kyūshū. Hiten was named after Japanese angels who fly alongside Buddha, while Hagoromo was named after the dress worn by such angels. The former, originally named Muses-A, was the nineteenth satellite of the Japanese Institute of Space and Astronautical Science (ISAS). Japan has had an official rocket programme since the 1950s, and the Japan Aerospace Exploration Agency began to launch satellites into Earth orbit in the early 1970s. The Hiten probe, weighing 200 kilograms, featured a simple design, reflecting its simple range of equipment. The probe carried a transponder and dust detector, though the mission's primary objective was for the Japanese scientists to gain a better understanding of the technologies involved. Hiten had a cylindrical body, 140 cm in diameter and 80 cm in height, and was completely clad with solar panels. Its round body featured a small number of recesses, which resembled irregularly placed windows. A medium-gain antenna protruded from the lower surface, forming part of the communications system and also marking the central axis of the spacecraft. Hagoromo, mounted on the opposite surface, was much smaller and had a polyhedral form. It was also clad with solar panels and was crowned with a small antenna mast. The subsatellite was put in lunar orbit during the first swing-by manoeuvre, which involved changing the direction of the two-probe system using Earth's gravity. Afterwards, the mother satellite began to test the aerobraking procedure while flying past the Earth at a distance of 126 kilometres. This part of the mission was successful, and the mother satellite continued to loop around the Earth and moon for three years. However, contact with the subsatellite was lost prematurely, which is why its location is unknown, though it probably crashed into the moon. Hiten was intentionally crashed into the moon once it had used up its fuel reserves.

The small probe goes on a piggyback ride on the mother probe to the moon.

NASA

Lunar Prospector
US

Launch: 7 January 1998, 2:28 am
Landing: 31 July 1999, 9:52 am
Last contact: 31 July 1999, 9:52 am
Launch site: Cape Canaveral
Crash location: 87.7°S 42.1°O

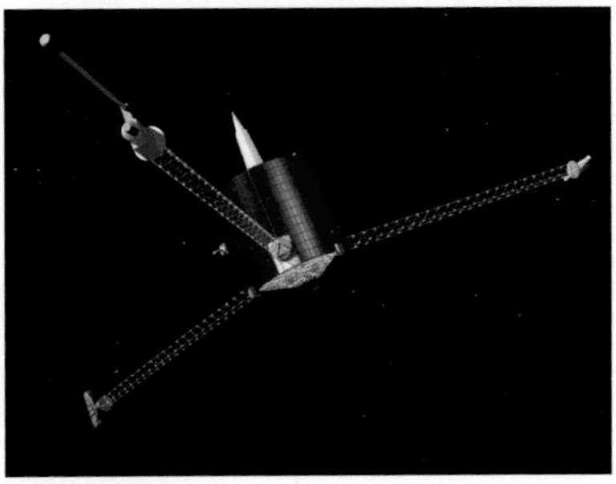

The Lunar Prospector mission saw the US land a spacecraft on the moon for the first time in 25 years. The probe's development had taken 13 years and had been marked by technical setbacks as well as successes. Lunar Prospector's design highly resembled that of the Japanese Hiten probe – with a barrel-shaped body clad with solar panels – though it was twice as tall. Three masts, arranged along the bottom edge, 120 degrees apart, protruded into space with instruments fitted on the tips. The prospector, loaded with six experiments and cutting-edge technology, had one primary objective: to find evidence of water ice on the lunar surface. It was therefore in many ways the follow-up probe to Clementine, which had brought back signs of water ice from the lunar south pole four years earlier. At the same time, Lunar Prospector captured data with which to create a map of thorium on the moon, the chemical being a radioactive substance that could be used to generate power for future lunar missions. It was still assumed in the late 1990s that NASA's lunar programme would be revived within the next 20 years. But the enormous costs required as well as delays destroyed this dream, and heavy budget cuts were eventually announced. The scientists' aim at the end of the mission was to steer the probe into a site, permanently in shadow, in the southern region of the moon in order to release water ice with the collision. Although this part of the mission was not successful, the probe's measurements proved that billions of tonnes of water, essential for life, exists at both poles of the moon. However, it is mixed together with lunar dust.

NASA

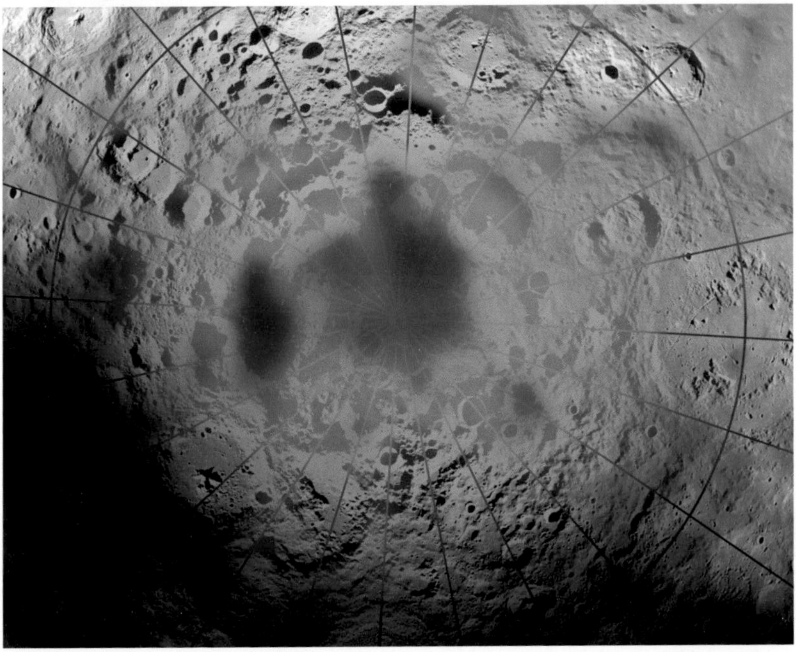

The Lunar Prospector mission proved the existence of water ice in the lunar south pole.

NASA

The probe doesn't look particularly big next to its engineers.

NASA

Rendering of the first US lunar mission after 25 years

NASA

A white head and a blue body: engineers and probe in matching colours

The Guiana Space Centre, a spaceport shared by France and the European Space Agency (ESA), is located in French Guyana, a country with a population of 260,000 people.

Corvaja / ESA

esa
vega

eutelsat
SMART 1
arianespace
esa
esa
cnes
cnes

SMART-1

Europe

Launch: 27 September 2003, 11:14 pm
Landing: 3 September 2006, 5:24 am
Last contact: 3 Sept. 2006, 5:24 am
Launch site: Kourou
Crash location: 34.26°S 46.19°W

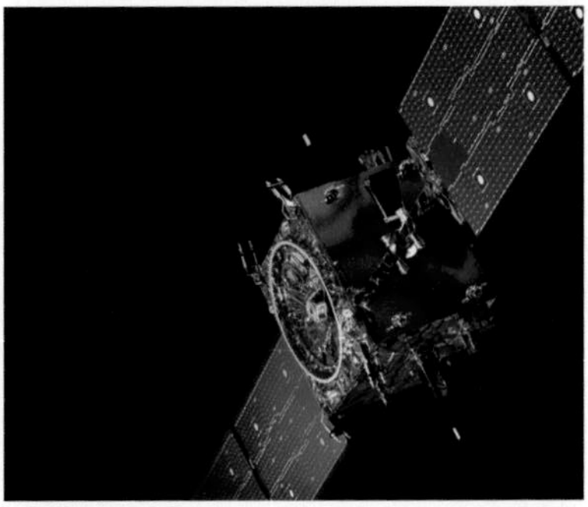

The European Space Agency (ESA) sent its first and only probe to the moon in the dawn of the new millennium. The probe was launched from the Guiana Space Centre in French Guiana. The French had started using the launch site in 1964, and ESA also began to use the launch pad in 1975, the year it was founded. SMART-1 (Small Missions for Advanced Research in Technology) aimed to test a solar-powered ion thruster as well as new navigation and communications technologies. Like older orbiters, the probe was composed of a core body and solar-panel wings mounted on the sides. The central body contained, among other things, the ion thruster and communications system, while the smaller instruments were arranged around the housing. The two wings, each comprising three solar panels, were folded up for the journey on the Ariane 5 launch vehicle, which remains the highest-capacity European rocket to this day. The wings unfolded in outer space to reach a full length of 14 metres. The probe, based on an efficient design, took four years to construct and is among the fastest-built probes in the world. During its elliptical lunar orbits, SMART-1 analysed the moon's chemical composition and incidence of water ice. The ion thruster's high capacity enabled an extension of the mission by a year. And the ample fuel reserves made it possible to crash the SMART-1 probe into the earth-facing side of the moon, which in turn enabled scientists to analyse the changes to the lunar surface caused by the collision of the probe, weighing almost 300 kilograms.

ESA – CNES – Arianespace

SMART-1's launch (left) and its trajectory and intentional crash (above)

ESA – CNES – Arianespace

ESA – CNES – Arianespace

Journey before the journey: arrival at the Cayenne-Rochambeau Airport and subsequent installation at the Guiana Space Centre

ESA – CNES – Arianespace

India's Journey to the Moon

Gurbir Singh

The sun and the moon dominated the lives of people living in the earliest civilisations on Earth. The moon played a central role in the creation stories of Aboriginal Australian peoples, Egyptian Pharos, and the myths of ancient Greece. So it was for the Indus Valley Civilisation in India which probably predates all others. Its people understood the key role of the sun and the moon in sustaining life on Earth but also in feeding a higher need to connect with the cosmos. The Vedas are the oldest religious texts from the ancient Indian subcontinent. Amongst the 350 million gods in those texts, the moon is represented in Sanskrit as Chandra and Soma with 120 dedicated hymns in ancient scriptures. In Hindi it is known simply as Chaand or Chandra and seen by children as a mother figure known as Chandamama.

The moon still plays a key role in Indian mythology, astrology, and ritual. Whilst many people from around the globe travelled to India in 1995 to experience the total solar eclipse, many in India stayed indoors. The belief was that to watch the moon during the eclipse (and other Hindu festivals) would be harmful. In Hindu mythology Chandravanshi are the people belonging to the 'Lunar Dynasty'. The first king of that dynasty was the all-powerful legendary emperor of India, Bharat. Today, Bharat is the official name of the Republic of India.

The space race of the 1960s and 1970s was the product of the Cold War. The celebrated Soviet space engineer Boris Chertok once claimed that without Gagarin there would have been no Armstrong. Meanwhile, India's main competition is China. China launched its first mission to the moon, Change' 1, in 2007. It had been designed to operate in lunar orbit for a year but it was so successful that it was extended to two. To date China has completed four lunar missions, one of which included a lunar lander and rover. It is currently working on its next mission, which is to land a rover on the far side of the moon by early 2019. India launched its first mission into lunar orbit in 2008. Following a series of delays, it will launch its second, Chandrayaan-2, which is to land on the lunar surface, in early 2019.

ISRO

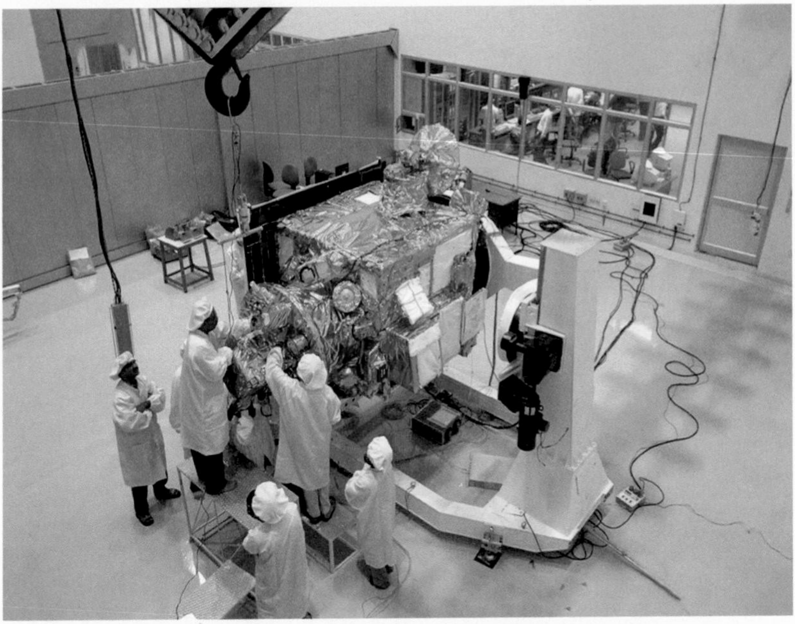

Putting the finishing touches before the probe is transported to the launch site

Destination moon

At the outset, India's space programme explicitly excluded any exploration of planets or the moon. Its clear mandate was to deliver socio-economic benefits to the ordinary people of India. By the turn of the millennium, ISRO had established a reliable track record in designing, building, launching, and operating satellites in Earth orbit. As one of a handful of nations, India had developed a sophisticated constellation of satellites for remote sensing, meteorology, and communication.

India's mission to the moon emerged from a proposal paper presented in 1999. At the time, India only had one operational launcher, the Polar Satellite Launch Vehicle (PSLV), with a limited capability. Once calculations confirmed that a PSLV could be used for a mission to the moon, planning for such a mission began. On 15 August 2003, India's prime minister made the formal announcement of the nation's intention to launch a mission to the moon, Chandrayaan-1.

ISRO was already familiar with designing, building, launching, and operating satellites in Earth orbit. Operating a spacecraft in lunar orbit was not very different conceptually, but the vast distance was a challenge. In addition, if India could get to the moon before China, then it would be the fifth country to do so, after the USSR (1959), the US (1959), Japan (1990), and ESA (2003). Krishnaswamy Kasturirangan, head of ISRO at the time, had originally planned a single lunar mission, initially called Somayaan, but Prime Minister Vajpayee wanted the mission to be a series rather than a one-off event and include other planets and not just the moon. Based on advice from Sanskrit scholars, Vajpayee also changed the name of the mission from Somayaan to Chandrayaan (in Sanskrit, chandra = moon and yaan = vehicle). As the first of a series of missions, it became Chandrayaan-1, which arrived on the moon a year after China's Chang'e-1. By the time the Chandrayaan-1 mission was complete, it had become an international, award-winning mission.

Chandrayaan-1
India

Launch: 22 October 2008, 00:52 am
Last contact: 28 August 2009, 8:00 pm
Launch site: Satish Dhawan Space Centre

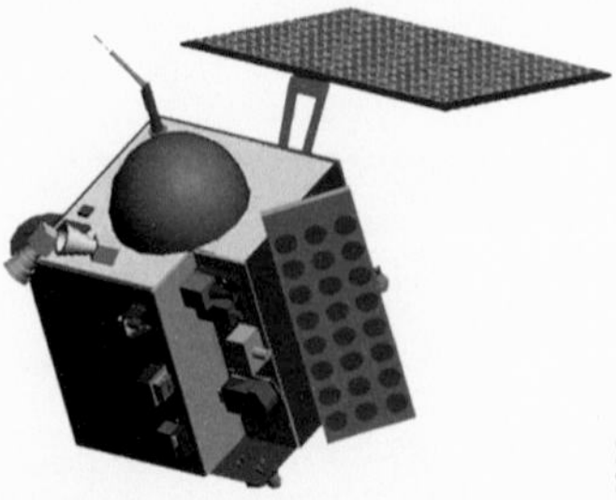

ISRO

Ostensibly, Chandrayaan-1's primary objectives were scientific, but the unstated motivation was national prestige and the desire to get to the moon before China. The spacecraft and launcher design evolved over time into Chandrayaan-1. It would be a cuboid, weighing 440 kg (524 kg with propellant) and each side measuring 1.5 m. It would be powered by one solar panel and lithium-ion batteries. The mission was designed to operate for two years. It would be launched by a PSLV and placed in a 100-km circular polar orbit around the moon. It would use S-band uplink for remote command, S-band downlink for telemetry, and X-band for data transmission from the moon to Earth. The single solar array would be canted at 30 degrees to provide 750 W peak power and use latest solid state memory devices to record data as it was captured before transmission back to Earth. The spacecraft's initial design was heavily revised twice. Calculations had shown that a carrying capacity of 10 kg and power capacity of 10 W were still available even after accommodating all six of ISRO's instruments. ISRO issued an Announcement of Opportunity, an invitation to scientists, engineers, as well as national and international institutions to make use of this excess capacity. ISRO received an unexpectedly large international response, which prompted the first design change. Two instruments proposed by NASA – M3 and Miniature Synthetic Aperture Radar (Mini-SAR) – looked promising, but together they exceeded the 10 kg allowance. To accommodate both, ISRO redesigned the spacecraft and increased the available payload capacity from 10 kg to 25 kg. Now, ISRO could accommodate even more than two additional instruments. Eventually, six instruments from international partners were incorporated. These were: the Moon Mineralogy Mapper (M3) from NASA; the Mini-SAR from the US and UK; the Near-Infrared Spectrometer (SIR-2) from ESA; the Sub keV Atom Reflecting Analyser (SARA) from ESA, India, Japan and Sweden; the Radiation Dose Monitor (RADOM) from the Bulgarian Space Science Institute; and the Chandrayaan-1 X-ray Spectrometer (C1XS), a collaboration funded by ESA and the UK. The second significant design change came when a Moon Impact Probe (MIP) was added. The MIP was designed to detach from Chandrayaan-1 and impact the lunar surface, delivering the Indian tricolour to the surface of Earth's nearest neighbour. Abdul Kalam, at the time the president of India and a highly respected former ISRO engineer, had advocated the MIP. His elevated political position ensured his proposal was taken seriously. Moreover, the delivery of the Indian flag to the surface of the moon was a source of national pride and would attract additional government commitment. The MIP, weighing 35 kg, called for a major redesign to optimise the subsystems shared by payloads – for power and data storage for example – and reduce built-in redundancies. The number of star sensors was reduced from four to two; the twelve Reaction Control Systems were reduced to eight; and two tanks, each with a capacity of 35 litres, used to pressurise fuel, were replaced with one with a capacity of 67 litres. To minimise cost overrun and reduce the delay these major redesigns caused, ISRO chose to dispense with the traditional practice of building three models (structural, engineering, and flight) and instead went straight to building a single-flight model.

The Chandrayaan-1 probe is placed inside a vacuum tank for testing.

ISRO

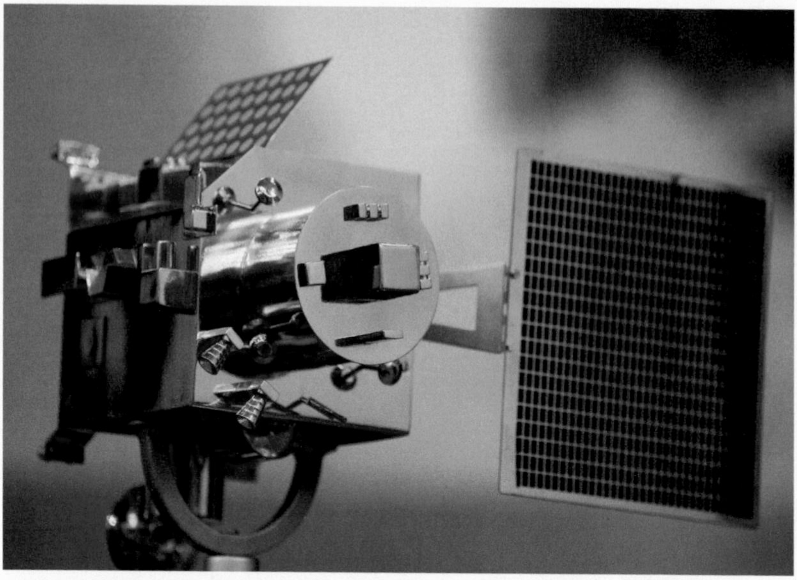

Model of Chandrayaan-1 at the Satish Dhawan Space Centre in Sriharikota (2008); PSLV carrier rocket shortly before its launch (right)

From the Earth to the moon

All of the spacecraft that have travelled from the Earth to the moon – around 130 – have used the Lunar Transfer Trajectory. While the US's Saturn 5 rocket could bring a 35,000 kg payload into lunar orbit, Chandrayaan-1 was launched with ISRO's carrier rocket Polar Satellite Launch Vehicle (PSLV), which could only deliver 675 kg. This limited Chandrayaan-1's maximum science payload to 105 kg and required five ever-increasing Earth orbits before heading out to the moon. When close enough to be within the moon's gravitational influence, the spacecraft would fire its engine to slow down and be captured in a lunar orbit. To minimise the fuel required for orbit entry, Chandrayaan-1 orbited the moon four times with an ever-decreasing altitude. Chandrayaan-1 was launched on PSLV-C11 at 10:19 UT from Sriharikota on 22 October 2008 and released 20 minutes later into a transfer orbit around the Earth at 254 × 22,932 km with an inclination of 17.9 degrees. Following five Earth orbits, Chandrayaan-1 arrived in lunar orbit just over two weeks later. During its first four days in lunar orbit, Chandrayaan-1 circularised its orbit from 507 × 7,510 km to 101 × 102 km through a series of engine burns.

First international moon mission

Several nations had sent spacecraft to the moon but Chandrayaan-1 was arguably the first international spacecraft to enter lunar orbit. Chandrayaan-1 was the product of a collaboration between nine nations and ESA, which itself is a collaborative effort between over 20 nations. During the 1960s and 1960s, the US's Apollo programme had taken 24 men to the moon, twelve of whom visited the surface during six missions. Both the US and the USSR brought back samples of rocks and soil, 382 kg by the US during six Apollo missions and 326 grams by the USSR with three unmanned sample return missions. In addition, both nations had conducted extensive experiments from lunar orbit. What science could be done by an Indian mission to the Moon that had not already been done by the numerous missions already completed?

ISRO scientists set three key goals for Chandrayaan-1. Deploying scientific instruments not used during the lunar missions of the 1960s and 1970s, Chandrayaan-1 was to: (i) prepare a three-dimensional atlas of the entire surface of the Moon with a resolution of 5–10 m; (ii) identify the location and distribution of lunar minerals as well as light

ISRO

and heavy elements; and (iii) search for and map the distribution of water, if it existed. Chandrayaan-1 was equipped with an array of eleven instruments (see p. 256). Although 98 per cent of the lunar surface had already been imaged in the past, only specific areas, such as potential landing sites, had been targeted for high-resolution imaging. Two of the 11 instruments – RADOM and TMC – were activated while Chandrayaan-1 was still in Earth orbit. The remaining nine instruments were commissioned gradually between 16 November and 9 December 2008. The moon was known to be extremely dry. The Earth's axis of rotation is inclined at 23° with respect to the plane of the orbit, so both poles get some sunlight over a year. The moon's orbit is inclined at 1.5°, so craters at its poles get almost none. If water existed, it would be trapped and frozen in the deep craters at the lunar poles which are never illuminated by sunlight. Chandrayaan-1's unusual pole-to-pole orbit allowed it to look directly down into these areas during each orbit. If water was present, Chandrayaan-1 had a good chance of detecting it.

Touching the moon

The mission's initial objective was to observe the moon from lunar orbit. But introducing the Moon Impact Probe (MIP) would also satisfy a deeper political goal. As a rising economic power, India wanted to stake a claim to the potentially vast amount of natural resources on the moon for future commercial operations. The MIP, weighing 35 kg, is best considered as a separate spacecraft designed to detach from Chandrayaan-1 upon reaching lunar orbit. It would then descend to and impact the surface, thereby spreading the colours of the Indian flag in a predefined location.

Chandrayaan-1 arrived in lunar orbit on 8 November, and the orbit was circularised by 12 November. MIP separated from Chandrayaan-1 two days later, on 14 November, at 20:06 IST. The date had been selected to coincide with the birthday of India's first prime minister Jawaharlal Nehru. MIP then started its journey to the surface, impacting at a speed of 1.75 km/s near the lunar south pole (latitude 89° south), close to the crater Shackleton at 20:31 IST. The impact site was named Jawahar Sthal (Jawahar site). The MIP had three key scientific objectives: to observe the moon close up; to test technologies for navigating to a predefined point on the lunar surface; and to prepare for future soft-landing missions.

Scientific findings

The primary objectives of Chandrayaan-1 were to create a high-resolution three-dimensional atlas of both the near and far side of the moon, including the polar regions, and to conduct a chemical and mineralogical mapping of the entire lunar surface. Before the termination of the mission, the M3 instruments had covered 90 per cent of the lunar surface. The TMC had only managed around 50 per cent, though most of that included the important polar regions not covered in detail before. The relative concentrations and locations of aluminium, magnesium, and silicon, along with iron-bearing minerals such as pyroxene, were integrated into a 3D map by combining data collected by multiple instruments.

Chandrayaan-1's data revised three key assumptions about the moon:

• The moon was not dry. Data, collected primarily by M3 but also by MIP and Mini-SAR, provided evidence for the existence of water and hydroxyl molecules, whose quantities grew ever larger towards the two poles.

• Solar wind protons were not absorbed by the moon. These protons were integral to the process of making water on the moon.

• The moon was geologically active. Hydrated magma indicated episodic explosive volcanic events. The TMC and HySI instruments identified uncollapsed lava tubes, which could be used for human habitation in the future.

The planned two-year mission was prematurely declared over on 29 August

2009, after 312 days of operation, when Chandrayaan-1 finally lost its ability to transmit data. The early termination of the mission was caused by the spacecraft's inability to regulate its internal temperature. On arrival in lunar orbit, Chandrayaan-1 found itself in a temperature regime of +100 °C on the day side and -100 °C on the night side of the moon. Designed to operate at 40°C or lower, Chandrayaan-1 was frequently in excess of 50 °C, and at one stage, 80 °C. To move to a lower-temperature environment, ISRO decided on 21 May 2009 to increase the orbit's altitude from 100 km to 200 km. This decision reduced the spatial resolution of the data collected by some of the instruments.

ISRO concluded that Chandrayaan-1 had completed 95 per cent of its mission objectives. ISRO engineers calculated that the moon's tenuous atmosphere would decay Chandrayaan-1's orbit and that it would crash into the Moon, like the MIP, by around 2012. But it did not. Surprisingly, Chandrayaan-1 was discovered to still be in lunar orbit in 2017. Though no longer operational, it was still using Earth-based radar. Apart from meeting its chief scientific objectives, Chandrayaan-1 will be remembered for three other key achievements: the strong international collaborative spirit it engendered, the wealth of experience many young ISRO engineers gained in operating a spacecraft in lunar orbit, and the recognition that ISRO had the capability to successfully undertake complex space science missions beyond Earth orbit.

Mission launch on 22 October 2008

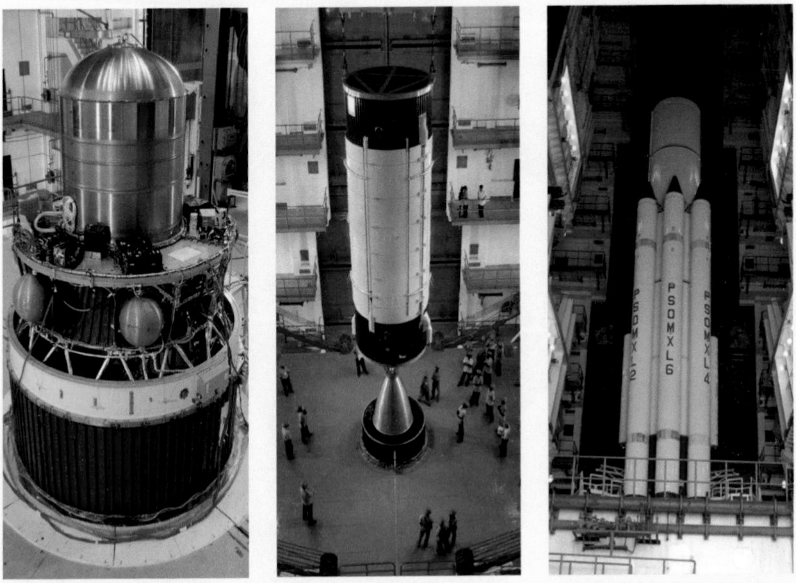

ISRO

Components of the PSLV carrier rocket: head, middle section, main engine (left to right)

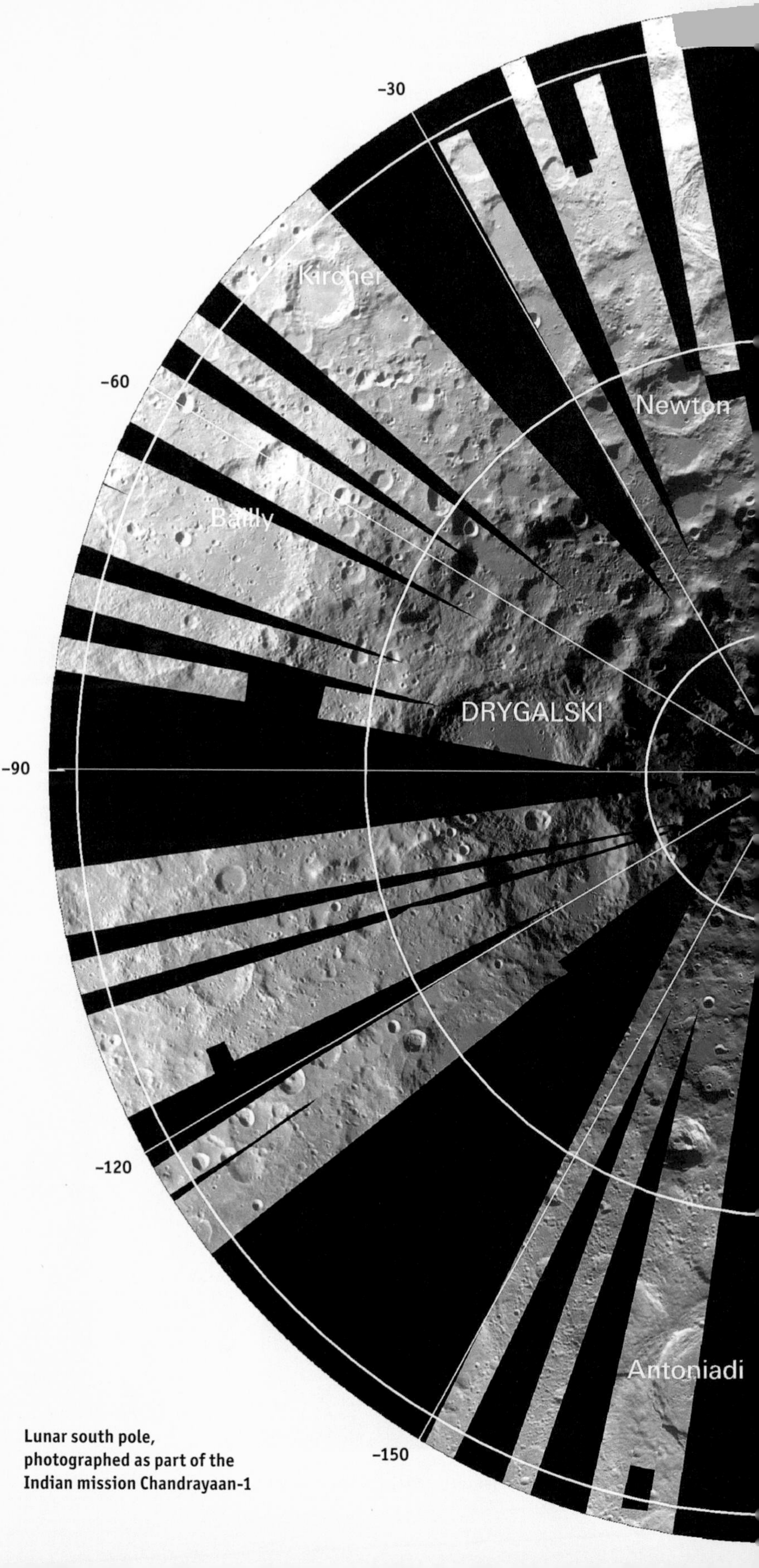

Lunar south pole, photographed as part of the Indian mission Chandrayaan-1

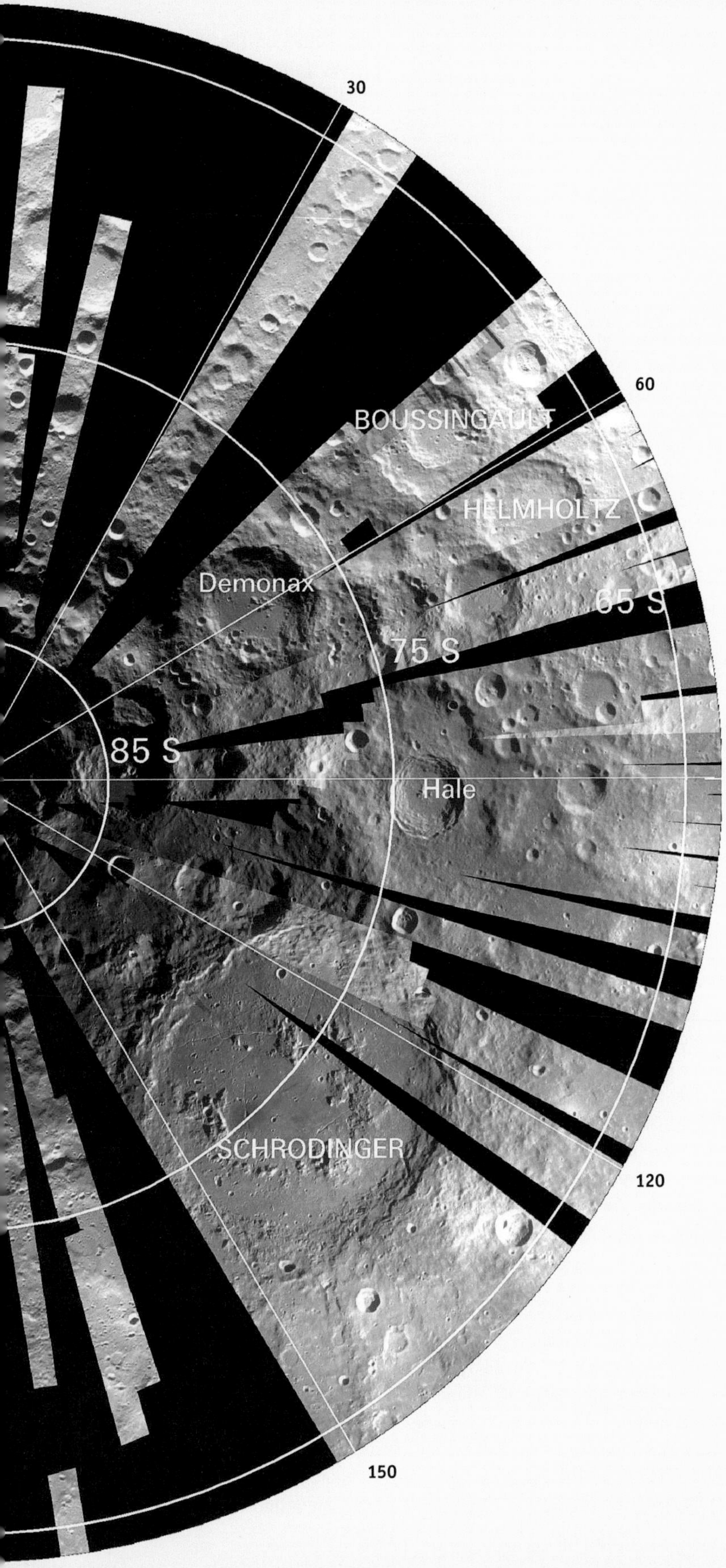

30
60
BOUSSINGAULT
HELMHOLTZ
Demonax
65 S
75 S
85 S
Hale
90
SCHRODINGER
120
150

ISRO

Chandrayaan-2

India

Under development

On 12 November 2007, a year before Chandrayaan-1 arrived at the moon, ISRO signed an agreement with Russia for Chandrayaan-2, a joint mission to the moon in 2011 or 2012. It would have three elements: an orbiter, lander, and rover. Russia had agreed to contribute the lander and rover. India would supply the PSLV launch vehicle and the orbiter. In November 2011, the Russian Zenit-2SB rocket, intended to carry to Mars two spacecraft – one Russian and the other Chinese – failed to leave Earth orbit, leading to the loss of both spacecraft. The resulting agency-wide review forced Roscosmos to first delay and then withdraw from the planned joint mission. Initially, ISRO sought to find alternative partners in NASA or ESA but then decided to go solo, making it a wholly Indian mission.

Since then the mission and spacecraft design have evolved and the initial launch date of 2016 and subsequent dates have been missed. A mission review in March 2018 recommended changes to the lander and to the mission flight profile, which necessitated a further redesign and delay. The provisional mission design consists of a lander, weighing 1.25 tonnes, with four instruments, and a rover, weighing 20 kg, with two instruments. The mission launch window was pushed to early 2019. The lander and rover will operate for at least two weeks but the orbiter is designed with a lifetime of about two years.

Each of the three components – orbiter, lander, and rover – has a distinct combination of science packages designed to make the best use of its local environment. Unlike Chandrayaan-1, Chandrayaan-2 does not involve international collaboration. The orbiter has a weight of 2,379 kg and carries five instruments. Two are enhanced versions of the those carried by Chandrayaan-1 and three will be entirely new. Following the March 2018 mission review, further redundancy was added, which increased the weight of the lander from 1,250 kg to about 1,471 kg and of the six-wheeled rover from 20 kg to 25 kg. Typically, for an increase of 1 kg in the spacecraft, the launch vehicle requires 4 kg of additional propellent. Initially ISRO planned to use a GSLV MkII launch vehicle, but that has now changed in favour of the more powerful GSLV Mk III.

Orbiter

The orbiter is designed to operate for a year in a 100 km lunar orbit. It has five instruments, all of which have been produced by ISRO. The Orbiter High Resolution Camera (OHRC) will continue the work of the TMC- 1 in Chandrayaan-1 and fill in the gaps from that mission to provide a complete 3D map of the lunar surface. The Imaging Infra-Red Spectrometer will scan the lunar surface using a wider wavelength range (extended from 3 microns in Chandrayaan-1 to 5 microns in Chandrayaan-2) to once again map the distribution of minerals, water molecules, and hydroxyl on the lunar surface.

Synthetic Aperture Radar (SAR), operating in a dual frequency, is expected to detect further evidence of water ice in the first few tens of metres below the lunar surface and in the regions permanently in shadow. CHACE- 2 (Chandra's Atmospheric Composition Explorer-2) will look at the composition of the tenuous lunar atmosphere during its descent. This is similar to the instrument carried aboard the MIP that detached from Chandrayaan-1 and captured data as it descended to its lunar surface impact.

Lander

The lander, Vikram, named after Vikram Sarabhai, the founder of India's space programme, will represent ISRO's first attempt to soft-land on an extra-terrestrial body. To facilitate the landing, it is equipped with an altimeter, velocity meter, and accelerometer. Developing this technology and associated software has been major hurdle for ISRO. The lander is

The planned lunar mission Chandrayaan-2

The Indian space probe Chandrayaan-2 will try to land a brief case-shaped rover on the lunar surface, 600 km from the lunar south pole. An orbiter flying above the moon will search for water – the prerequisite for future crewed missions.

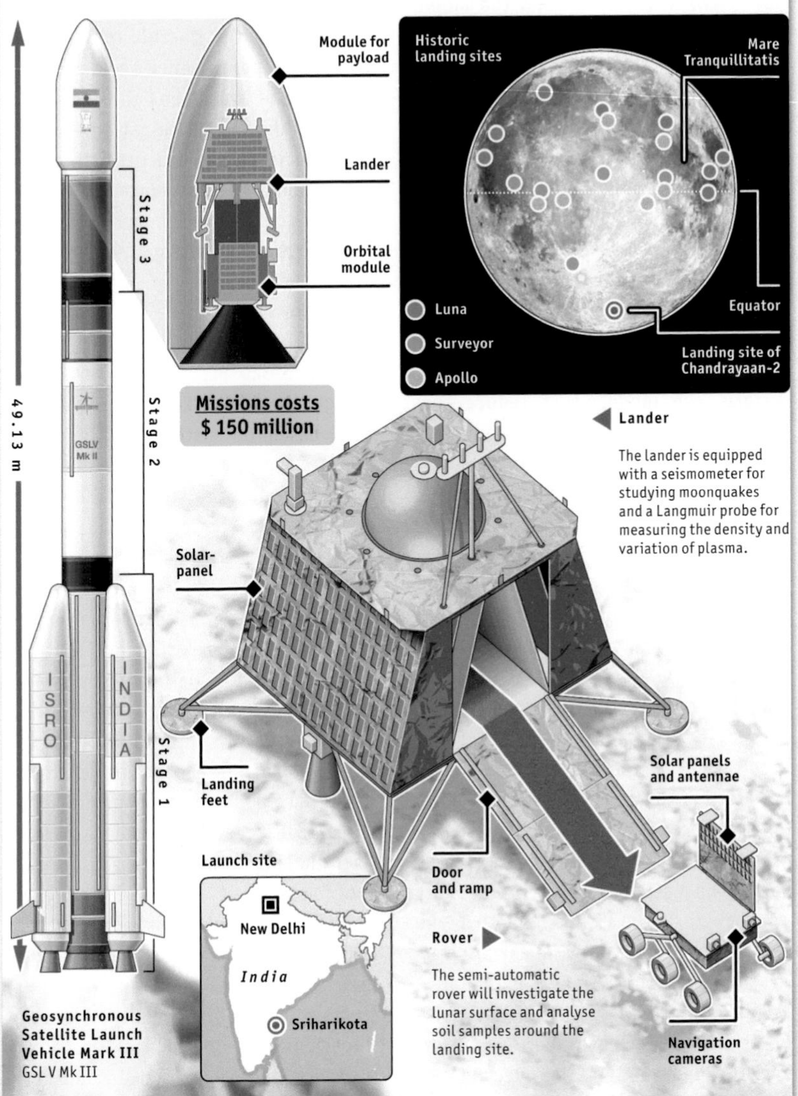

Graphic New/ISRO

ISRO

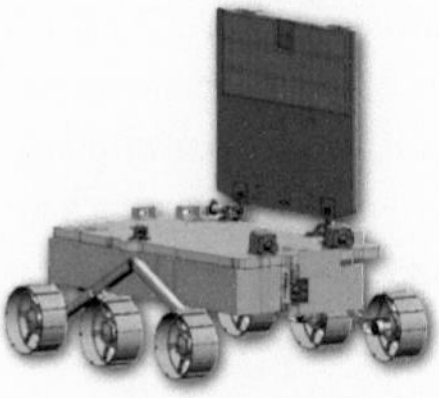

Rover

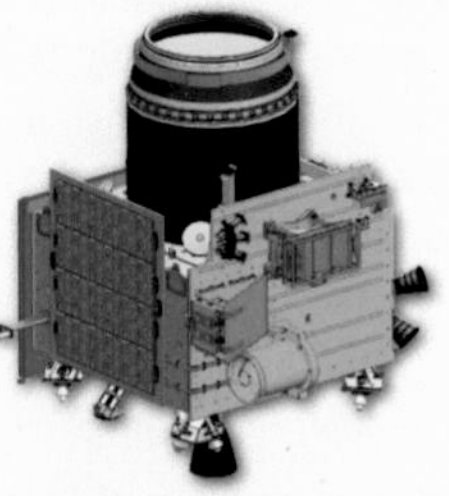

Orbiter

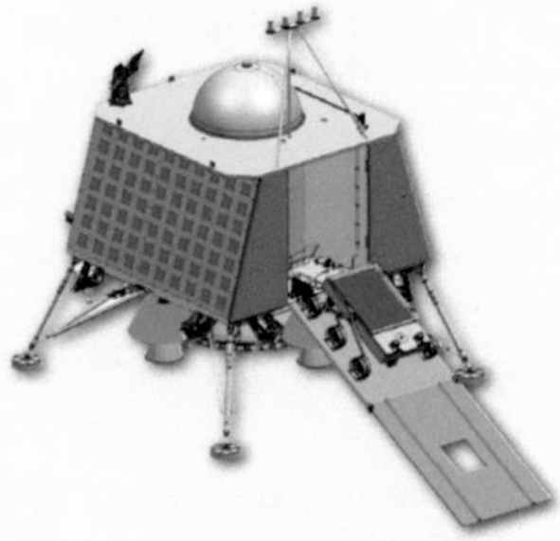

Lander

equipped with a high-resolution descent camera and a throttleable liquid-fuel main engine for a soft landing.
Given the time delay in the communication between the Earth and the moon, the lander is designed to land using real-time data from the altimeter and descent camera without intervention from Earth. Since the moon is devoid of any substantial atmosphere, parachutes cannot work. The review in March 2018 recommended strengthening the landing legs and increasing the number of the lander's engines from four to five. Prior to separation the orbiter will be moving at 1.6 km per second in a lunar orbit 100 km above the moon. During its journey to the lunar surface, the lander must reduce that speed to virtually zero for a safe landing. A preliminary orbit of 100 km by 30 km has been added to the mission profile to allow the lander engines to be tested prior to touchdown.
The lander's onboard science payload consists of a seismometer for studying moonquakes, a thermal probe (a thermometer) to measure the temperature at the landing site during the mission, and two experiments (Langmuir Probe and Radio Occultation) to detect electrons emanating from a few centimetres above the lunar surface. On Earth, the high-energy radiation from the sun is deflected by the Earth's magnetic field or absorbed by its atmosphere. In the absence of both on the moon, this high-energy radiation from the sun reaches its surface, and the interaction of the radiation with dust particles causes atoms to release electrons, resulting in a thin layer of plasma over the sunlit side of the moon.

Rover

The rover, weighing 27 kg, has a limited payload. The onboard cameras, antennae, and communication systems are powered by a battery, which is charged by a single solar panel. It will communicate with Earth via the orbiter, situated 100 km above the moon, and not via the lander just a few metres away. It has six wheels controlled by 10 motors. It will be guided by a pair of stereoscopic cameras, which provide a 3D view of its environment. To test the traction of its wheels and driving performance on Earth, a helium balloon is attached to the rover to simulate the reduced lunar gravity. The rover's science payload consists of two instruments. A Laser Induced Breakdown Spectroscope and a Langmuir probe will examine the lunar plasma.

Scientific temper

In around 320 BC, Alexander the Great brought the Greeks' knowledge of astronomy, based on mathematics, precision measurements, and systematic observations, to India. The structured approach of the Greeks, however, could not uproot the entrenched culture of astrology that had persisted in India for centuries. At first, observational astronomy served only to support the existing ancient traditions rather than ushering in a new way of thinking based on the scientific method. Over the next few centuries, Indian scientists continued to engage with the natural world through the language of science, albeit still steeped in ancient (and non-scientific) Indian rituals and traditions.
The founders of modern India recognised the value of science for the newest and largest democracy on the globe. Article 51A in the constitution requires that every citizen of India shall 'develop the scientific temper, humanism and the spirit of inquiry and reform'. In addition to a mission to the moon, India has a space observatory in Earth orbit and a spacecraft orbiting Mars. Both are operational and provide scientific data. India has plans to return to Mars and to visit Venus. Recently, the country announced its plans to launch human missions into space before 2022. As India stretches outward into the solar system, it is reaching back to meet its ancient roots.

A GSLV-MKIII rocket is being transported from a hall in the Satish Dhawan Space Centre in Sriharikota to its launch site. The Chandrayaan-2 mission will also be launched into space from the Indian space port, 50 kilometres to the north of Chennai.

ISRO

Not only James Bond villains like their rocket launch pads by the beach. The Japanese spaceport gives legendary film sets a run for their money.

Ingalls/NASA

Tanegashima Space Center

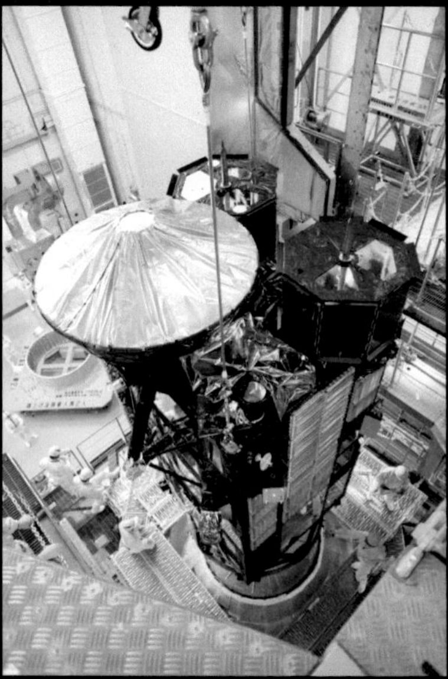

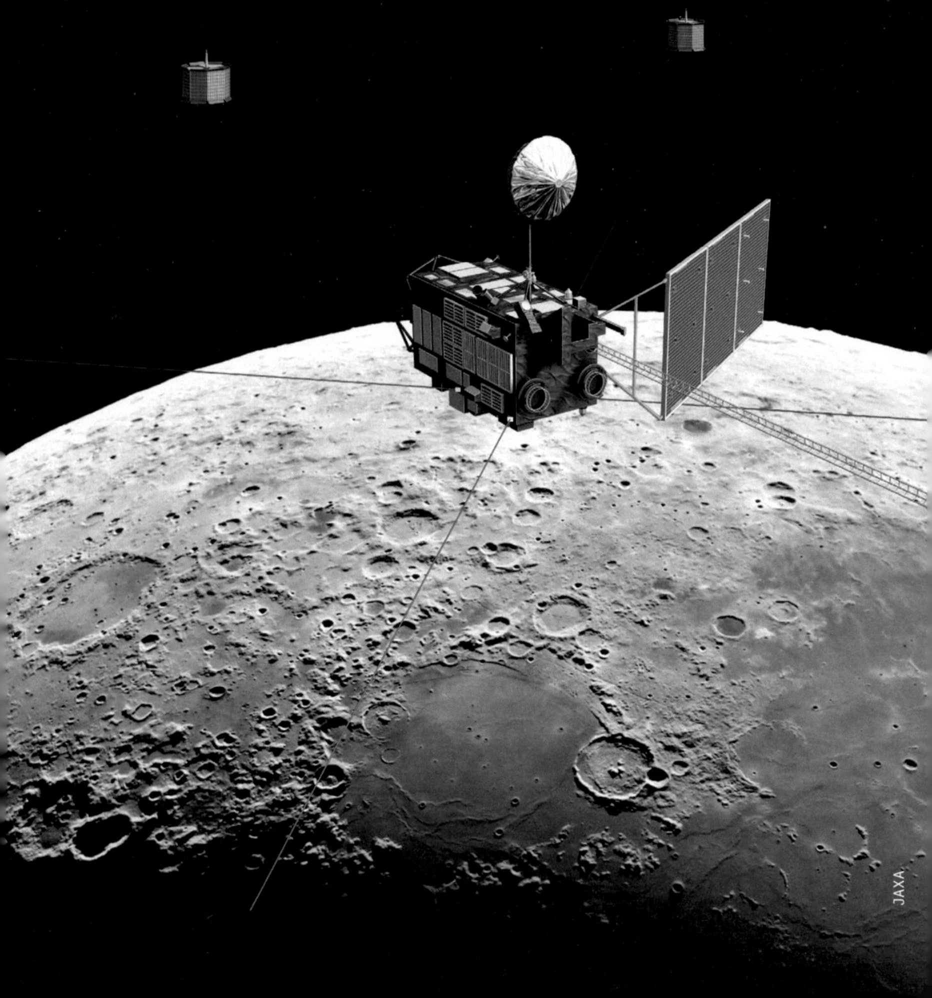

JAXA

Kaguya/SELENE
Japan

Launch: 14 September 2007, 10:31 am
Landing: 10 June 2009, 6:25 pm
Last contact: 10 June 2009, 6:25 pm
Launch site: Tanegashima Space Center
Crash location: 80.4°O 65.5°S

Small and large: the mother probe with its two small impactor probes

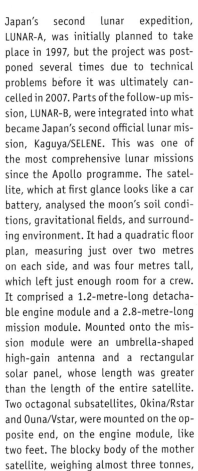

Okina/Rstar *(subsatellite)*
Japan

Release: 9 October 2007, 00:36 am
Landing: 12 February 2009, 7:46 pm
Last contact: 12 February 2009, 7:46 pm

Ouna/Vstar *(subsatellite)*
Japan

Release: 12 October 2007, 04:28 am
Landing: currently still in orbit
Last contact: unknown

Japan's second lunar expedition, LUNAR-A, was initially planned to take place in 1997, but the project was postponed several times due to technical problems before it was ultimately cancelled in 2007. Parts of the follow-up mission, LUNAR-B, were integrated into what became Japan's second official lunar mission, Kaguya/SELENE. This was one of the most comprehensive lunar missions since the Apollo programme. The satellite, which at first glance looks like a car battery, analysed the moon's soil conditions, gravitational fields, and surrounding environment. It had a quadratic floor plan, measuring just over two metres on each side, and was four metres tall, which left just enough room for a crew. It comprised a 1.2-metre-long detachable engine module and a 2.8-metre-long mission module. Mounted onto the mission module were an umbrella-shaped high-gain antenna and a rectangular solar panel, whose length was greater than the length of the entire satellite. Two octagonal subsatellites, Okina/Rstar and Ouna/Vstar, were mounted on the opposite end, on the engine module, like two feet. The blocky body of the mother satellite, weighing almost three tonnes, therefore recalled brutalist architecture. The mirror-reflected panels, mounted on the outer walls, looked like windows. Two panels, in which the names of and messages from 410,000 Japanese citizens were engraved, were also attached to the outer body. After the three satellites separated in lunar orbit, they entered into elliptical orbits of different sizes. This combination of simultaneous orbits made it possible for the first time to measure the gravitational fields of the far side of the moon and obtain more precise measurements and data in general. The triumvirate worked together for a year. The relay satellite Okina withstood the moon's gravitational force longer than expected and did not crash until over a year later. Kaguya followed suit four months later, after it had depleted its energy reserves, and was intentionally crashed into the moon at a speed of 6,000 km per hour. This successful mission's achievements included not only high-resolution 3D images of the lunar surface and a map of the uranium, thorium, and potassium on the lunar surface, but also the finding that there are caves on the moon, for example, a former lava tube measuring 50 km in length and 100 m in width.

A parade on Tiananmen Square in Beijing on 1 October 2009: the Chinese astronaut Zhai Zhigang waves the national flag from a spaceship module while his colleagues Jing Haipeng, Liu Boming, Nie Haisheng, and Fei Junlong stand before him.

China's Lunar Programme

Brian Harvey

'The great Asian space race' was the media title given to the extraordinary events in 2007 and 2008, when no less than three Asian countries sent spacecraft to the Moon. Japan won, sending Kaguya (launch: 14 September 2007) to orbit the moon with two mini-satellites, and India came in last, with Chandrayan-1 (launch: 22 October 2008). In reality, China, which came second, was the more important participant in this race. China's probe, Chang'e-1 (pronounced 'Chung urr') was the first of a series of probes on a schedule that would lead to a crewed lunar landing. This was not the case with its two rivals. Overlooked by the media, Japan already won the Asian moon race when it sent two small probes, Hiten and Hagoromo, to the moon in 1990. No other Japanese spacecraft has followed Kaguya, while India's second moon probe, Chandrayaan-2, is still due. These can therefore be seen as important but occasional missions, while China's Chang'e can be seen as the start of a more wide-reaching programme.

China's space programme started on 8 October 1956, a year before Sputnik, and in 1962, Nanjing University carried out the first studies for a lunar probe. Delayed by the Great Leap Forward and thereafter the Cultural Revolution, China did not launch its first satellite until 1970. In the following twenty years, China built scientific, recoverable communications satellites, but the launch rate was very slow: typically only one or two launches per year, and there were many years when there were none.

This changed in 1992 with the government decision to substantially expand its space programme, the flagship project being crewed flight. At this stage, what might be called the 'lunar underground' emerged. There had always been scientists in China who wanted a moon mission, and now they saw their chance. The most prominent figure was Ouyang Ziyuan, geologist, meteor expert, and the man who had studied the 500 g of Apollo moon rock donated to China by US President Carter in 1978. The lunar underground campaigned for a moon mission, held conferences and symposia in the universities, received funding to study such missions, gave press interviews, and cheekily suggested postponing the crewed space-flight programme in favour of a lunar probe. The underground published 67 papers and argued for a coherent programme with three stages: orbiting, soft landing, and sample return.

The Chinese government approved the moon programme on 28 February 2003. It allocated ¥ 1.4bn (€ 140 m) to the project and put its chief evangelist, Ouyang Ziyuan, in charge. A Lunar Exploration and Engineering Centre followed in 2005. The scientists were told to keep costs down by using existing technology, in this case an early generation drum-shaped Dong Fang Hong (DFH) 3 communications satellite.

Chang'e-1
China

Launch: 24 October 2007, 10:05 am
Landing: 1 March 2009, 8:03 am
Launch site: Xichang
Landing site: 1.50°S 52.36°0

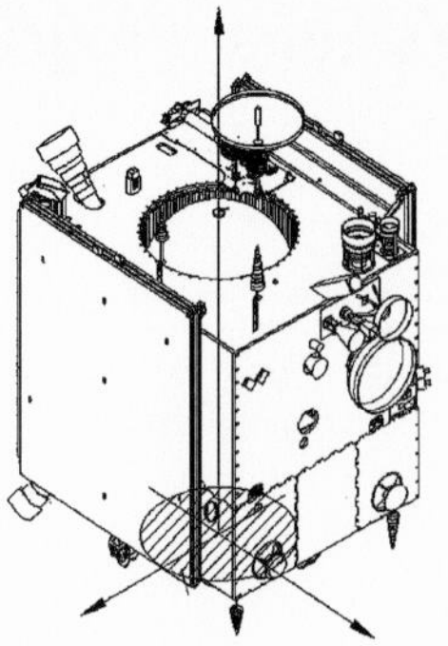

The moon probe was named Chang'e-1 after a beautiful fairy from Chinese mythology that flies to the moon. It was fitted with nine instruments: a stereo camera, an ultraviolet imager, an imaging spectrometer, a laser altimeter, a gamma-ray spectrometer, a microwave radiometer, a high-energy particle detector, a solar-wind-ion detector, and an X-ray spectrometer. It weighed 2,350 kg, half of which was taken up by fuel. It had solar cells generating 161 W of power and a transmission rate of 3 MB/s. Chang'e-1 took off on a Long March CZ-3A rocket from Xichang in the scenic hills of Sichuan on 24 October 2007. It received a wave of publicity, being covered live on television and watched by 2,500 event-paying spectators. The mission pushed the rocket to the limits of its capacity: Chang'e-1 was placed in a highly elliptical orbit, which was raised four times over the following two weeks, before the orbiter eventually reached the moon on 5 November. It took another three manoeuvres to reach an operational lunar orbit at a distance of around 200 km. China celebrated this outcome on 21 November, when Prime Minister Wen Jiabao formally unveiled the first of Chang'e-1's pictures of the moon while a band played the national anthem. Media hype or not, the Chang'e-1 mission was of great scientific value. It enabled China to make its own lunar map, at a resolution of 120 m, becoming the third country do so after the Soviet Union and the US. Moreover, it was able to create a topographic map, thanks to the 9 m measurements taken by the altimeter. A chemical map also followed, highlighting uranium, iron, titanium, and KREEP (Potassium (K), Rare Earth Elements (REE), and Phosphorus (P)). Chang'e-1 found new features, remapped mascons, recalculated the radius of the moon (1,737.013 km), measured the volume of Helium 3 (658,000 tonnes on the near side and 286,000 tonnes on the far side), measured the depth of the regolith across the surface, and detected solar plasma as it reached the Earth-moon system. On 1 March 2009, Chang'e-1 was intentionally crashed into the Moon, carrying a gift of 30 recorded songs on board.

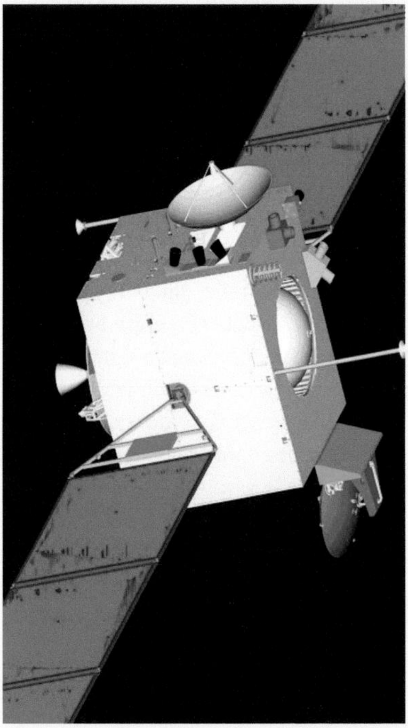
CNSA

3D drawing of Chang'e-1

Chang'e-2
China

Launch: 1 October 2010, 10:59 am
Launch site: Xichang
Landing site: currently in space

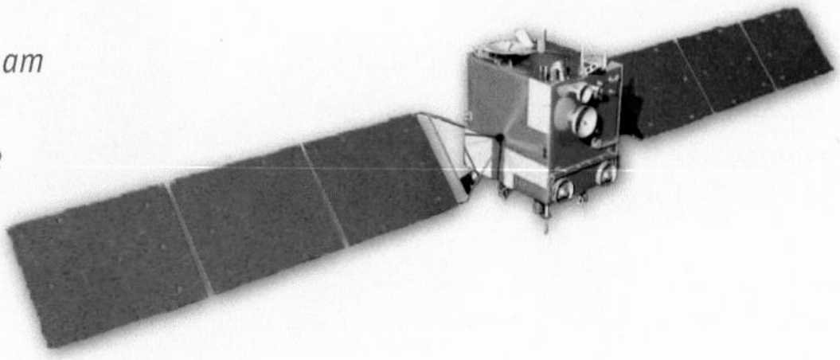

Although China could have proceeded directly to a lander project, it decided to launch its flight-qualified backup on what became one of the least publicised and most over-performing lunar missions ever. Chang'e-2 left Xi Chang on the more powerful and recently introduced CZ-3C rocket on 1 October 2010, the National Day of the PRC, hitting the single-second launch window exactly on the mark. This made possible a shorter, five-day transit to the moon, and Chang'e-2 reached its operational 100-km-high orbit on 9 November. Its first task was to take photographs of prospective touchdown sites for a lander. To this end, Chang'e-2 began on 26 October a series of dives down to a depth of 15 km, using a technique developed by the Soviet Union's Luna 19 and 22 in the 1970s. This is where the mission began to be different: Chang'e-2's cameras had a resolution of 1.2 m, a ten-fold improvement over that of the Soviet cameras. The first map compiled was of Sinus Iridum, the lander's prospective target. As was the case with the first Chang'e, there was a huge scientific return: a new and much more detailed map as well as new interpretations of lunar history. For the Chinese scientists, it was crucial to work with their own country's data rather than using data relayed by others. The following summer, Chang'e-2 still had 520 kg of fuel remaining. It had been hinted when it was launched that there might be an extended mission, though no one had paid much notice at the time. On 8 June 2011, Chang'e-2 fired out of lunar orbit and on 25 August reached the L2 point of gravitational stability in the Earth-moon system where its particle detectors began a continuous analysis of solar and high-energy particles. It was the furthest distance any Chinese space probe had travelled to from Earth, one requiring the construction of a network of new, large 50 m ground-tracking dishes. The best was yet to come. As sharp-eyed amateur observers noted, Chang'e-2 left L2 the following year, in April 2012. China later announced that Chang'e-2 was on course to intercept Toutatis, an asteroid, discovered in 1989, that orbits the sun every 4.02 years. When China remained silent on the interception date, 13 December, the mission was presumed to have failed. But several days later, China presented the outcome to an astonished scientific community. Chang'e-2 passed the asteroid at a distance of just 3.2 km, at a speed of 10.73 km/s, some 7 km away from Earth. It used tiny 358 g cameras to take 300 images of Toutatis with a resolution of 10 m as it whizzed by at a distance of 85 km. Toutatis proved to have the shape of a dog's bone, with craters, lumps, mounds, and boulders: an object that had been battered by collisions, spin, fragmentation, and recombination. Chang'e-2 is still in solar orbit but will be back in the Earth's vicinity in 2027. The most remarkable feature of the interception was that it was done by a country without 'autonav' – an automatic navigation technology, developed by the Soviet Union in 1971, that allows spacecraft to independently steer themselves towards a target. The Toutatis interception was achieved by precision aiming alone.

Chang'e-3 *Lander*
China

Launch: 1 December 2013, 5:30 pm
Landing: 14 December 2013, 1:11 pm
Launch site: Xichang
Landing site: 44.12°N 19.51°W

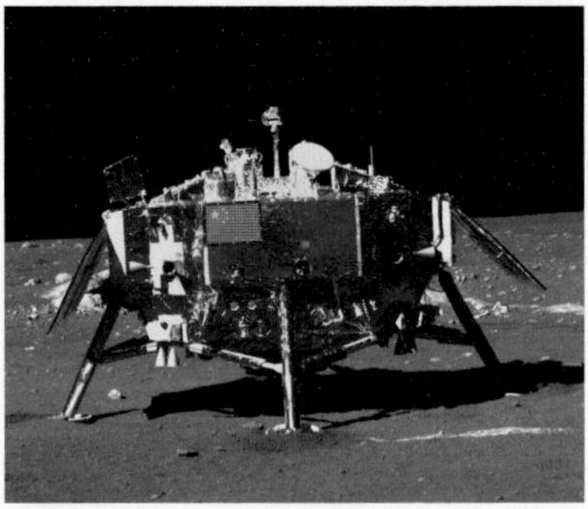
CNSA

Now China could proceed to the second stage of its lunar programme: the landing. This required a much larger spacecraft, a lander weighing 3,780 kg, which in turn required what was then the most powerful rocket in China's fleet, the Long March CZ-3B. The decision was made to send a small rover (140 kg) – the closest analogue being the American Sojourner which roved Mars in 1997 – rather than the car-sized Soviet Lunokhods. It was 1.5 m in length, 1 m in width, and 1.1 m in height. The most challenging engineering aspects were the development of a 7,500 kN throttleable engine and a landing guidance system. Here, engineers developed a system for computerising Chang'e-2's images and matching them against the terrain detected by the landing radar so that the lander could be steered into a smooth spot. Chang'e-3 was launched from Xi Chang on 1 December 2013. It entered lunar orbit six days later. Not only was the launch broadcast live on Chinese television, with rocket cameras (rocketcams) showing the Earth recede in the distance; the landing itself was also televised, a world first. Viewers could see the lander pitch over, descend to a smooth point, stir up a cloud of particles, and then come to a steady halt. It was the first soft landing on the moon since Luna 24 in 1976. The landing site was not Sinus Iridum as originally planned, but the nearby Mare Imbrium. The rover, called Yutu, or 'jade rabbit', was soon deployed. Unlike Lunokhod, which rolled down a ramp, Yutu drove out on a horizontal ramp which was then lowered like a platform on the back of a delivery truck. Yutu drove 21 m around the lander, as the two vehicles photographed one another. Yutu's next set of journeys had to await the next lunar day the following month. Yutu began a wider clockwise circle back toward the lander, but on 15 January the wheels seized up some 15 m from the lander. This was a disappointment, for the first Soviet Lunokhod had lasted eleven months, and the American rover

First photograph captured by the time-lapse camera aboard Chang'e-3

Chang'e-3 *Yutu*
China

Last contact: March 2015
Launch site: Xichang
Landing site: 4.0°S 11.0°W

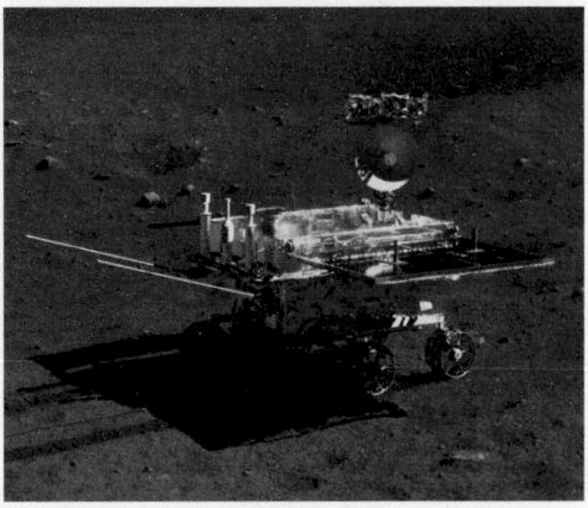

CNSA

Opportunity has been roving on Mars for over 14 years. The total distance travelled was 115 m (compared to Sojourner's 100 m). Although the failure was initially blamed on gravel getting stuck in the wheels, it transpired that there was an electronic problem in the power system. Yutu continued signalling until 3 August 2016 as the longest-lasting, but not moving, lunar rover. Nonetheless, both the lander and the rover were able to perform substantial scientific experiments, returning 2.86 TB of data to Earth by 2016. During eight traverses, Yutu's active particle X-ray spectrometer made a chemical analysis, while its lunar penetrating radar made a subsurface map at a depth of up to 140 m, making it possible to draw a cross section of the regolith. The Chinese scientists concluded that the spacecraft had landed on a new type of mare basalt never previously sampled. It represented some of the youngest volcanism on the moon, rich in olivine and ilmenite. There was a big crater, Zi Wei, and a large rock, Loong, nearby, and both were analysed. For the first time in many years, a new theory of lunar history could be formulated. The experiment that perhaps caught the most attention was the extreme ultraviolet telescope on the lander, which is still sending back images of galaxies. It has shown that the Earth is surrounded by bulging plasma and binary stars and has enabled the creation of new star maps.

CNSA

The Chang'e-3 lander on the moon

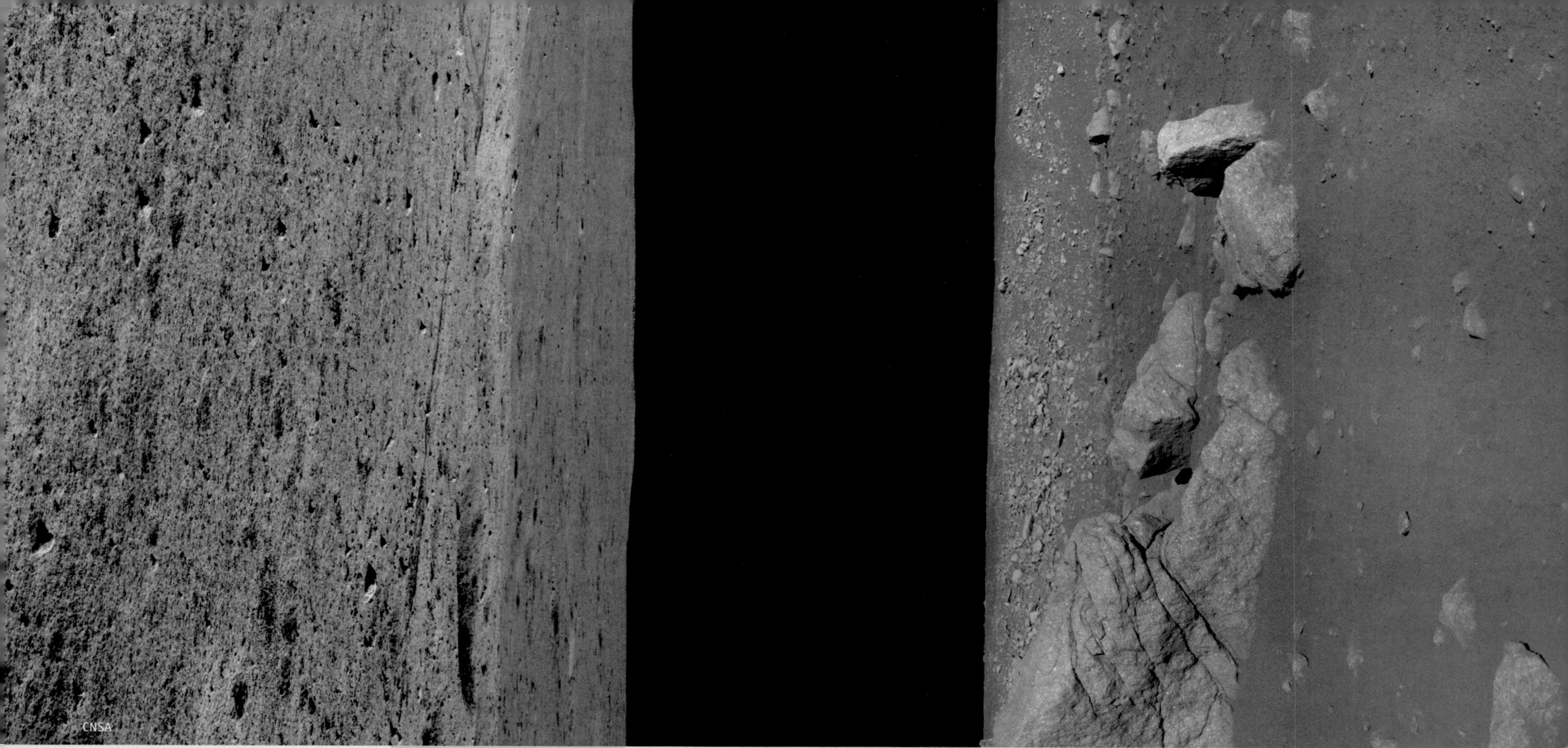
CNSA

Chang'e-5T1
China

Launch: 24 October 2014, 6:00 pm
Launch site: Xichang
Landing site: currently still in orbit

Chang'e-3 was considered so successful that its flight-ready backup was placed in the hanger soon afterwards and launched for a sample-return mission. To this end, China followed the same approach as the Soviet Luna missions, but with one critical difference. The Luna return cabin had flown directly from the moon back to Earth, using a carefully calculated trajectory, whereby a single but precise launch from the moon let it fall into Earth's gravity and return to Earth without the need for a course correction. China instead chose to carry out a lunar orbit rendezvous: the cabin housing the lunar sample would be launched into lunar orbit, dock with an orbiter, and transfer the sample, and the orbiter would return to the Earth, as did the American Apollo spacecraft. A test mission – Chang'e-5 Test 1 (5TI) – was deemed necessary. Chang'e-5T1 again used the small DFH-3 comsat spacecraft and was launched on the CZ-3B on 24 October 2014. It was expected that it would fly around the moon and return to Earth, like the Soviet Zond tests of 1968–1970. Chang'e-5T1 duly rounded the moon, took remarkable photographs of the Earth-moon system, and, as it returned to Earth, released a sub-scale model of the return spacecraft, Xiaofei ('little flier'). Like Zond, Xiaofei made a double-skip re-entry to reduce heat and speed and touched down on the same landing site used by the crewed Shenzhou spacecraft, in the grasslands of inner Mongolia. This appeared to be the end of an effortlessly successful mission, but as it approached Earth, the mother craft made an evasion manoeuvre with some of its 800 kg fuel remaining. Chang'e-5T1 flew back to the moon, where it made a gravity-assisted manoeuvre to fly onto a halo orbit at L2. Next, Chang'e-5T1 fired out of L2 and braked into a 200 km lunar orbit in January 2015. In February, it began a series of manoeuvres as the mother ship to rendezvous with a phantom target, the following month diving to 18 km to play the reverse role of the ascent stage. Its cameras sent back 1 m resolution images of prospective landing sites for Chang'e-5.

Chang'e-5

This impressive tour de force cleared the way for the Chang'e-5 lunar sample return mission, set for November 2017. Then came the first setback in the programme. The sample return required such a large spacecraft that the new Long March CZ-5 rocket was needed, and the latter failed on its second launch attempt on 2 July 2017. A turbine broke in one of its two engines, starving it of fuel, and it crashed into the Pacific. It was grounded for two years while the problem was resolved. The principal engineering challenges for Chang'e-5 were: the new engine for the orbiter and ascent stage, each with a capacity of 3,000 kN; radar systems for lunar orbit rendezvous, from which radars from the Shenzhou crewed programme were adapted; and a drill to recover lunar rock. Systems for transferring samples from one spacecraft to another using remote arms were tested in Earth orbit. In 2017, a landing site was announced: 150 km north-east of the dome of Mons Rümker in the north-west of the moon. The mission was expected to acquire samples of the youngest volcanic rocks. If successful, it will be followed by Chang'e-6P1 (2023) and 6P2 (2023), which will go to the lunar poles.

CNSA

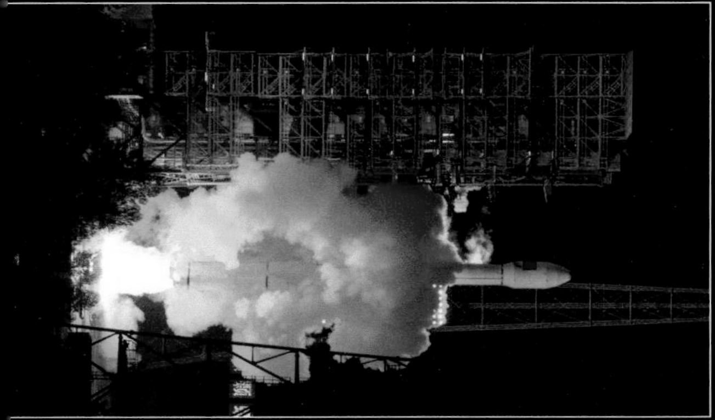

CNSA

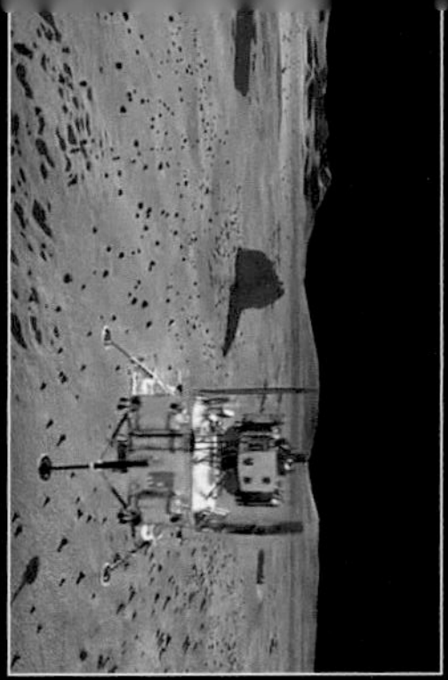

CNSA

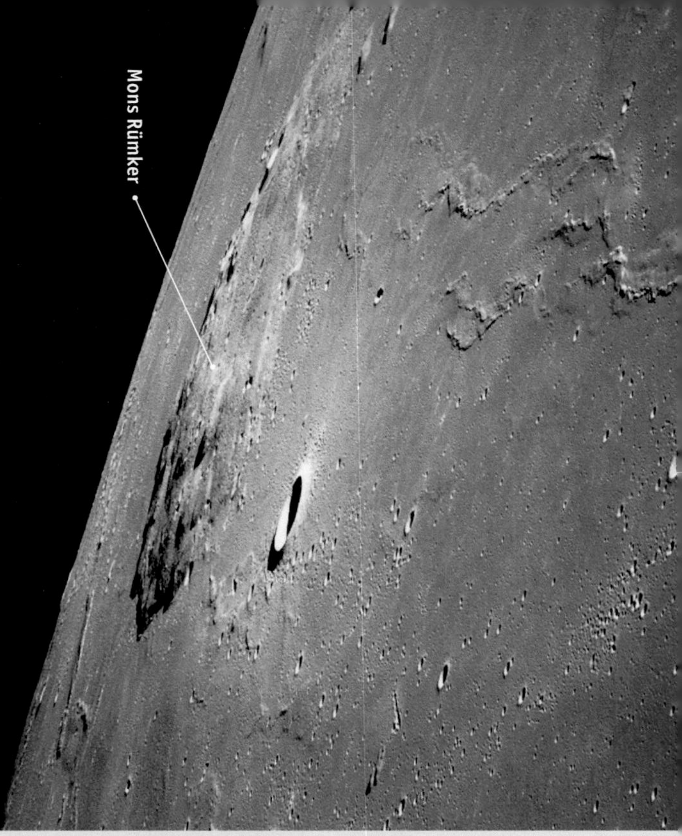

Investigating the uncrewed landing capsule after its impact

CNSA

Chang'e-4
China

Launch: 7 December 2018, 6:23 pm
Landing: 3 January 2019, 2:26 am
Launch site: Xichang
Landing site: 44.8°S 175.9°

Queqiao
China

Launch: 21 May 2018
Launch site: Xichang
Landing site: currently still in orbit

With Chang'e-5 delayed, the problem of what to do with the Chang'e-3 backup resolved itself. It would indeed repeat the mission of Chang'e-3, but in a daringly different location and as part of an international cooperation. China selected the far side of the moon as the target, where no spacecraft had ever soft-landed, though such a mission had been sketched by the Soviet Union in the 1970s. Landing on the far side of the moon would, to say the least, be difficult, for it would by definition be out of sight of ground control. Not only that, but the far side was much rockier than the near side with only two flat parts, the crater Tsiolkovsky and Mare Moskve. The Chinese scientists selected another landing site altogether, the crater von Karman in Mare Ingenii, since redefined as the South Pole Aitken Basin. On the other hand, a far-side landing offered opportunities to examine the oldest parts of the moon and its deepest crustal features. Foreign contributions came from Germany (neutron dosimeter for finding water), the Netherlands (a radio interferometer), Saudi Arabia (cameras), and Sweden (particle analysers). The most attention-grabbing experiment was housed in the top of the lander: a mini-ecosystem of silkworm eggs, arabidopsis seeds, and potato seeds in a nutrient solution, stored in an aluminium box weighing 3 kg and 18 cm in height. It was developed by Chongqing University with 28 other universities and became he first-ever biological lab to land on the moon. The idea was that once the first rays of the sun flooded into the box, the silkworms would hatch and produce carbon dioxide. The potato and arabidopsis seeds would then absorb the carbon dioxide and emit oxygen, which would then help the silkworms. All of this would be livestreamed by a camera for television and the internet. The first plant to be grown on the moon would be the humble potato. The first task, though, was to send up a relay satellite to enable communication between the descending (and thereafter descended) lander with Earth. A relay satellite was launched on 21 May on the small Long March CZ-4C rocket from Xi Chang. After a public naming competition, it was called Queqiao, or 'bridge of magpies', based on a folklore story of lovers reunited across the Milky Way with the help of magpies making a bridge. Queqiao also acquired two passengers – two mini-satellites called Longjiang ('Dragon River'). Queqiao passed the moon on 25 May, dropped off the two mini-satellites into lunar orbit (Longjiang 2 was lost), and entered a halo orbit of 14 days, 65,000 km from the moon. The orbit had the form of an irregular 3D curve behind and above the moon with a view of Earth and the lunar far side.

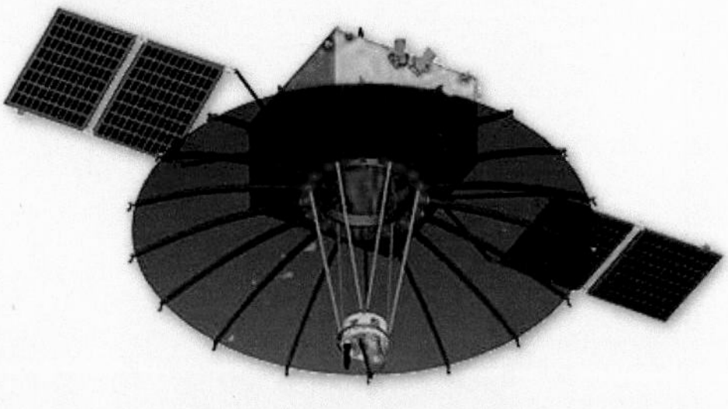

Queqiao

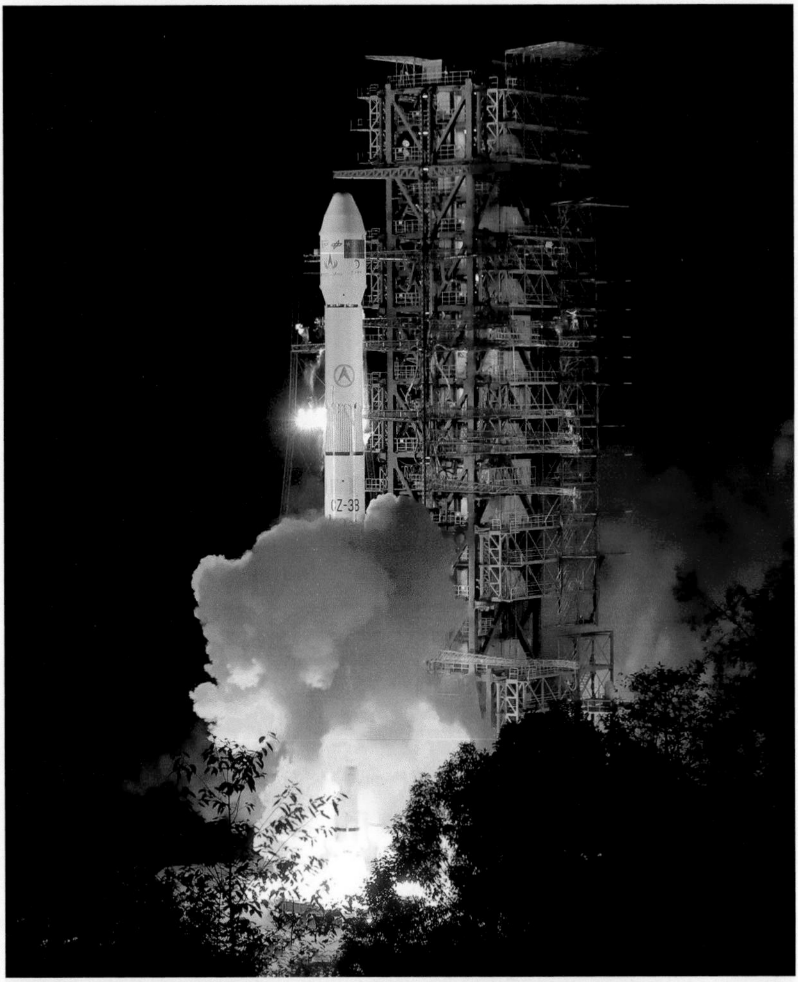

CNSA

Launch of Chang'e-4 on 7 December 2018 with the CZ-3B carrier rocket

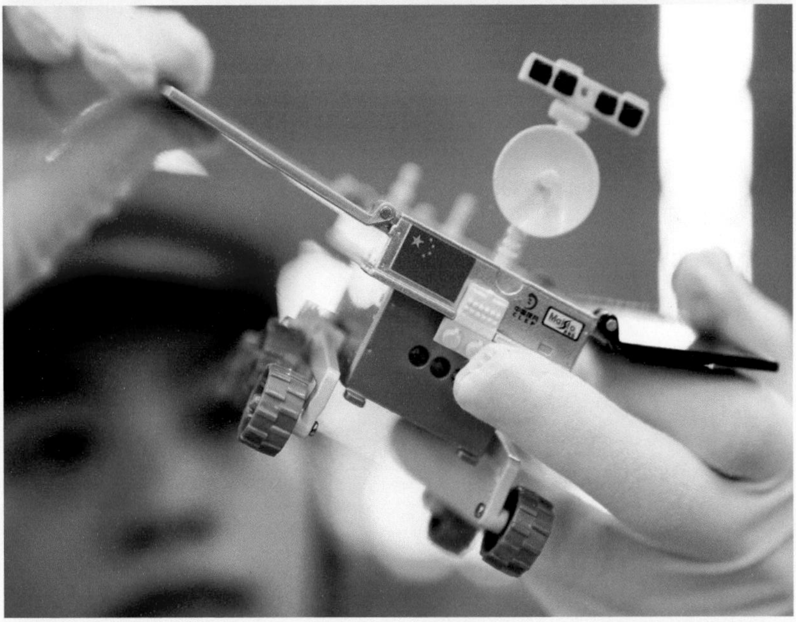

Reuters / Stringer

A toy manufacturer employee with a model of the Chang'e-4 rover during a quality check

CNSA

CNSA

CNSA

Finding a parking spot is not a problem: the Chinese rover exploring the far side of the moon

…hang e-4 is the first mission to involve … landing on the far side of the moon. …On 3 January 2019, the probe landed in …he crater Von Kármán, one of the largest …known impact craters in the solar system …diameter: 2,600 kilometres).

The first photo taken by Chang'e-4 after its landing

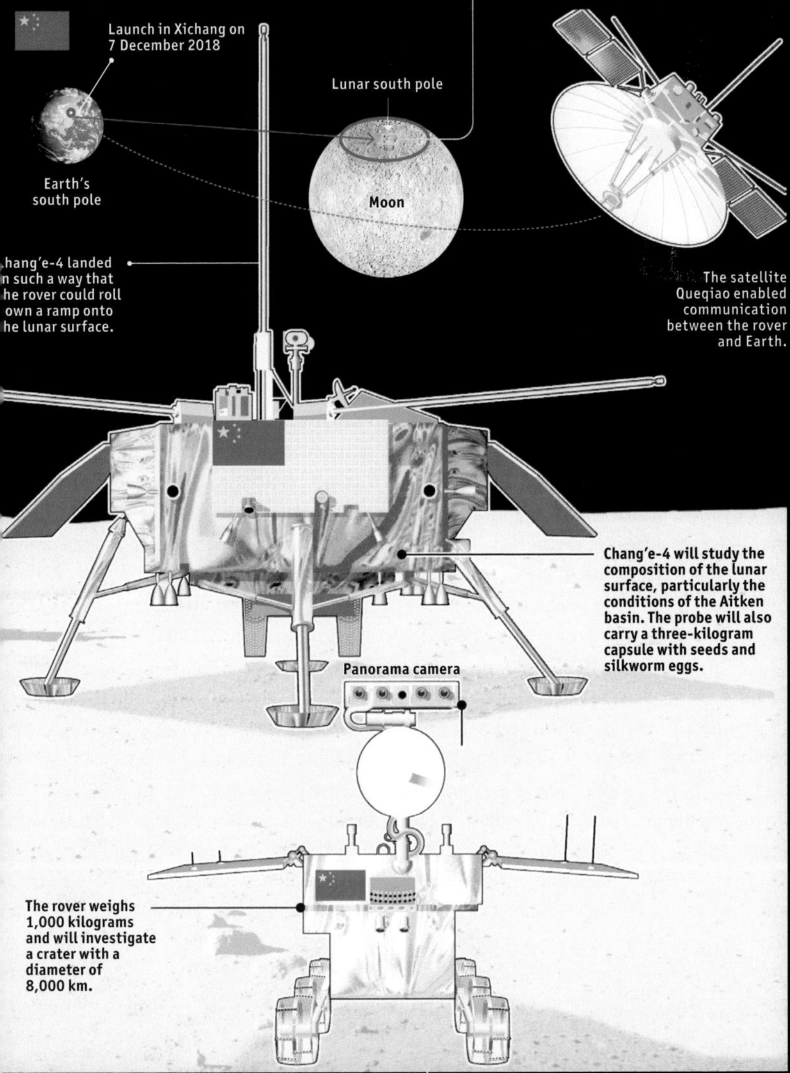
Launch in Xichang on
7 December 2018
Lunar south pole
Earth's
south pole
Moon
…hang'e-4 landed
…n such a way that
…he rover could roll
…own a ramp onto
…he lunar surface.
The satellite
Queqiao enabled
communication
between the rover
and Earth.
Chang'e-4 will study the
composition of the lunar
surface, particularly the
conditions of the Aitken
basin. The probe will also
carry a three-kilogram
capsule with seeds and
silkworm eggs.
Panorama camera
The rover weighs
1,000 kilograms
and will investigate
a crater with a
diameter of
8,000 km.

China's Lunar Base

Brian Harvey

The ultimate goal of the Chang'e programme is to set up a base on the moon. 'Roadmap 2050', a prospectus for China's scientific progress, identified a lunar base as an objective to reach by 2040. A Chinese crewed moon landing is not yet a formal government objective, although enthusiastic interviews given by Ouyang Ziyuan might encourage the media to think so. Nevertheless, several preparatory steps have been made to bring that lunar base closer to reality.

First, work has begun on a large launcher, the Long March CZ-9, intended to land Chinese astronauts (yuhangyuan in Mandarin) on the moon, with a first launch date set for between 2028 and 2030. Although this is not yet a government-approved project, development of the powerful YF-650 and YF-220 rocket engines has been approved and begun (the Chinese and Russian philosophy is always to start with the engines). The CZ-9 has been in the design stage since 2005, relentlessly pushed forward by the top Chinese rocket designer, Long Lehao. The design was finalised in 2018, and it will have a payload of 140 tonnes, height of 93 m, diameter of 10 m, weight of 4,000 tonnes, and thrust of 6,000 tonnes, making it bigger and more powerful than the American Saturn V and Space Launch System or the Russian N-1 or Energia. A rocket of this size will enable China to achieve not the kind of land-and-quickly-return mission, as were the Apollo missions, but will instead start with lunar surface missions lasting a month or longer, as had been sketched by the Soviet Union in the 1980s. In 2017, Chinese engineers travelled to Dnipropetrovsk in the Ukraine. This is where the Yangel design bureau built and successfully tested the Soviet lunar module, one whose RD-843 engine now powers the European Vega rocket. Updated, improved, and fitted with new electronics not available back then, this will be the basis of the Chinese lunar lander.

Second, China has already begun to design a lunar base. In 2017 at the Global Space Exploration Conference (GLEX) in Beijing, Professor Guo Linli published its design, a habitat of rigid, inflatable, and 3D printed modules. She had published a design for a crewed lunar base as far back as 2013, but nobody had taken much notice then – but they did this time.

Third, China has built a prototype lunar base on Earth, the Yuegong ('lunar palace'). Developed by Liu Hong of Beihang University, it comprises three modules of 500 m^3 with living areas and a garden. It is intended to be self-sustaining, as a moon base must be, with yuhangyuan growing and eating their own food and recycling their own air and water. It has been tested by volunteers with experiments lasting 105 or 370 days (between 2014 and 2018), leading to important gains in knowledge of self-sustainability and the interactions of participants. A perhaps unexpected finding was the unsuitability of men for this type of project: women eat less, require less air and water, and may be much better candidates for China's lunar base in 2040.

CNSA

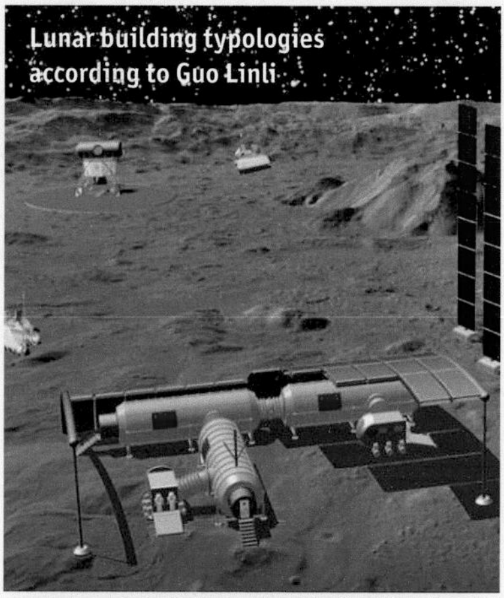

Type 1: rigid cabin structure

Type 2: inflatable structure

CNSA

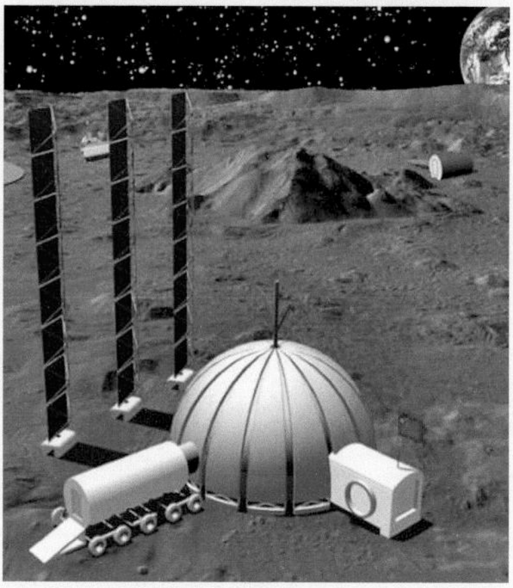

Type 3: rigid and flexible structure

Type 4: architectural structure

Model of the Yuegong-1 training station, where test subjects went through a feasibility study between 2017 and 2018.

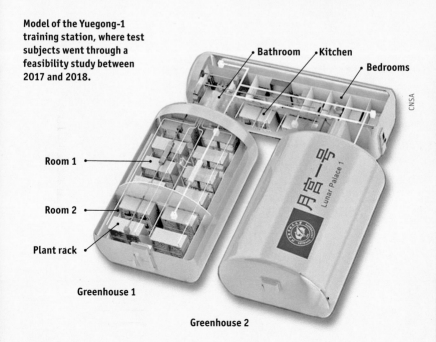

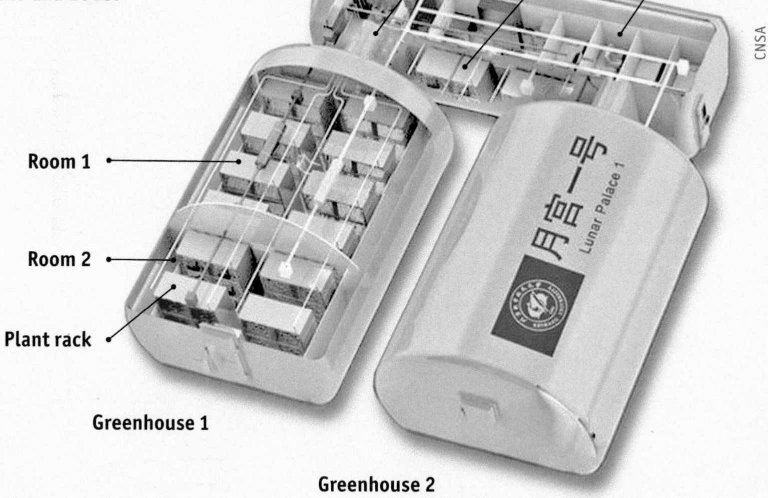

CNSA

CNSA

Yuegong-1, the Chinese Moon Palace, as a test station on Earth

LCROSS *Shepherding Spacecraft*
US

Launch: 18 June 2009, 9:32 pm
Landing: 9 October 2009, 11:37 am
Last contact: 9 October 2009, 11:37 am
Launch site: Cape Canaveral
Crash location: 84.72°S 49.61°0

NASA

Centaur Upper Stage *EDUS*
US

Landing: 9 October 2009, 11:31 am
Landing site: 84.68°S 48.71°0

It had been over ten years since the US had last sent a probe to the moon. The LCROSS (Lunar Crater Observation and Sensing Satellite) mission was, alongside the missions to Mars, one of the few NASA missions that flew beyond Earth orbit. The mission's objective, as was the case with Lunar Prospector, was to find evidence of water ice on the moon. LCROSS achieved this objective by jettisoning the upper stage of its launch vehicle into a lunar crater – a strategy that had been used during the Apollo missions since Apollo 13. LCROSS therefore comprised two parts: the Shepherding Spacecraft (S-S/C) satellite and the Atlas V carrier rocket's Centaur Earth Departure Upper Stage (EDUS). This upper stage had the form of a long cylinder with truncated cones at the two ends, one of which featured a jet propulsion engine. Attached on the other end was a compact satellite, measuring only 10 cubic metres and comprising a cylinder with a fuel tank and engine on one end and a high-gain antenna on the other. Rectangular solar panels were attached around the round, barrel-shaped body, forming a hexagonal shell around it. All experiments and communications instruments were mounted on the rear side of the panels. The satellite and the upper stage remained attached until after they entered lunar orbit. They circled the moon together for almost four months, and it wasn't until and didn't separate until shortly before the end of the mission. The upper stage was intentionally crashed into a pre-selected crater in the lunar south pole, while the Shepherding Spacecraft satellite braked in order to create a distance from the crashing upper stage. After four minutes, the satellite flew through the cloud created by the collision, analysing the composition of the stirred-up lunar soil before impacting the moon itself. It was crucial that the upper stage's fuel reserves were nearly empty before the crash; otherwise, they could have distorted the measurements. The scientists used optical spectrometers to carry out the analysis. They confirmed the presence of hydroxyl fragments in the cloud. The compound, which by all appearances seems to be formed during the lunar day, led to the deduction that ice or water is present in the crater.

NASA

KOHLER

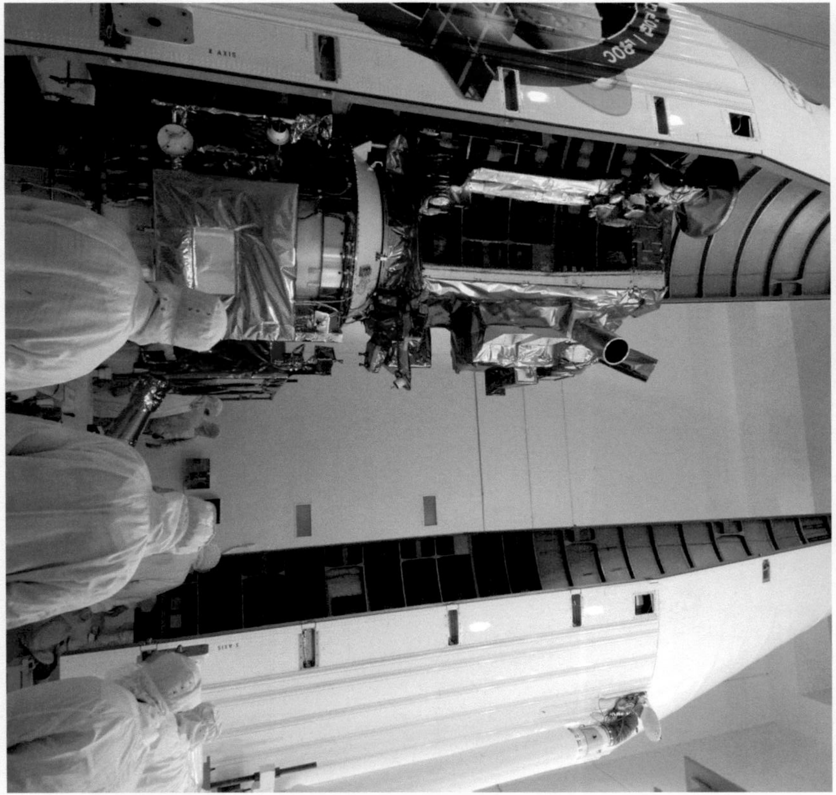

NASA

Lunar Reconnaissance Orbiter
US

Launch: 18 June 2009,9:32 pm
Launch site: Cape Canaveral
Crash location: currently still in orbit

NASA

The Lunar Reconnaissance Orbiter (LRO) was launched on an Atlas V rocket on 18 June 2009, at the same time as LCROSS. The probe, built in the Maryland-based Goddard Space Flight Center, was primarily equipped with instruments manufactured by private companies. Its seven experiments were designed not only to map the entire lunar surface at a high resolution but also to measure the cosmic radiation in the lunar environment and analyse its biological effects. The mission, like most lunar missions that took place in the new millennium, aimed to find evidence of water on the moon. The probe, wrapped in a grey foil at the Goddard Space Flight Center, unfolded in outer space to become a satellite resembling the Japanese Kaguya/SELENE probe: a cuboid with an engine, scientific instruments, a protruding antenna mounted on a mast, and a wing composed of three large solar panels. The probe circled the moon for 60 days in a polar orbit. During this time, all devices were calibrated and prepared. Afterwards, a one-year exploratory mission and two-year scientific mission took place. The probe documented all Apollo landing sites, where the remnants of landing modules can still be seen, and discovered new potential landing sites. It also determined the collision sites of failed missions and located where the mission of the Russian lunar rover Lunokhod 1 had ended. The third stage of the mission is still ongoing: the probe continues to collect data and transmit photographs. One day, the probe will be intentionally crashed into the moon.

NASA

Launch of the Atlas V rocket on 18 June 2009

NASA

Preparing to launch the LCROSS/LRO mission

NASA

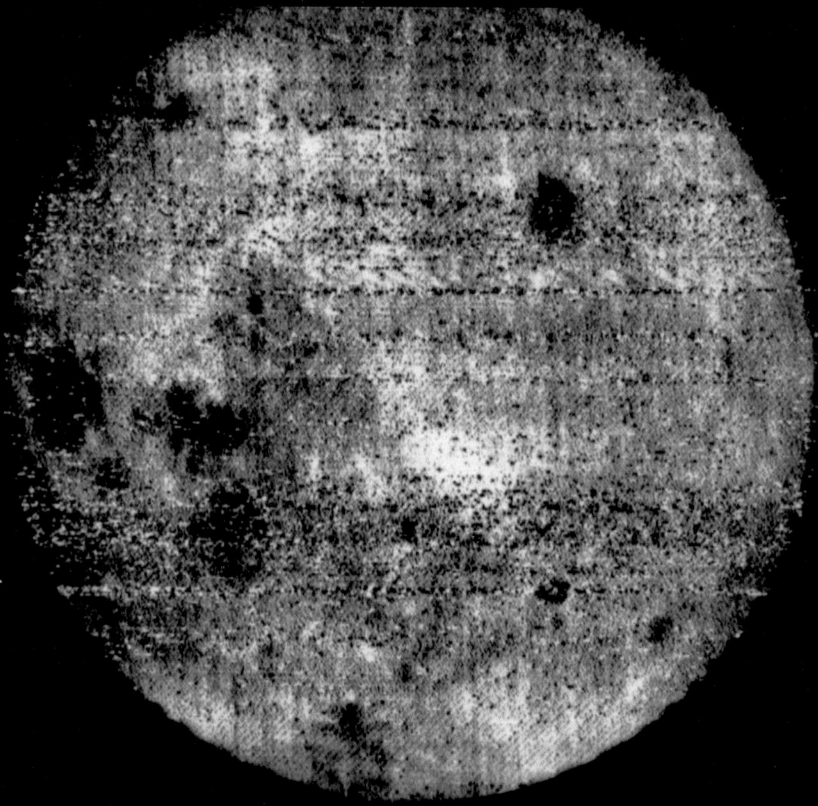

Moscow Polytech

The first photo of the far side of the moon by Luna 3 (left) looks rather grainy compared to the one taken by its American successor LRO 50 years later (right).

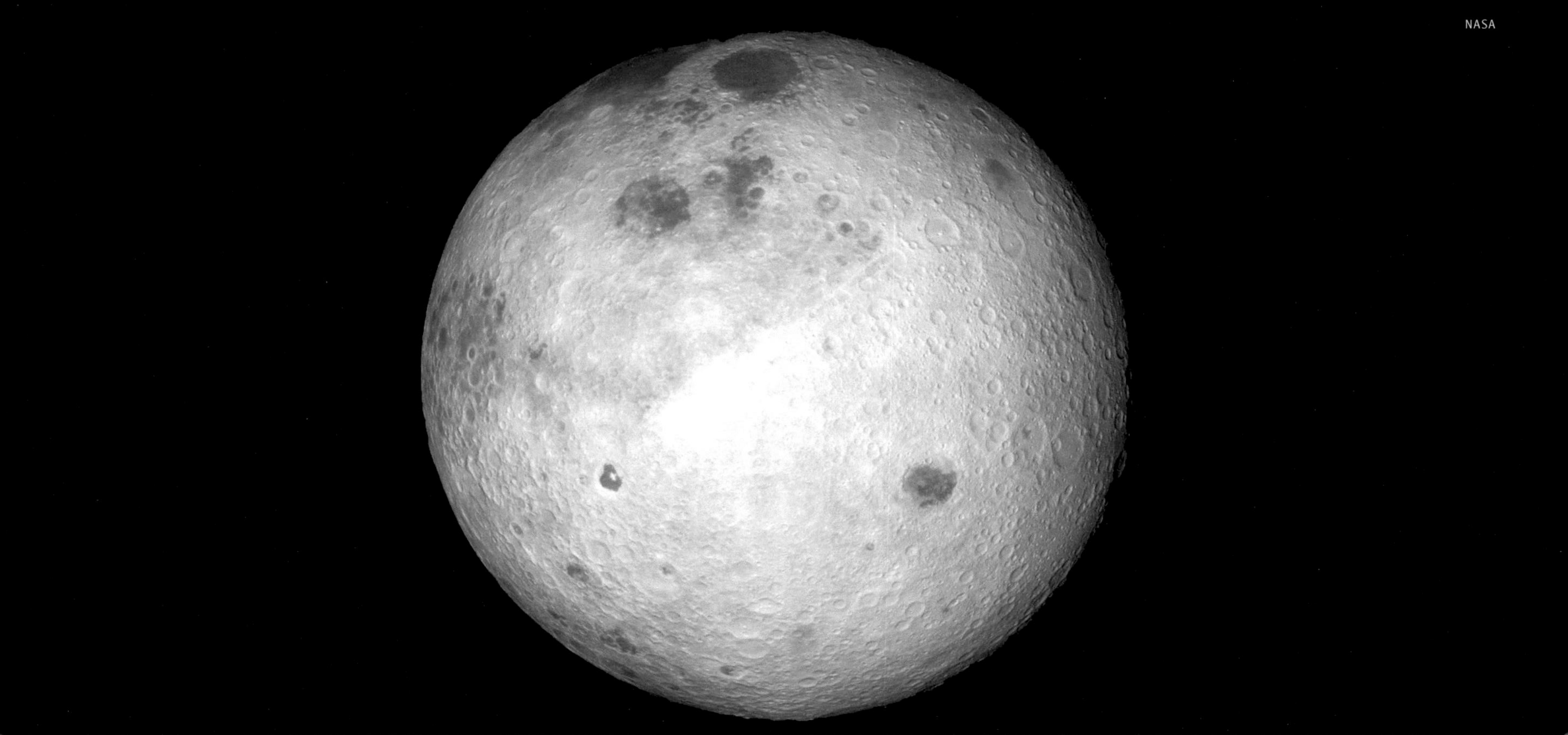

NASA

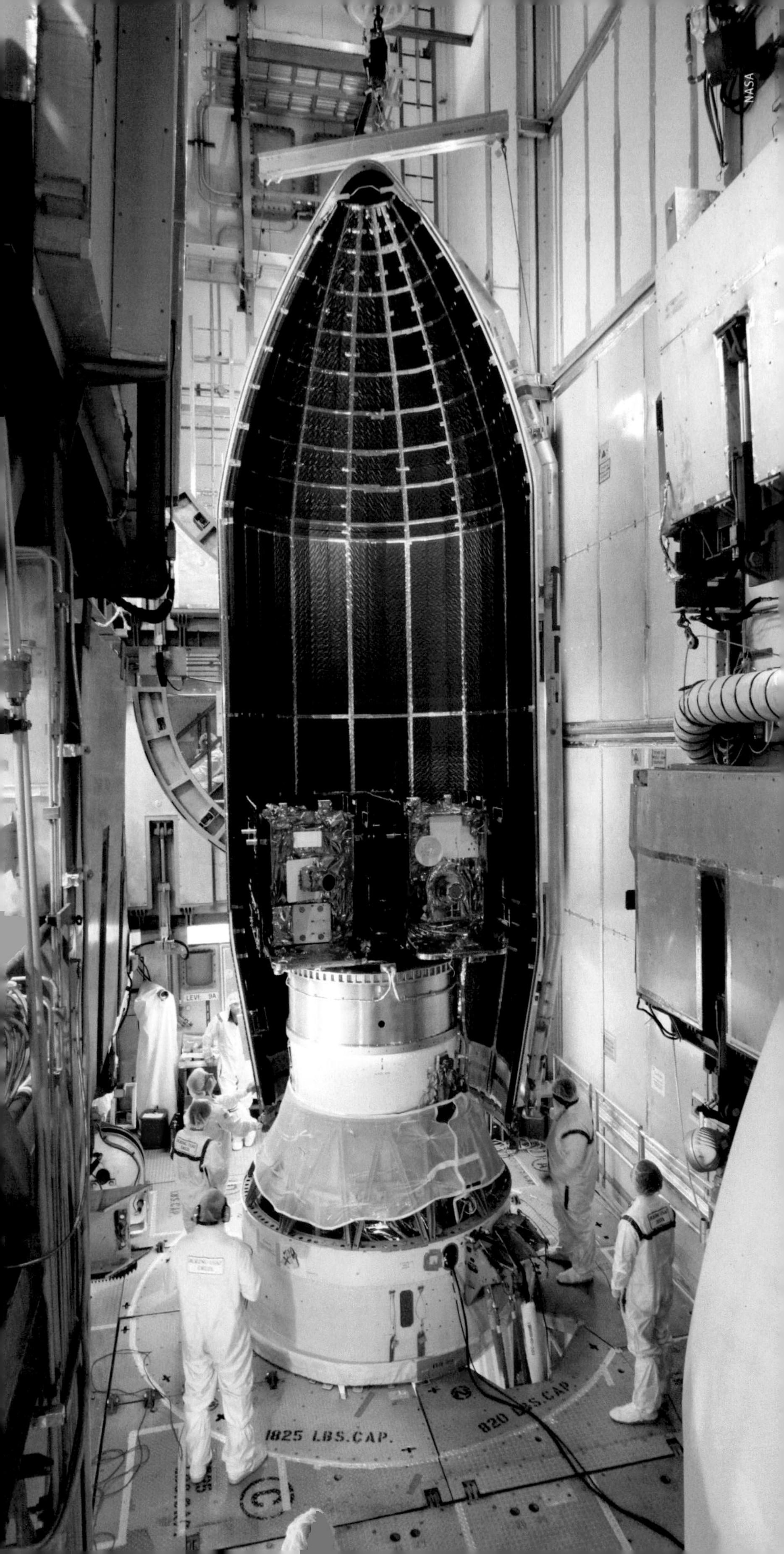
NASA
1825 LBS.CAP.
820 LBS.CAP

NASA

GRAIL A Ebb
US

Launch: 10 September 2011, 1:08 pm
Landing: 17 December 2012, 10:28 pm
Last contact: 17 December 2012
Launch site: Cape Canaveral
Landing site: 75.61°N 33.40°O

GRAIL B Flow
US

Launch: 10 September 2011, 1:08 pm
Landing: 17 December 2012, 10:28 pm
Last contact: 17 December 2012
Launch site: Cape Canaveral
Landing site: 75.65°N 33.16°O

The Gravity Recovery and Interior Laboratory (GRAIL) was part of NASA's Discovery Program, a series of short-term, low-cost, and highly specialised space missions. GRAIL comprised two satellites, GRAIL A Ebb and GRAIL B Flow, which created new maps of the moon's gravitational field. These maps in turn enabled scientists to analyse the composition of the moon's interior structure. The two satellites were virtually identical in form. Each comprised a flat cuboid – where the instruments and communications devices were housed – fully covered by a solar panel on both sides. The solar panels unfolded like wings during the journey through space. The twin satellites separated from the Delta II rocket one after the other, with an interval of eight minutes, but their entry into lunar orbit took place 25 hours apart. They remained in the same orbit as a pair, just 50 km from the moon. Reaching the orbit required a three-month flight and very precise preparatory work, which in fact delayed the launch four times. It wasn't until March 2012, half a year after the carrier rocket's launch, that the scientific measurements began. The main data was collected after 82 successful mission days. Afterwards, it was decided to take more precise measurements in an orbit 23 km above the lunar surface. The satellites were intentionally crashed into the northern region of the moon 14 months after the mission's launch.

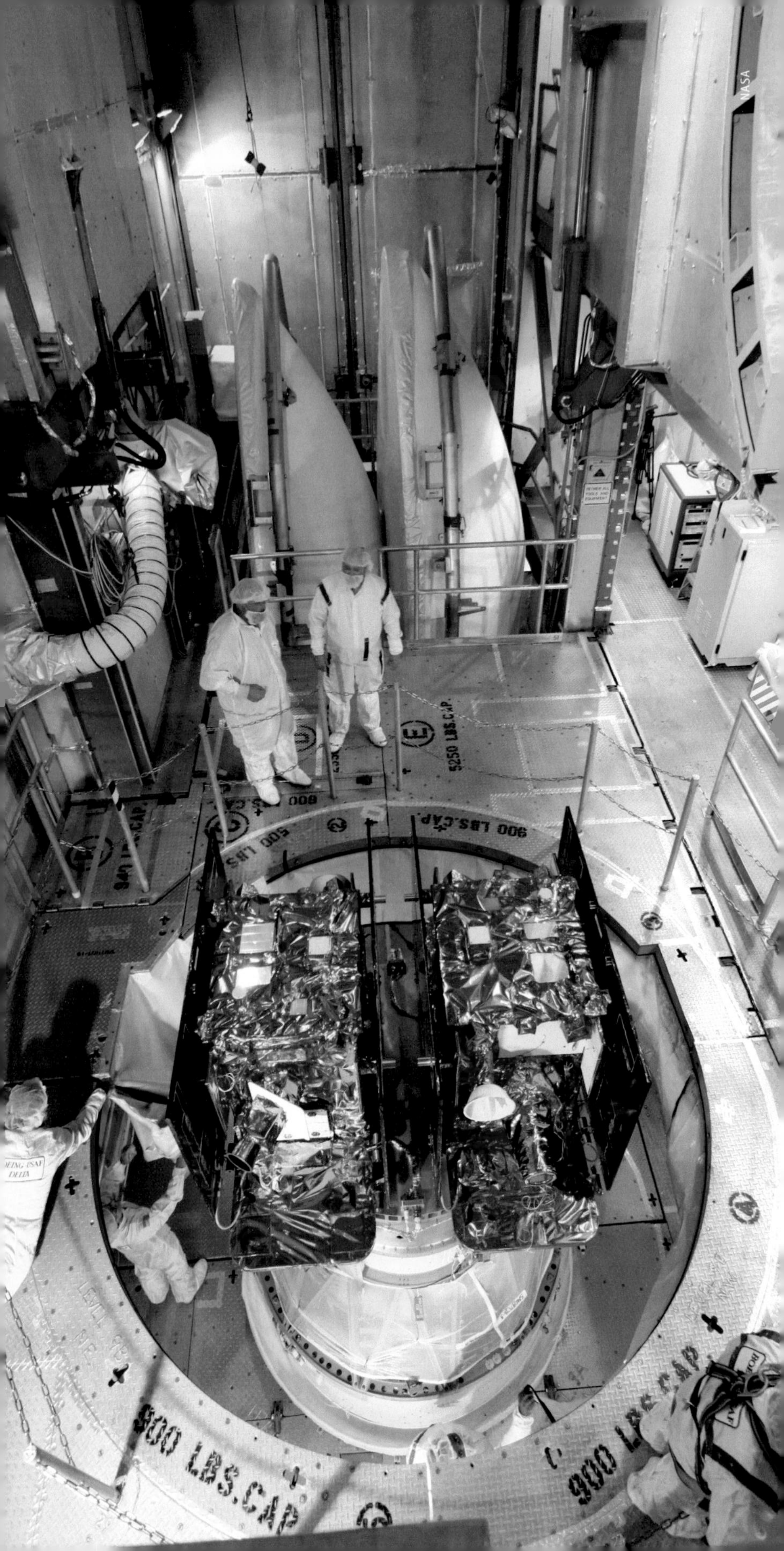
NASA
5250 LBS.CAP.
900 LBS.CAP.
900 LBS.CAP

NASA

NASA

Dust-free and meticulously dimensioned: final assembly of the lunar probe GRAIL

The Minotaur 5 rocket, with LADEE on board, being launched from Cape Canaveral

NASA EDGE

NASA

Speed check: LADEE on the test bench

LADEE
US

Launch: 7 September 2013, 3:27 am
Landing: 18 April 2014, 6:59 am
Last contact: 18 April 2014, 6:59 am
Launch site: Wallops Flight Facility
Crash location: 11.85°N 266.75°O

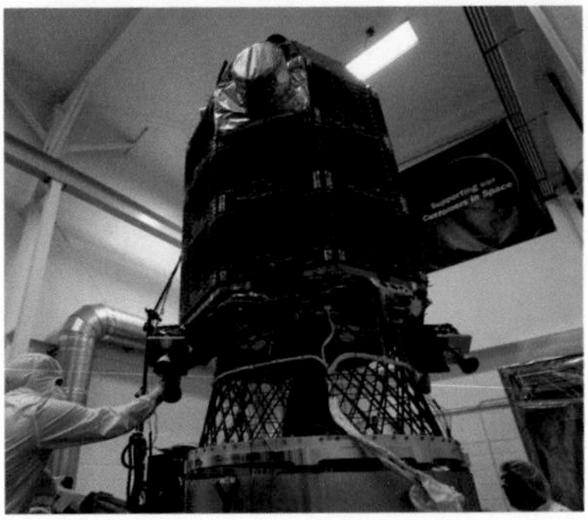

NASA

In the late 1960s, scientists on Earth and astronauts on the moon began to notice bright vapours, or smoke clouds, rising from the lunar surface. It is unclear what causes these clouds – meteorite collisions are one possibility – but it is certain that there is a thin atmosphere above the moon. The Lunar Atmosphere and Dust Environment Explorer (LADEE) was launched on a Minotaur 5 rocket with the aim of analysing this atmosphere and the moon's surrounding environment. The initial plan was to launch the mission together with the two GRAIL probes, but the proposed carrier rocket did not have enough room to house the spacious spacecraft. LADEE was therefore launched, as an exception, from the Wallops Flight Facility in Virginia two years later. The launch site is among the oldest in America. The satellite itself resembled a rocket with an octagonal cylinder measuring two metres in height and one metre in diameter. It was completely clad with solar panels. One large propelling nozzle and four smaller ones were mounted on the lower end of the satellite. The instruments were housed either in the top section or around the sides and included a mass spectrometer, a UV spectrometer, and a dust detector. The probe was built to an innovative design: the body could be fitted with different instruments, depending the respective mission's objectives. Once the probe, weighing almost 400 kg, entered lunar orbit, the scientists tested all instruments for 30 days before attempting, and achieving, a descent to a lower orbit at a distance of just 50 km from the moon. The successful mission's duration was even extended by a month. During its time in orbit, LADEE withstood not only the lunar nights but also a lunar eclipse, when the satellite did not receive any light, and hence no energy, for a longer period of time. The probe was intentionally crashed into the far side of the moon after the mission's completion.

NASA

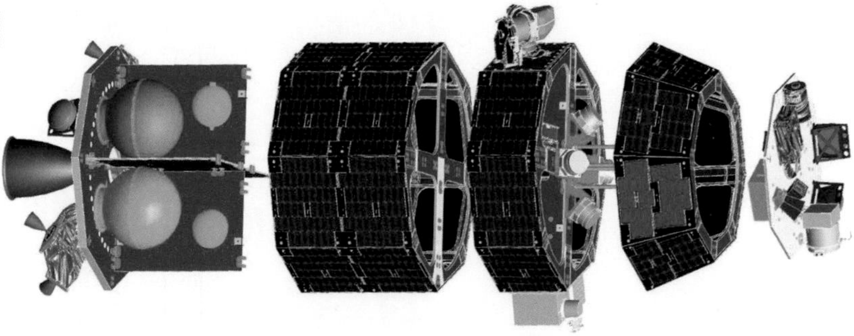

Composition of the LADEE probe

NASA

The packaging artist Christo was not involved in this work: transport cover for LADEE

NASA

Like the tip of a skyscraper: the nose cone of the Minotaur 5 rocket

Beresheet-1

Israel

Launch: 22 February 2019, 1:45 am
Planned landing: 11 April 2019, 7:30 pm
Last contact: 11 April 2019, 7:23 pm
Launch site: Cape Canaveral
Planned landing site: 33.0°N 17.0°O

SpaceIL, an Israeli lunar exploration agency, was founded in order to participate in the Google Lunar XPRIZE competition (see p. 318). The Beresheet-1 mission – 'Beresheet' is Hebrew for 'Book of Genesis' – was the first lunar exploration project to be completely privately financed. By the time of the launch, SpaceIL, a non-profit organisation, had incurred 100 million euros in costs, which were largely covered by the Israeli billionaire Morris Kahn. The relatively light probe, weighing 600 kg, was also the smallest ever developed for a lunar mission. However, it lost contact with the control station on Earth a few minutes before its scheduled landing and crashed into Mare Serenitatis: the braking procedure could not be initiated via wireless contact as planned. The Beresheet lander resembles an insect. The probe's body, like that of all other landing probes, resembles a kind of bowl filled with instruments. It features a cylindrical propulsion unit at the bottom and rests on arched buttresses with round feet. SpaceIL has been working on a follow-up mission since the failure of Beresheet-1.

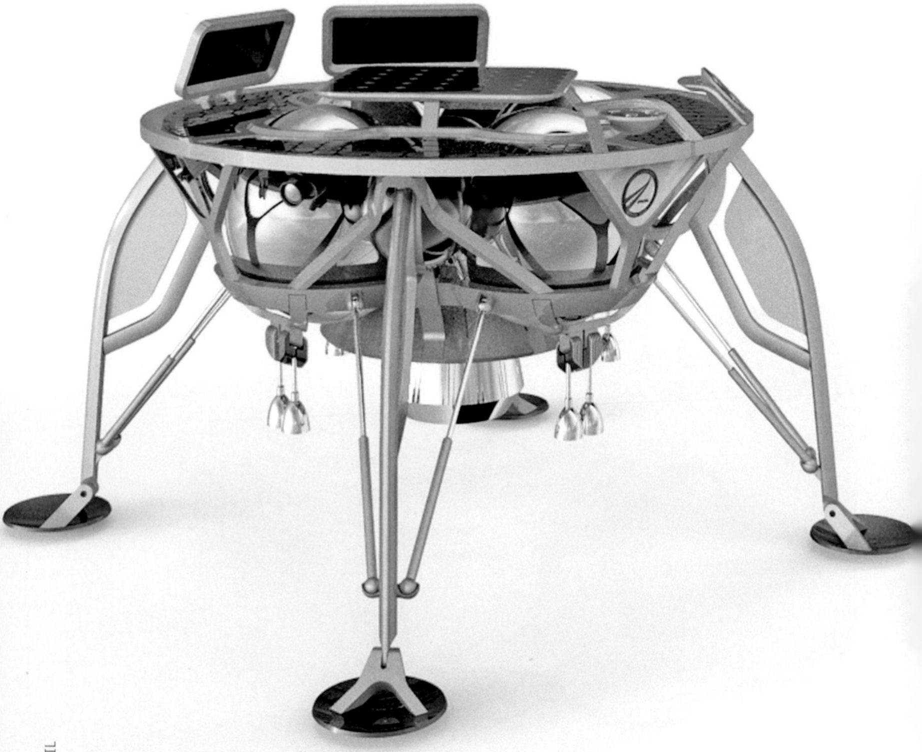

SpaceIL

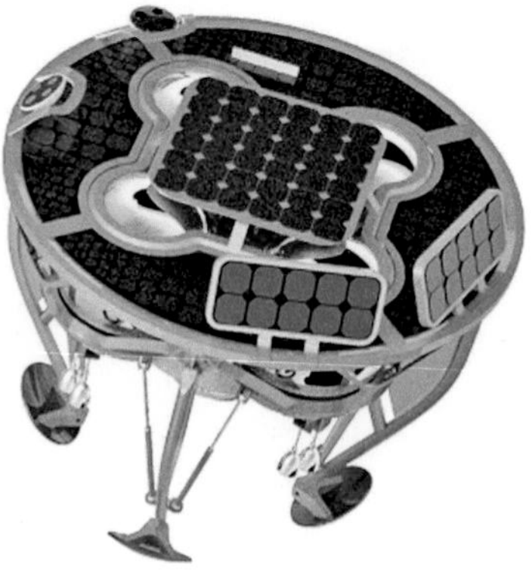

SpaceIL

SpaceIL

SpaceIL

Moon lander Beresheet-1 on the test bench

3D rendering of moon lander Beresheet-1

SpaceIL

Gold-wrapped probe headed for the regolith desert: selfie shortly before impact

SpaceIL

Beresheet-1 at a handy scale: Opher Doron (Israel Aerospace Industries) and Ido Anteby (SpaceIL) comment on the landing live on Israeli television.

Famous drawing with a tragic fate: *Moon Landscape* was discovered posthumously, after the artist, Petr Ginz, had perished at the Auschwitz concentration camp. The young Czech artist was a talented draughtsman who also wrote his first novel when he was 14. The Israeli astronaut Ilan Ramon took a copy of the drawing with him on the Columbia space shuttle to commemorate those who died in Nazi death camps. On 1 February 2003 – the 75th birthday of Ginz – Columbia disintegrated while re-entering Earth's atmosphere. All crew members died, and the image, which Ginz had drawn in the ghetto of Theresienstadt, became widely known across the world. Beresheet-1 later also carried a copy of the drawing on board, along with the Israeli Declaration of Independence of 1948, the Israeli flag, and a time capsule with 30 million pages of data about humanity.

SpaceX / SpaceIL

The Beresheet-1 lander, 1.50 m in height, in front of a Falcon 9 B5 rocket nose cone

Commercialisation of the Lunar Programme
Planned Missions

Although huge sums are still required for research and development, flights to the moon are becoming increasingly affordable. And today, private companies play a key role in journeys to outer space. On the one hand, different countries are competing to develop new technologies in order to attain supremacy of the still unclaimed space, as they did in colonial times. On the other hand, the profit-driven economy lusts after power and reputation with perhaps equally questionable motives. The first person to walk on the moon was an American, who marked his country's territory with a colourful flag. But a reversal of the world order might soon take place.

Whether on the moon or on Mars: the present and future of humanity can no longer be defined solely by economics, wars, and social and environmental problems on Earth, but by what happens in the upcoming years and decades away from the blue planet.

The upcoming lunar missions belong to an equally important chapter of the history of the moon as the crash landings of the Luna and Ranger probes. All future flights to the moon share a common idea, of setting up a permanently crewed base on the moon – a kind of international space station that, rather than circling the Earth at a distance of 400 km, at a speed of up to 27,6000 km/hr, within 1.5 hours, is fixed to a specific place at a distance of almost 400,000 km.

How will humans live on the moon in the future? Cylindrical modules and roofed craters are still merely futuristic visions. Will humanity realise them in the upcoming decades? Or will these missions in outer space only represent a stepping stone towards a journey to Mars?

Google

Google Lunar XPRIZE

'Welcome to the New Space Race'. This is the message that greets visitors to the home page of the XPRIZE competition, initiated by Google in 2015. The global player has now joined the ranks of lunar explorers. The goal of the competition is nothing less than to achieve a soft landing on the moon, after which a rover, built to a proprietary design, is to travel a distance of at least 500 metres, capturing high-resolution photographs and videos of its surroundings and transmitting them to Earth. The StreetView project, carried out since 2007, will thus have a counterpart in outer space. The competition was endowed with 30 million dollars of funds and was aimed at privately financed engineering teams from across the world. Until now, no participating team has managed to bring a spacecraft of its own design to the moon. But solid designs were developed as part of the competition. In fact, five teams reached the final round of the competition: Moon Express from the US (1); SpaceIL from Israel (2), whose mission already took place (unsuccessfully) in April 2019 (see pp. 312–315); HAKUTO from Japan (3); and Synergy Moon (4) and TeamIndus (5), both from India. Though the competition is over, and no team achieved a lunar landing by the deadline, all teams have been continuing to work independently towards this objective.

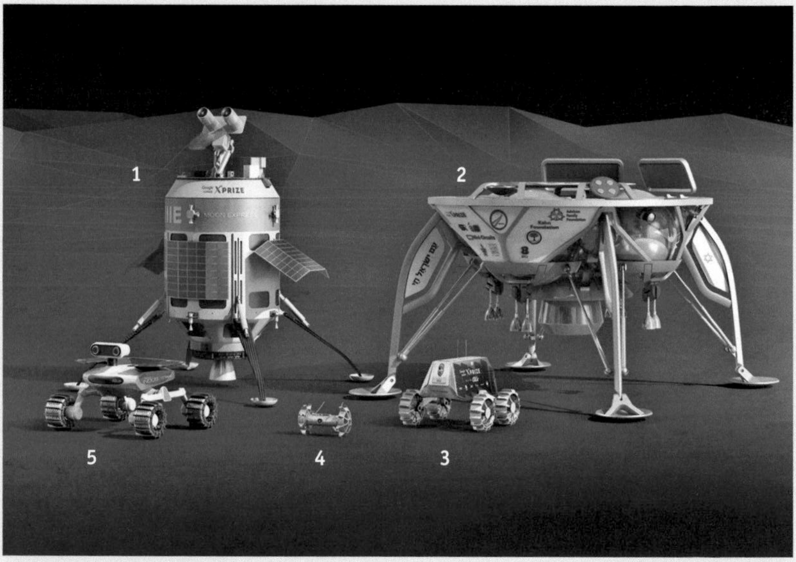

National Geographic

Five final-round candidates on a test flight to the moon

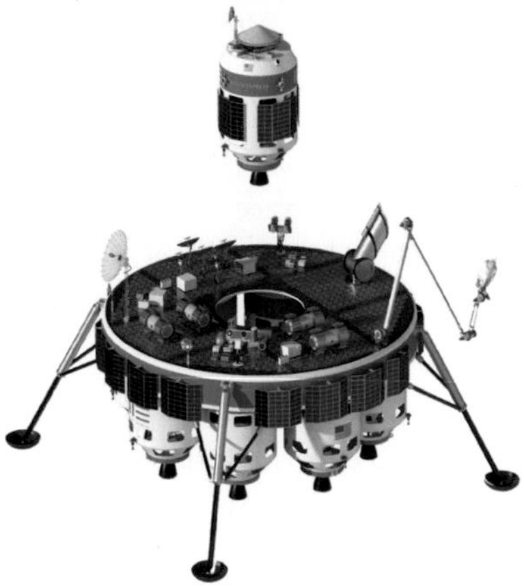

Moon Express
US

Under development

The developers of the team Moon Express was awarded a launch contract alongside their Israeli counterparts. They have an agreement with Rocket Lab, a private American company that provides access to launch pads across the world, to carry out up to three launches. As part of the competition, the team was awarded 'milestone prizes' in the imaging and landing categories, which secured for them additional funds amounting to 1.25 million dollars for a launch. The motivation for the project is to lay the foundation for commercial space travel and to exploit the wide range of potential resources on the moon. The launch of the MX-1 probe is planned for 2019 and will be followed by the launch of MX-2 one year later. In 2020, as part of the Harvest Moon mission, the first lunar soil samples for commercial purposes will be brought to Earth. The probes feature a corporate design that marks them as part of an ensemble. The recurring element of the design is a rocket-shaped barrel that is connected to a modestly designed engine and is decorated with turquoise rings. The circular ring element is even found on the landing module. A wide ring-shaped ceiling holds the barrels together. Various instruments and antennae are arranged on it, and the overall construction stands on four legs with round feet. Finally, the barrels of the return probes are crowned with a flattened sphere and clad with solar panels. The Lunar Scout is a further variation of the design. It features four landing feet, with which it will bounce along the lunar surface, propelled by a jet engine, as part of an exploratory mission. It thereby takes on the function of a rover.

Moon Express

Promising prospects for space flight once again, thanks to Google

TeamIndus
India

Under development

The Indian team, Indus, which also participated in the Google Lunar XPRIZE, was founded in 2010. Originally headquartered in Delhi, it soon moved to Bangalore to be closer to the Indian Space Research Organisation (ISRO). The mission's launch is planned to take place in 2019. It will last 30 days, ten of which will take place on the lunar surface. A probe, based on the Apollo spacecraft, will be loaded with a rocket engine, 16 thrusters, and fuel tanks and will thus be able to carry scientific equipment and commercial freight to the moon. Also on board is the central protagonist of the mission: Ek Choti Si Asha (ECA), the world's smallest rover developed to date. It features four legs on wheels and recalls the Lunokhod models, though with a futuristic robotic design with soft white edges. It carries small wing-like solar panels on its back. What looks like a flat head with two eyes set far apart is the camera system, which will take high-resolution images of the lunar surface. After the probe's landing, it will roll down the steep ramp of the landing module, which looks like a dematerialised Aztec pyramid.

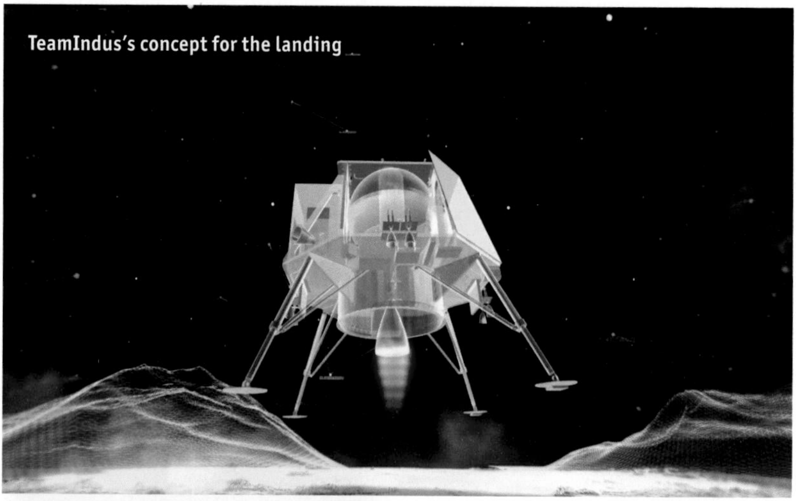
TeamIndus's concept for the landing

Team Indus

Design of a moonbase

Lander

Hakuto
Japan

Under development

The central element of the Japanese team's contribution to the Google competition is the compact moon rover Hakuto ('white rabbit'), weighing four kilograms and roughly the size of a rabbit. Though it appears black in an image of the model, in reality, it is a shiny silver. The vehicle's design follows the principle of reduction when it comes to its form as well as its equipment and power supply. The trapezoidal cuboid with sloping, canted sides is crowned by just two antennae, like a rabbit's ears. All other devices are integrated into the body and are therefore hardly or not at all visible. The solar panels are not mounted onto the vehicle but embedded in the outer shell. Four high-resolution cameras, which are currently being designed to take panoramic photos of the surroundings, present a particular challenge for the team's precision engineers. The mission will be the outcome of 15 years of research carried out by a team with Takeshi Hakamada and Kazuya Yoshida at the helm. Yoshida is currently also developing other designs for the Japanese space programme. Around 20 sponsors financed the project, whose funds were boosted by 500,000 dollars of prize money.

Hakuto

The 'white rabbit' on a test drive through a sandy terrain on Earth

Autonomous Landing and Navigation Module (ALINA)

PTScientists/Germany

Under development

The private space company PTScientists, founded in 2009, aims to reduce the cost of space exploration and to democratise access to the moon. The researchers and developers are working on space vehicles that will be available to a wide range of users. The ALINA landing module has a conventional design, resembling a four-legged insect, though it is somewhat shorter than the older American prototypes of the Apollo missions and is designed somewhat more appealingly. It is a preliminary design, and the hardware will be modified before each launch, depending on the mission's objectives. The body is composed of different elements of various grey hues: dark-grey struts forming the framework, which rests on four plate-like feet, along with eight light-grey tanks in the centre. Shimmering silver and gold details add colour accents. The landing module measures 2.5 m in length, 2 m in width, and less than 2 m in height, which means the distance between two diagonally opposite feet is merely 3.5 m. There

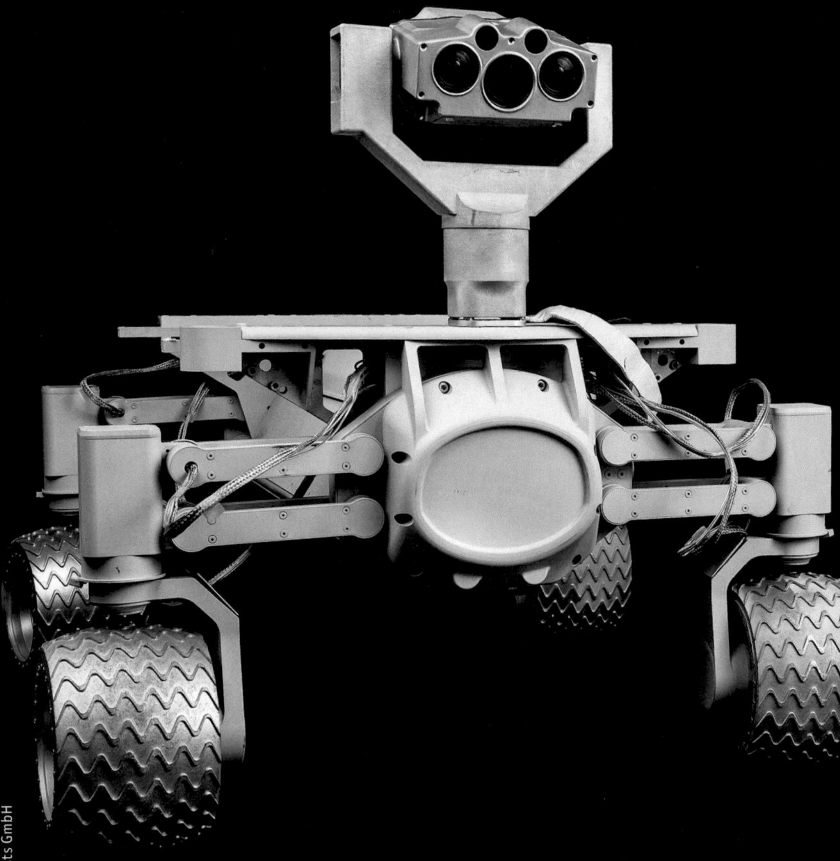

PTScientists GmbH

are eight small propelling nozzles arranged in a circle at the bottom, while on top, there is an 'attic floor', demarcated by a round, open, truss-like section. Solar panels are mounted here on the sides, and two of them extend downwards to form a kind of curtain wall with triangular openings. ALINA is designed to land softly on the moon. Fitted with seven cameras, the landing module can navigate its surroundings with a high degree of independence. The body is compatible with all larger rocket types, allowing users to make decisions flexibly regarding the carrier and launch site. It has the capacity to carry a payload of 300 kilograms and two lunar rovers. At the same time, it is designed to function as a work of architecture on the moon. The company is cooperating with the mobile phone provider Vodafone, seeking to equip ALINA to serve as the first LTE base station on the moon. A further cooperation with the automotive company Audi made it possible to develop two lunar rovers for use on the lunar site: the Audi Lunar Quattro (ALQ). The rover, with two axles and four wheels, is completely powered by electricity, with its energy generated by a tiltable solar panel on its back. The vehicle, resembling a robust robotic dog, features a three-eyed camera head and an antenna and can transport a payload of five kilograms.

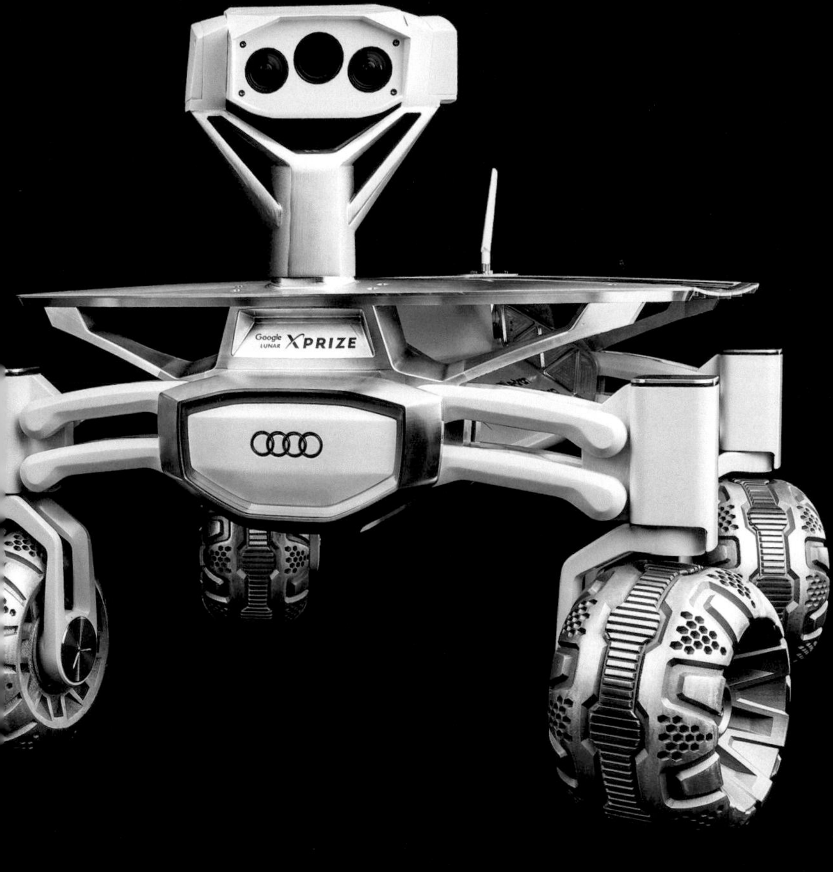

**PTScientists in cooperation with the automotive industry:
Audi Lunar Quattro will be the first off-road vehicle to go for a spin on the moon**

PTScientists GmbH

The ALINA lander in its prototypical design

PTScientists GmbH

ALINA's design details

Daniel Ziegenberg

Test rover Asimov Jr. R3 during a test drive along a slag heap in Austria (2012)

PTScientists GmbH

ALQ on an imaginary exploratory trip on the moon, where it encounters the LRV

PTScientists GmbH

Size comparison of the lander and rover

Lunar Exploration Programme
Europe

Under development

ESA

ESA's dream of lunar exploration has appeared to be in jeopardy several times in the last few years. Since 2012, the space agency has planned three projects that had to be abandoned due to a lack of funds. The German government was willing to fund 45 per cent of the costs, but not enough of the other ESA member states, such as France and the UK, wished to participate, nor did investors feel enough enthusiasm for a lunar project. However, ESA's mission to the lunar south pole has appeared very promising. The probe looks simple on the outside: a cylindrical round body fully clad with solar panels. But it was designed for big plans. The objective is not only to document the moon's surroundings and take measurements of ice water deposits but also to lay the foundation for astronautical missions in the region. As inconspicuous as the probe looks in the grey lunar wasteland, it overshadows some of its predecessors with its diameter of 3.5 metres. The antenna and camera mast extend a short distance from the flat, six-square-metre roof. A seemingly headless rocket, the probe remains to this day a symbol of the still uncompleted ESA mission to the moon. The new, ambitious aim is to land the first probes on the lunar surface from 2025 onwards and to thereafter produce oxygen and water on site. To this end, no new missions will be planned from the ground up. Rather, private missions will house the required devices and experiments.

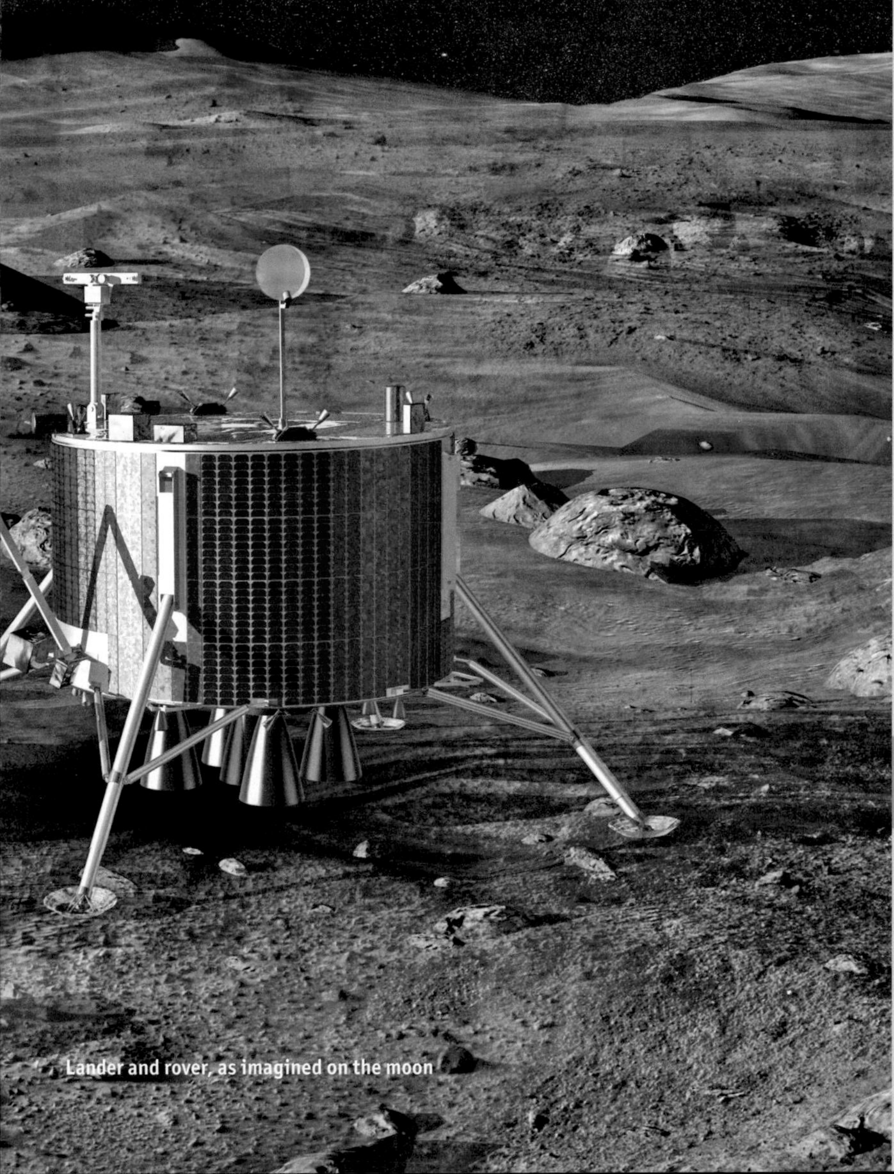

Lander and rover, as imagined on the moon

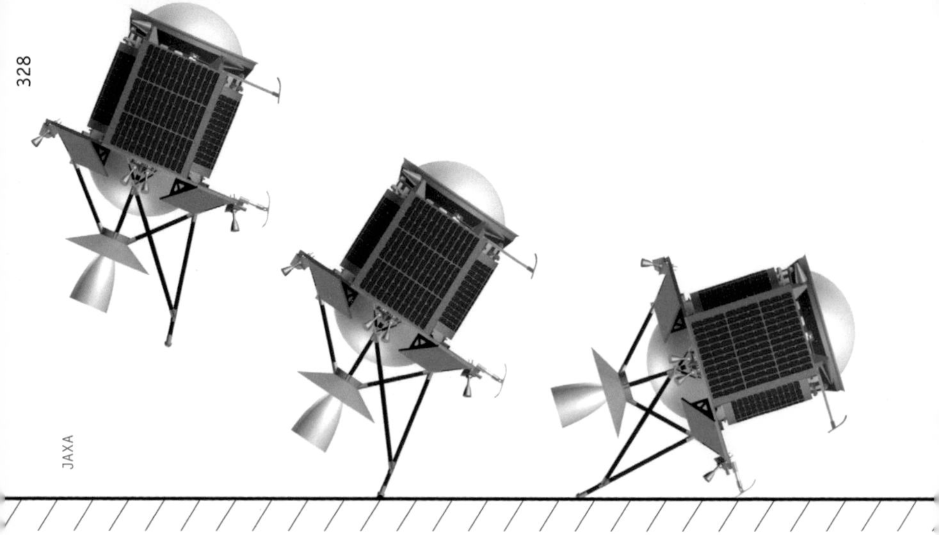

JAXA

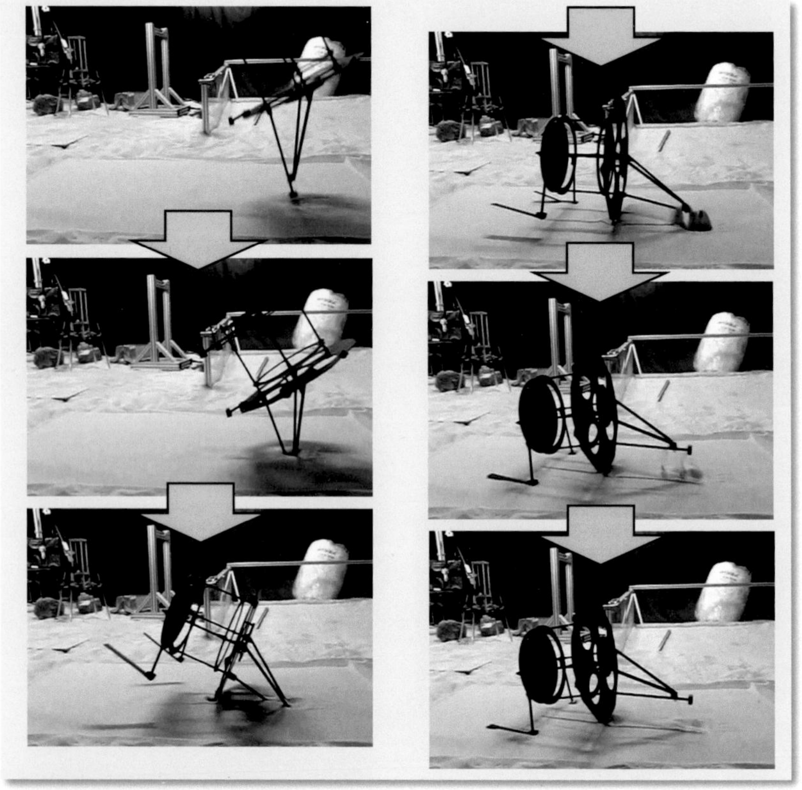

JAXA

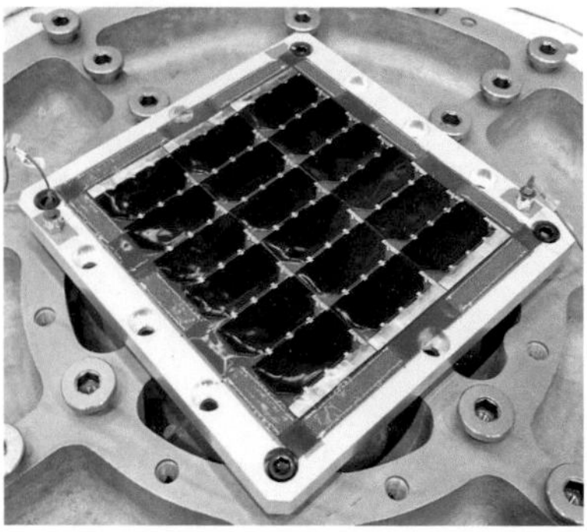

JAXA

Alternative landing techniques (above) are designed to prevent the SLIM probe from tipping over. Conventional braking jets are seen as a risky technology. Another technical challenge for the Japanese space engineers: thin-film solar cells (left)

Sophisticated landing manoeuvre that outsmarts both the moon's gravity and topography

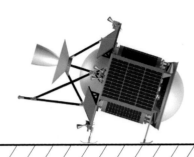

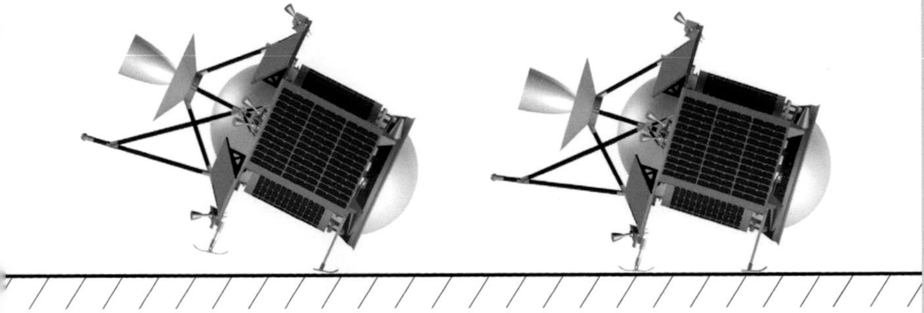

SLIM
Japan

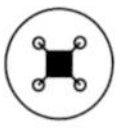

Under development

JAXA

Alternative design with a hexagonal body for the probe

JAXA

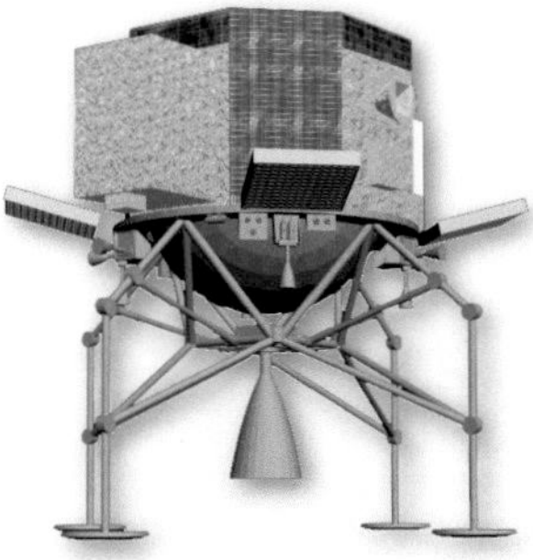

Alternative design with an octagonal body for the probe

Almost 20 years have past since Japan became the first nation after the US and the Soviet Union to fly to the moon. During this time, the Japan Aerospace Exploration Agency (JAXA) has sent two more multi-part satellites into lunar orbit. Its current plan is to achieve a soft landing with a landing module. The agency also plans to analyse useable moon rocks so that the findings can form the basis for future astronautical missions. Unlike older lunar landers, for which an optimal landing site was carefully selected in advance, the Japanese lander will be able to touch down at any location desired. The small stature of the probe in particular will enable this flexibility. One version of the design highly resembles the older Apollo probe's frame. A hexagonal body rests on four supports with round feet. The body joins a half-spherical section at the bottom, which finally leads to a propelling nozzle. A wide range of instruments for experiments are arranged around the central body, which is fully clad with solar panels. Another version of the design features an innovative support structure that enables a new soft-landing technique. It will allow JAXA to push further into outer space in future.

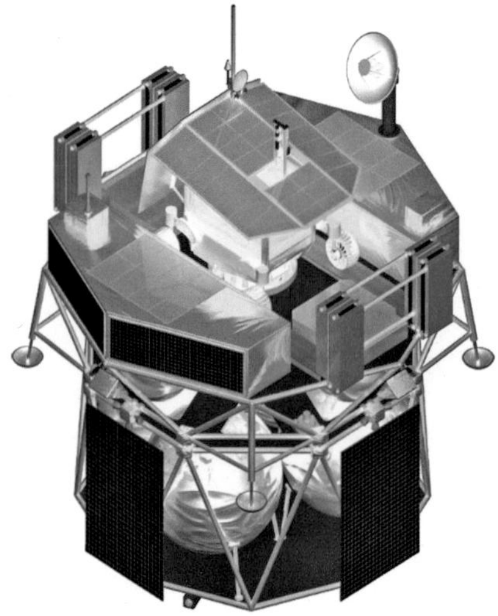

JAXA

Lander with ...

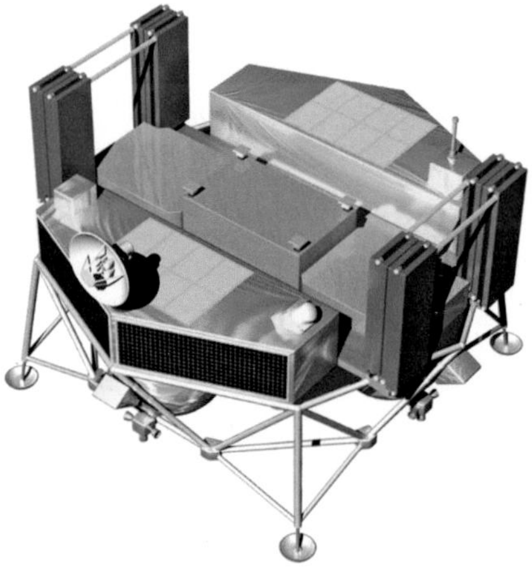

... and without the
Selene 2 lunar rover

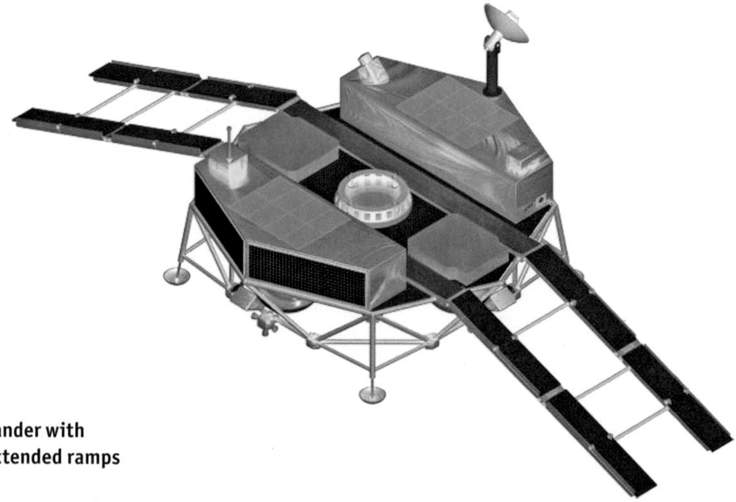

Lander with
extended ramps

JAXA

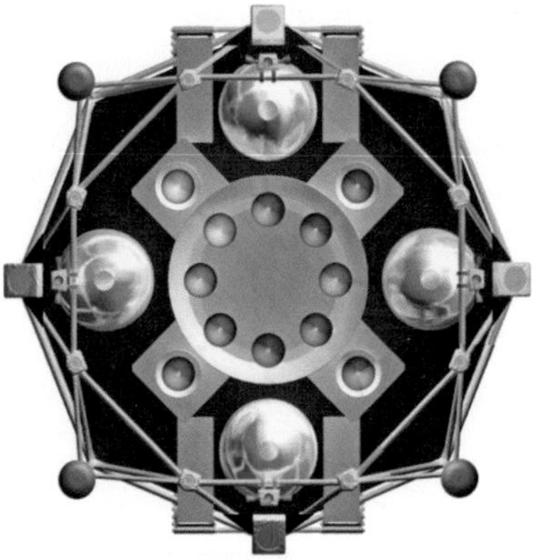

Underside of the lander

Section: lander in rocket's nose cone

JAXA

Selenological and Engineering Explorer 2: the Selene 2 programme was discontinued in 2007.

Roscosmos

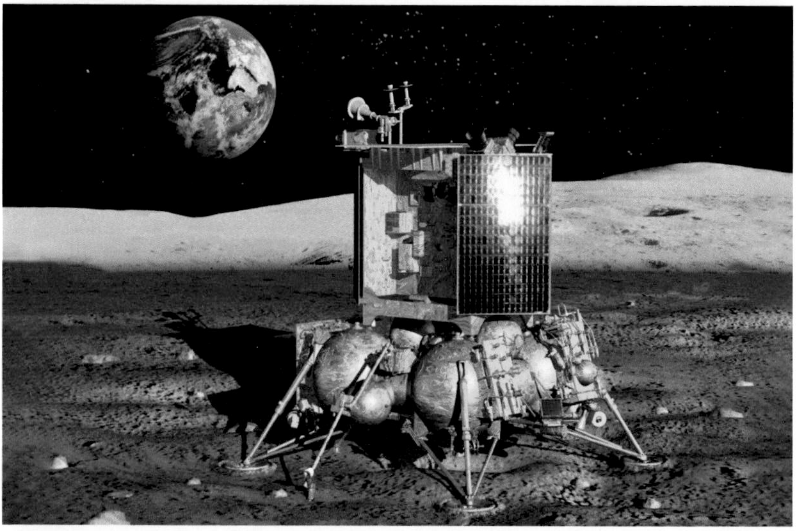

Roscosmos

Luna 25 keeps the Soviet tradition alive, displaying spherical fuel tanks as clearly identifiable design elements.

Roscosmos

Roscosmos

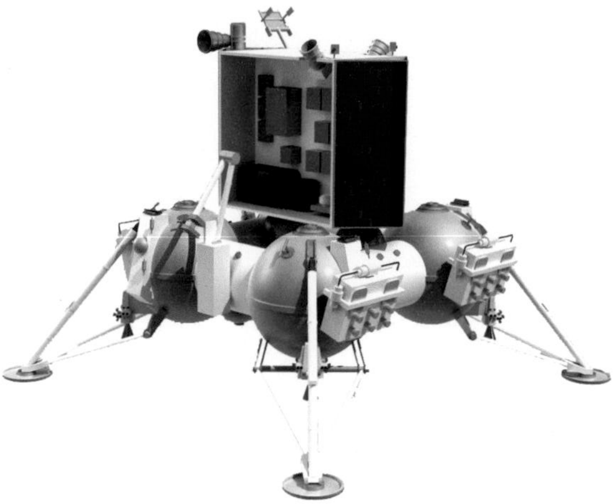

Luna 25
Russia

Under development

Luna 25 was unveiled in 1997. The lunar probe takes the name of the earlier Soviet moon exploration programme, and its design also borrows elements from the older Luna models. Four spherical capsules are connected by cylinders and supported by four struts with round feet. Standing on this pyramidal structure is an upright cuboid, fitted with instruments and flanked by solar panels. The cuboid is in fact an orbiter that will separate from the structure beneath it on the way to the moon so that it can enter a lunar orbit while the landing module touches down in the lunar south pole. The Soviet Union's interest in the moon dwindled after NASA's triumph in the race to the moon. And this in turn led to a number of budget cuts for Soviet space flights to the moon from the mid-1970s onwards. However, this trend seems likely to reverse over the next decade, as Russia once more engages in the new international race to the moon.

Roscosmos

Luna 25

Luna 26

Luna 27

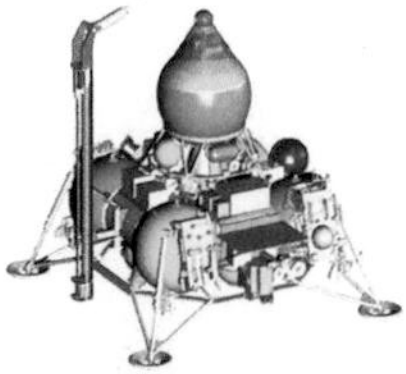

Luna 28 (Lander)

Luna 28 (Orbiter)

Luna 29

Typology of probes: future Russian uncrewed lunar missions (2020–2025)

Model of Luna 25 (true to scale) at an exhibition by the Russian manufacturer NPO Lavochkin in Khimki, near Moscow

SPUTNIK/Igor Lekarkin

SLS Block 1

SLS Block 1B Cargo

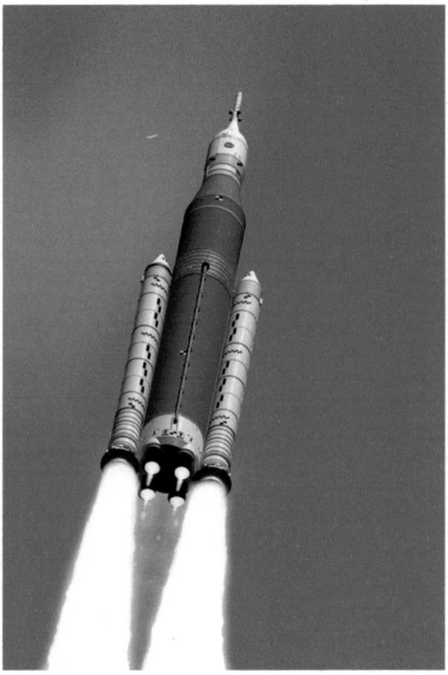

NASA/Marshall Space Flight Center

Space Launch System SLS
US

Under development

NASA is also taking part in the international race to achieve the next moon landing. The centrepiece of this endeavour is a new rocket, the Space Launch System (SLS), which is the world's most powerful rocket to date. It will not only reach the moon with ease, it will also head for destinations far beyond, specifically, Mars. The SLS rocket is capable of transporting a payload of 130 tonnes, which might also include the first human travellers to Mars. The design is based on the Ares V rocket, designed specifically for lunar missions but never realised. In contrast to previous models, which were kept white and bright grey with black accents, the new rocket type catches the eye with the auburn insulation layer on its shaft, which in turn makes the boosters and rocket head, in regular colours, stand out. It remains to be seen which space capsules will ultimately crown the rocket shaft; for the time being, development of the rocket design is still the focus, especially given the pressure of competing with SpaceX. Nonetheless, this project represents an important step for US space travel in the twenty-first century.

SLS Block 1B Crew

SLS Block 2 Cargo

NASA/MSFC/MAF/Jude Guidry

Basic version of the Space Launch System (SLS) with a space capsule standing on the launch pad (artistic rendering, 2015)

NASA/Marshall Space Flight Center

Modules currently being developed by private companies on behalf of NASA

Lockheed Martin

Workshop for repairing components

Northop Grumman

Transport module based on many years of experience

Bigelow Aerospace

Inflatable module for habitation and work

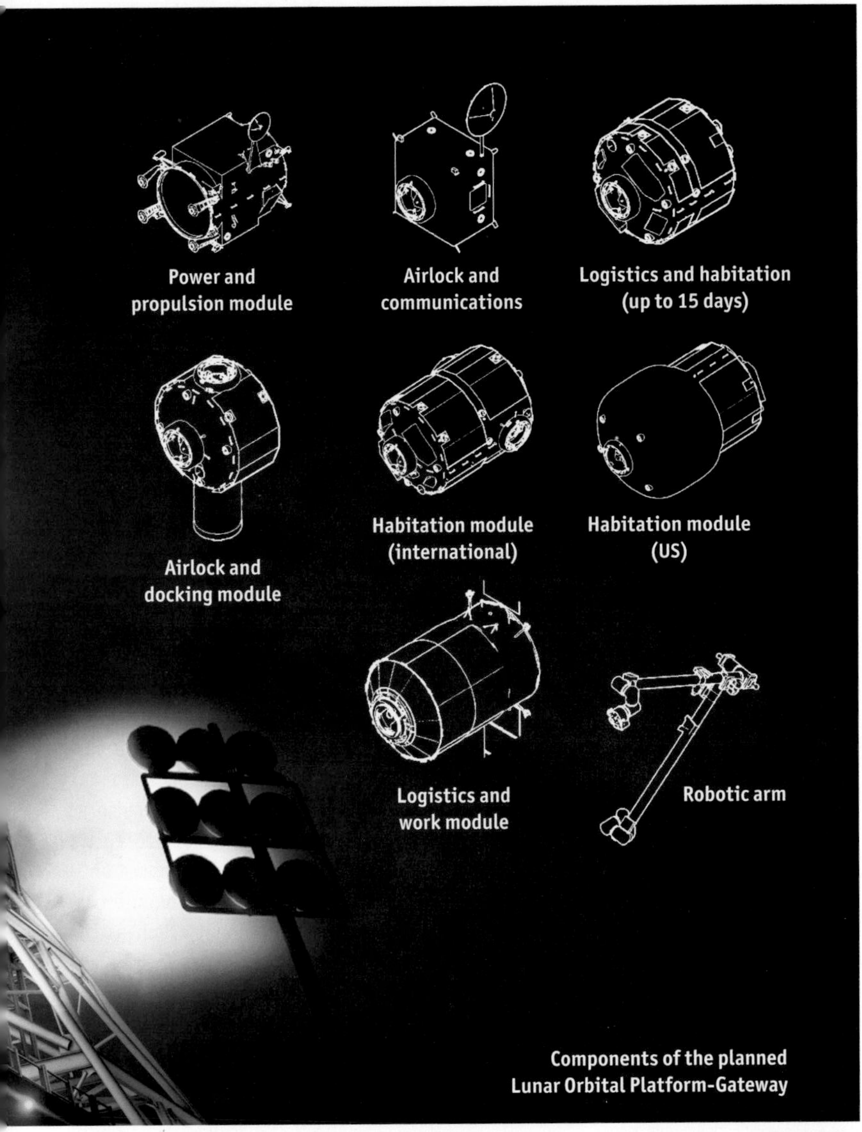

Components of the planned Lunar Orbital Platform-Gateway

Boeing

Development of a highly liveable space ship

Sierra Nevada

Habitation module based on Dream Chaser Cargo module

NanoRacks

Converted upper stages of Centaur rockets

Crew Dragon

SpaceX/US

Under development

SpaceX

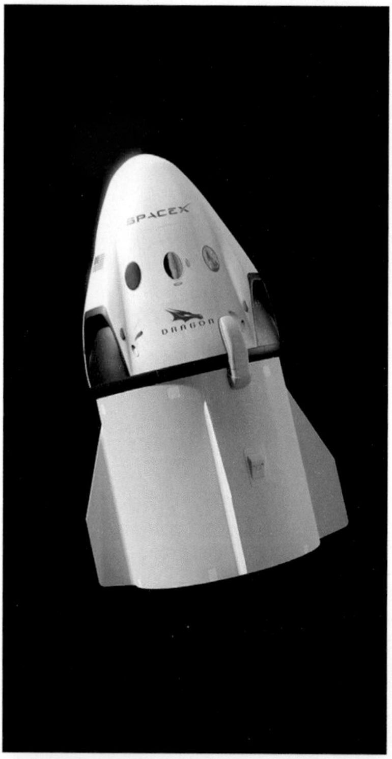

The private American company SpaceX was founded with the aim of colonising other planets. For this reason alone, this book must consider it as the potential future of lunar architecture. Though there are currently no plans for a moon landing, the spaceship Crew Dragon is being developed for journeys around the moon. Crew Dragon is a space capsule in the form of a stylised rocket. A conical housing with a rounded tip sits on a cylindrical propulsion engine. A row of openings are arranged around the midway point of the housing, providing the passengers a view of the surroundings in outer space. The capsule will enable two private citizens to make a trip around the moon and return back to Earth. This would make it not only the first mission operated by a private company, but also the first spacecraft, carrying private passengers, not intended for scientific purposes. SpaceX sees a market niche in space flights to the moon – a profitable business catered to the adventurous super rich. The mission will cost the company and customers no more than a trip to the International Space Station, located closer to Earth. However, the project has stagnated until now, since the company has been focusing on other developments such as the new Big Falcon Rocket.

SpaceX

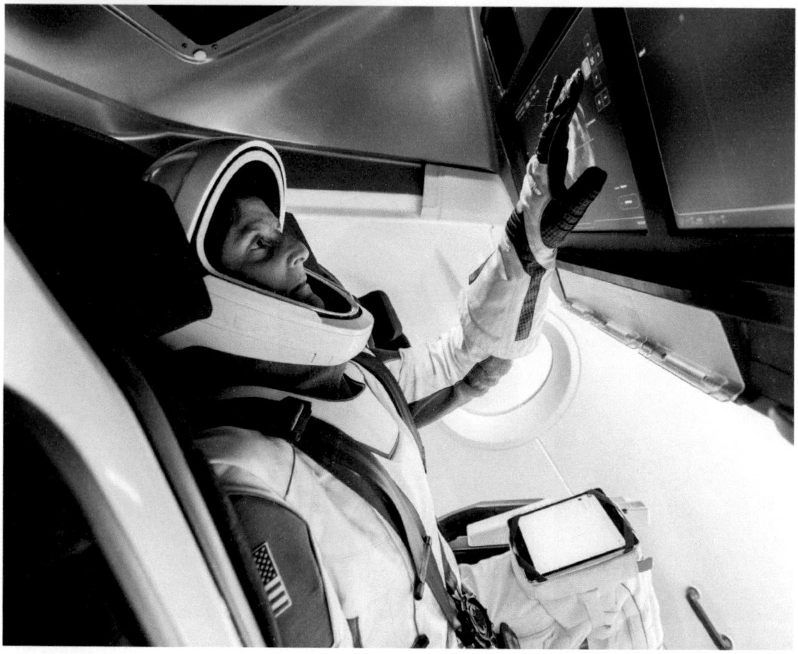

To the moon in a designer jet: SpaceX even allows the space engineers to consider design.

SpaceX

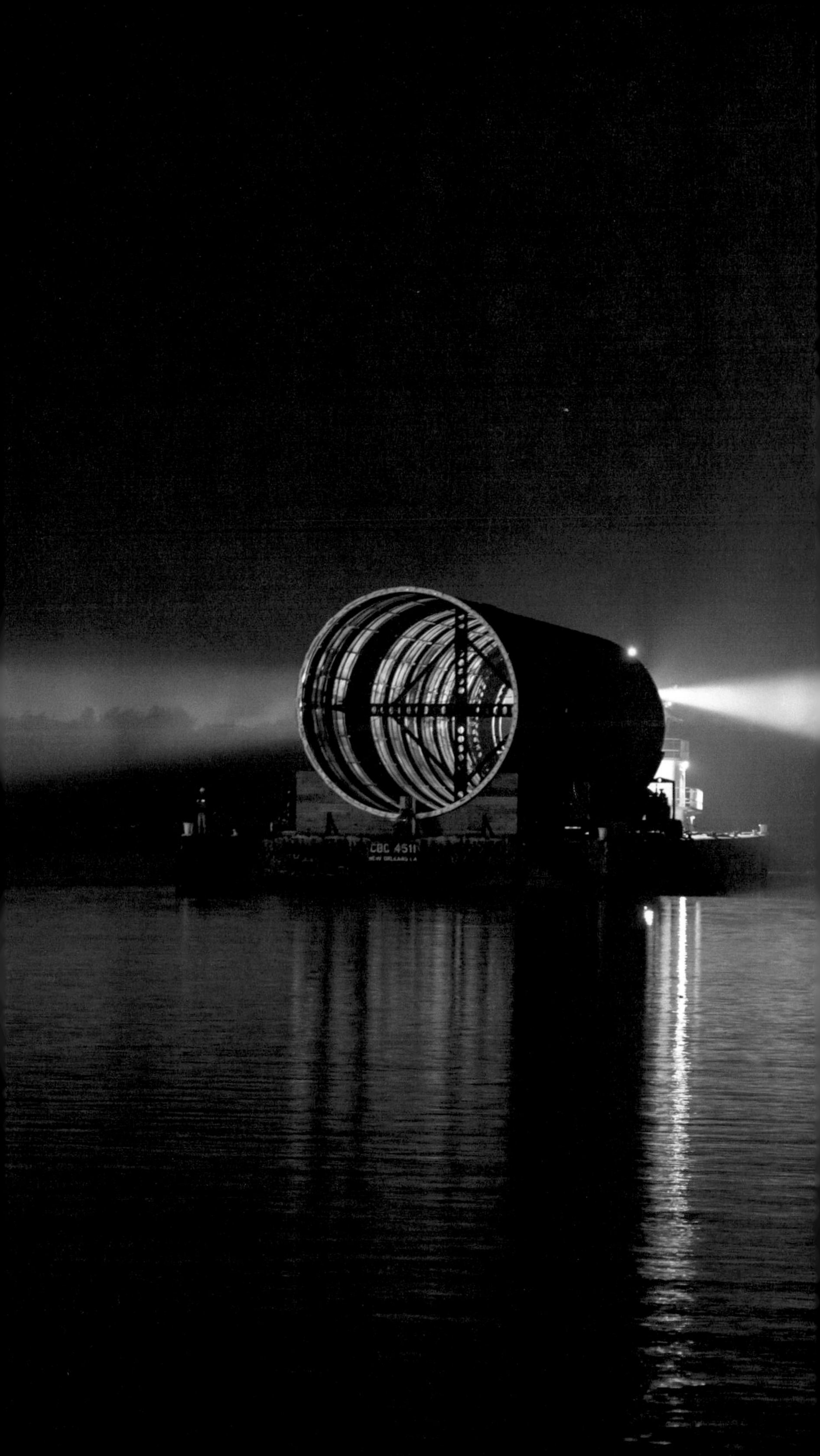

Towards the future: the Space Launch System core stage, a module measuring over 60 metres in length, arrives at the Michoud Assembly Facility, NASA's production and assembly hall near New Orleans, on the morning of 27 September 2017.

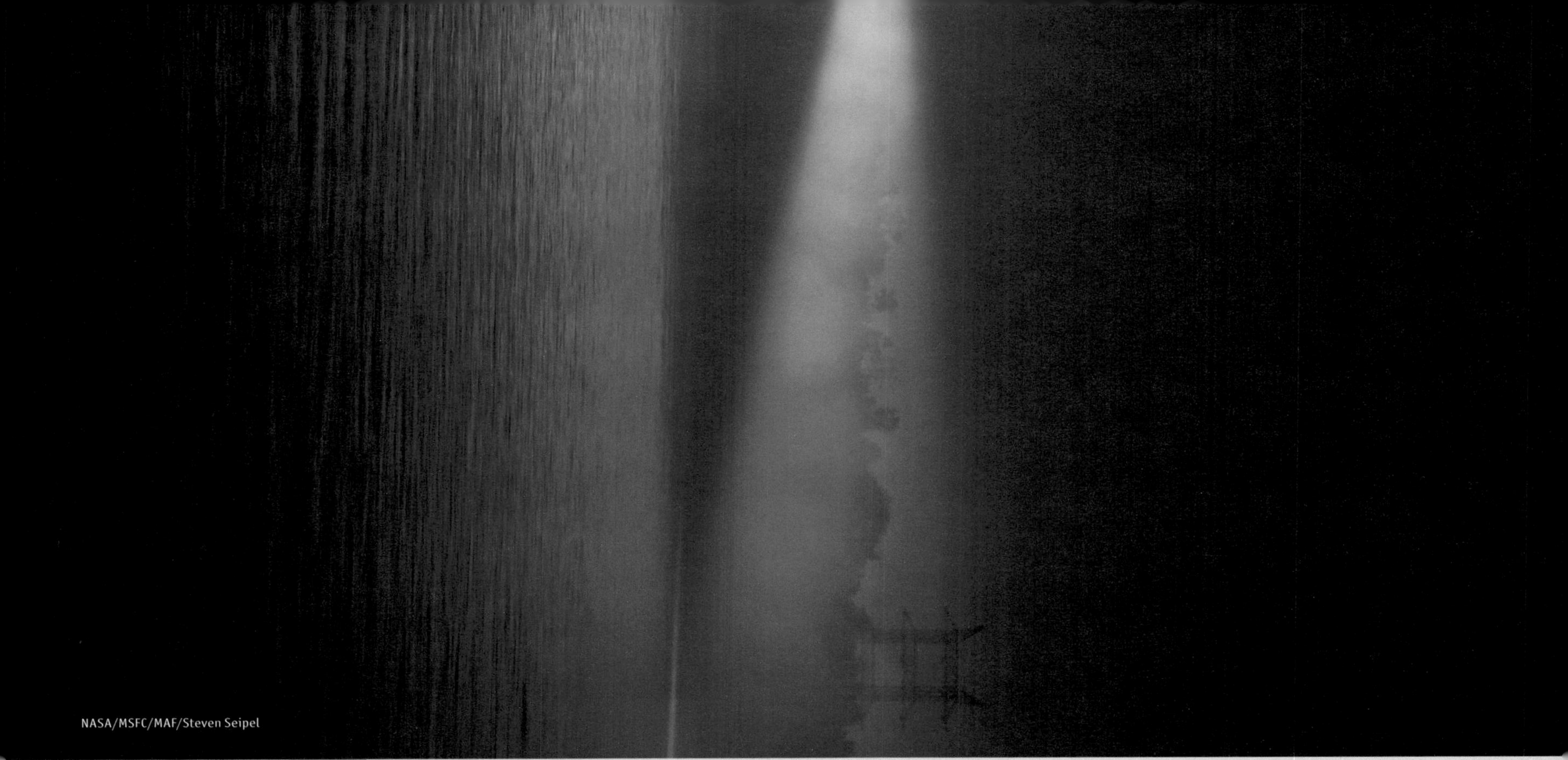

NASA/MSFC/MAF/Steven Seipel

Appendix

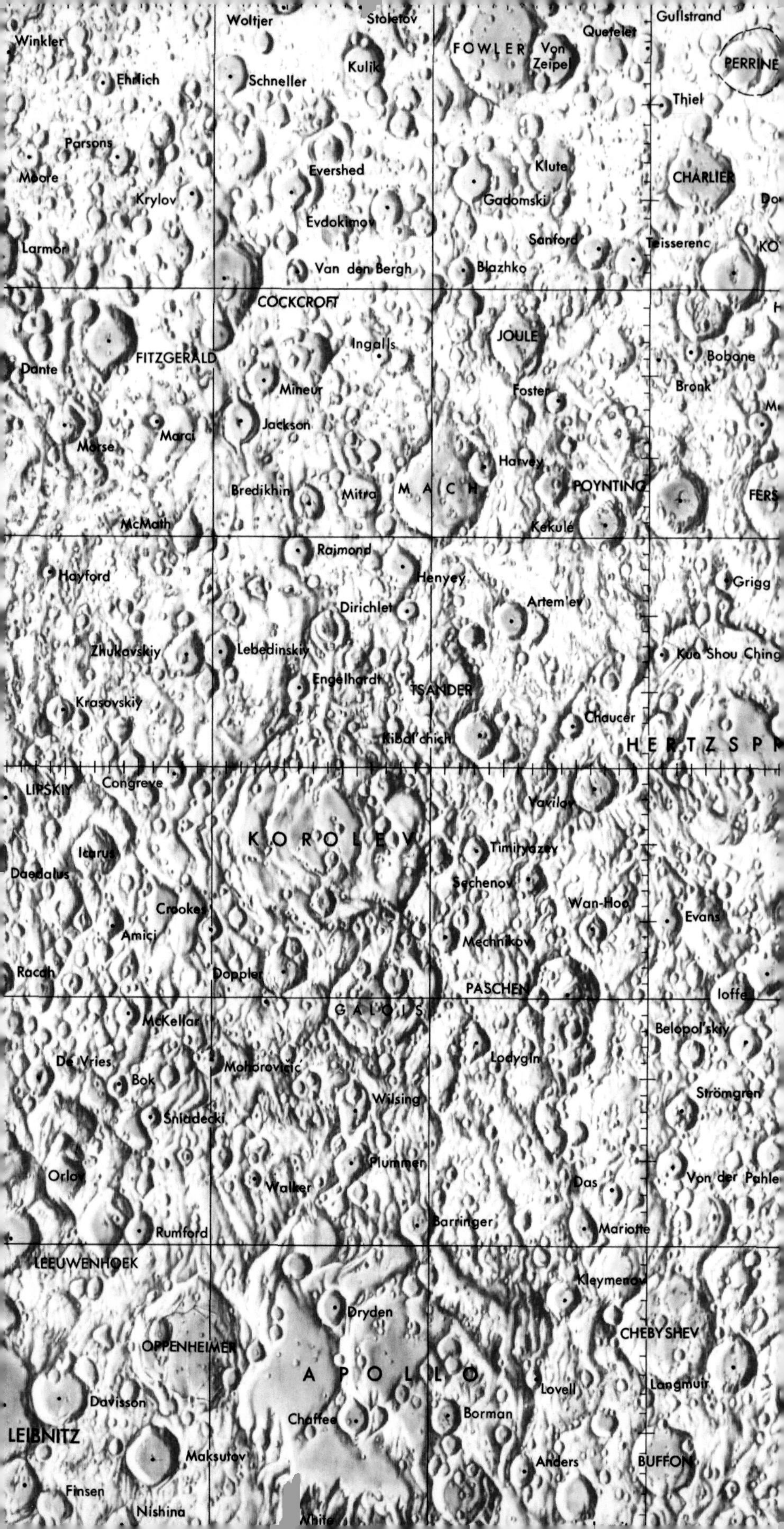

Winkler
Woltjer
Stoletov
Gullstrand
Quetelet
FOWLER
Von Zeipel
PERRINE
Ehrlich
Schneller
Kulik
Thiel
Parsons
Evershed
Klute
CHARLIER
Moore
Krylov
Gadomski
Evdokimov
Larmor
Sanford
Teisserenc
Van den Bergh
Blazhko
COCKCROFT
JOULE
Ingalls
FITZGERALD
Bobone
Dante
Mineur
Foster
Bronk
Jackson
Marci
Morse
Harvey
Bredikhin
Mitra
MACH
POYNTING
McMath
Kekulé
Raimond
Henyey
Hayford
Grigg
Dirichlet
Artem'ev
Zhukovskiy
Lebedinskiy
Kuo Shou Ching
Engelhardt
TSANDER
Krasovskiy
Chaucer
Kibal'chich
LIPSKIY
Congreve
Vavilov
KOROLEV
Timiryazev
Icarus
Daedalus
Sechenov
Wan-Hoo
Crookes
Evans
Amici
Mechnikov
Racah
Doppler
PASCHEN
Ioffe
GALOIS
McKellar
Belopol'skiy
Lodygin
De Vries
Mohorovičić
Bok
Strömgren
Wilsing
Sniadecki
Plummer
Orlov
Von der Pahlen
Walker
Das
Barringer
Rumford
Mariotte
LEEUWENHOEK
Kleymenov
Dryden
CHEBYSHEV
OPPENHEIMER
APOLLO
Lovell
Langmuir
Davisson
Chaffee
Borman
LEIBNITZ
Maksutov
Anders
BUFFON
Finsen
Nishina
White

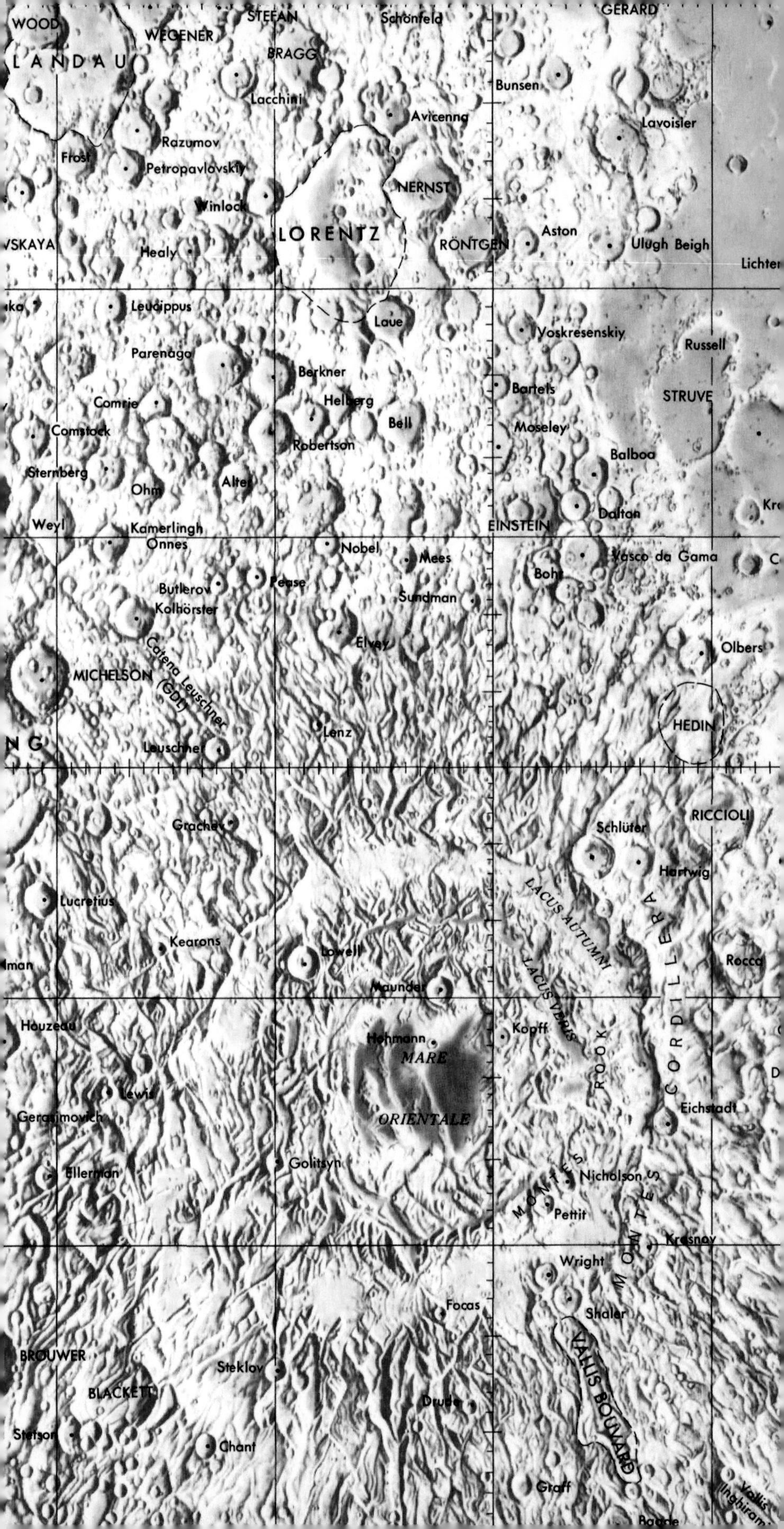
WOOD
LANDAU
WEGENER
STEFAN
Schönfeld
GERARD
BRAGG
Lacchini
Bunsen
Avicenna
Lavoisier
Razumov
Frost
Petropavlovskiy
NERNST
Winlock
LORENTZ
Aston
RÖNTGEN
Ulugh Beigh
VSKAYA
Healy
Lichter
Leucippus
Laue
Voskresenskiy
Russell
Parenago
Berkner
Bartels
STRUVE
Comrie
Helberg
Comstock
Bell
Moseley
Robertson
Balboa
Sternberg
Ohm
Alter
Dalton
Weyl
Kamerlingh
Onnes
EINSTEIN
Nobel
Mees
Vasco da Gama
Butlerov
Pease
Bohr
Kolhörster
Sundman
Elvey
Olbers
MICHELSON
Catena Leuschner
(GDL)
HEDIN
Lenz
Leuschner
Grachev
Schlüter
RICCIOLI
Hartwig
Lucretius
LACUS AUTUMNI
Kearons
Lowell
LACUS VERIS
Rocca
Maunder
CORDILLERA
Houzeau
Hohmann
Kopff
MARE
ORIENTALE
ROOK
Lewis
Eichstadt
Gerasimovich
Golitsyn
Ellerman
MONTES
Nicholson
Pettit
MONTES
Krasnov
Wright
Focas
Shaler
VALLIS BOUVARD
BROUWER
Steklov
BLACKETT
Drude
Stetson
Chant
Graff
Vallis Inghirami
Baade

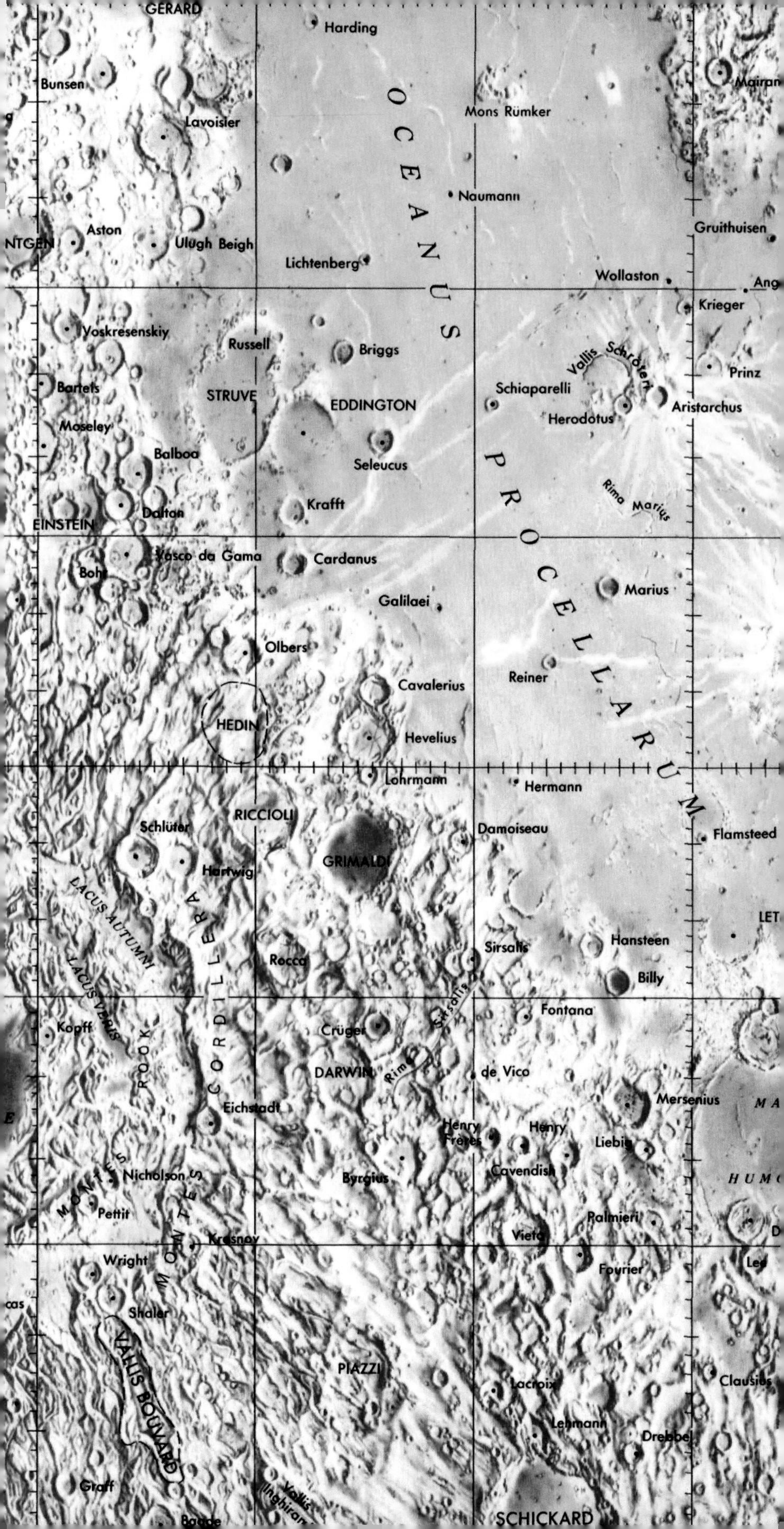
GERARD
Harding
Bunsen
Mairan
OCEANUS
Mons Rümker
Lavoisier
Naumann
Aston
Ulugh Beigh
Gruithuisen
Lichtenberg
Wollaston
Krieger
Voskresenskiy
Russell
Briggs
Vallis Schröteri
Prinz
Bartels
STRUVE
EDDINGTON
Schiaparelli
Herodotus
Aristarchus
Moseley
Balboa
Seleucus
PROCELLARUM
Rima Marius
Krafft
EINSTEIN
Dalton
Vasco da Gama
Bohr
Cardanus
Marius
Galilaei
Olbers
Reiner
Cavalerius
HEDIN
Hevelius
Lohrmann
Hermann
RICCIOLI
Schlüter
Damoiseau
Flamsteed
Hartwig
GRIMALDI
LACUS AUTUMNI
CORDILLERA
Sirsalis
Hansteen
Rocca
LACUS VERIS
Billy
Rima Sirsalis
Fontana
Kopff
ROOK
Crüger
DARWIN
Rima
de Vico
Eichstadt
Mersenius
Henry Frères
Henry
Liebig
MONTES
Nicholson
Byrgius
Cavendish
Pettit
Palmieri
Krasnov
Vieta
Wright
Fourier
Shaler
VALLIS BOUVARD
PIAZZI
Clausius
Lacroix
Lehmann
Drebbel
Graff
Vallis Inghirami
Baade
SCHICKARD

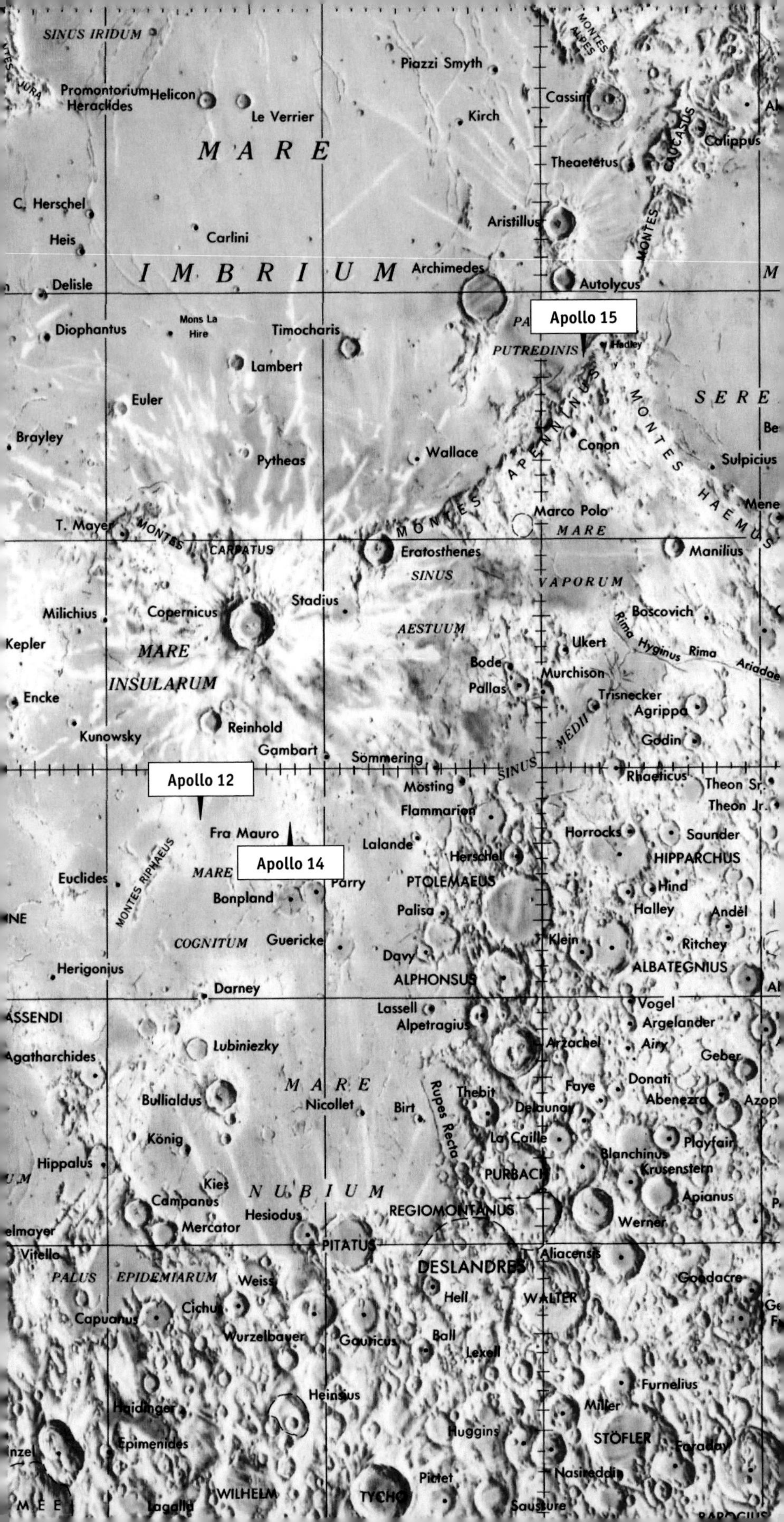

SINUS IRIDUM
Piazzi Smyth
MONTES ALPES
Promontorium Heraclides
Helicon
Le Verrier
Kirch
Cassini
CAUCASUS
Calippus
MARE
Theaetetus
C. Herschel
Aristillus
Heis
Carlini
MONTES
IMBRIUM
Archimedes
Autolycus
Delisle
Apollo 15
Diophantus
Mons La Hire
Timocharis
PUTREDINIS
Hadley
Lambert
SERE
Euler
MONTES
Brayley
APENNINUS
Conon
Pytheas
Wallace
Sulpicius
Marco Polo
Mene
T. Mayer
MONTES CARPATUS
MARE
HAEMUS
Eratosthenes
Manilius
SINUS
VAPORUM
Stadius
Milichius
Copernicus
Boscovich
AESTUUM
Rima Hyginus
Rima Ariadaeus
Kepler
MARE
Ukert
INSULARUM
Bode
Murchison
Pallas
Trisnecker
Encke
Agrippa
Reinhold
SINUS MEDII
Kunowsky
Godin
Gambart
Sömmering
Rhaeticus
Apollo 12
Mösting
Theon Sr.
Theon Jr.
Flammarion
Fra Mauro
Horrocks
Saunder
Lalande
Apollo 14
Herschel
HIPPARCHUS
MONTES RIPHAEUS
MARE
Euclides
Parry
PTOLEMAEUS
Hind
Bonpland
Palisa
Halley
Andĕl
COGNITUM
Guericke
Klein
Ritchey
Davy
Herigonius
ALPHONSUS
ALBATEGNIUS
Darney
Vogel
Lassell
ASSENDI
Argelander
Alpetragius
Arzachel
Airy
Lubiniezky
Agatharchides
Geber
MARE
Faye
Donati
Thebit
Bullialdus
Nicollet
Birt
Rupes Recta
Delaunay
Abenezra
Azop
König
La Caille
Playfair
Blanchinus
Hippalus
PURBACH
Krusenstern
Kies
NUBIUM
Campanus
Apianus
REGIOMONTANUS
Hesiodus
Mercator
Werner
PITATUS
Vitello
Aliacensis
DESLANDRES
PALUS EPIDEMIARUM
Weiss
Goodacre
WALTER
Hell
Cichus
Capuanus
Wurzelbauer
Gauricus
Ball
Lexell
Furnelius
Heinsius
Haidinger
Miller
Huggins
STÖFLER
Epimenides
Faraday
Nasireddin
Pictet
WILHELM
TYCHO
Lagalla
Saussure

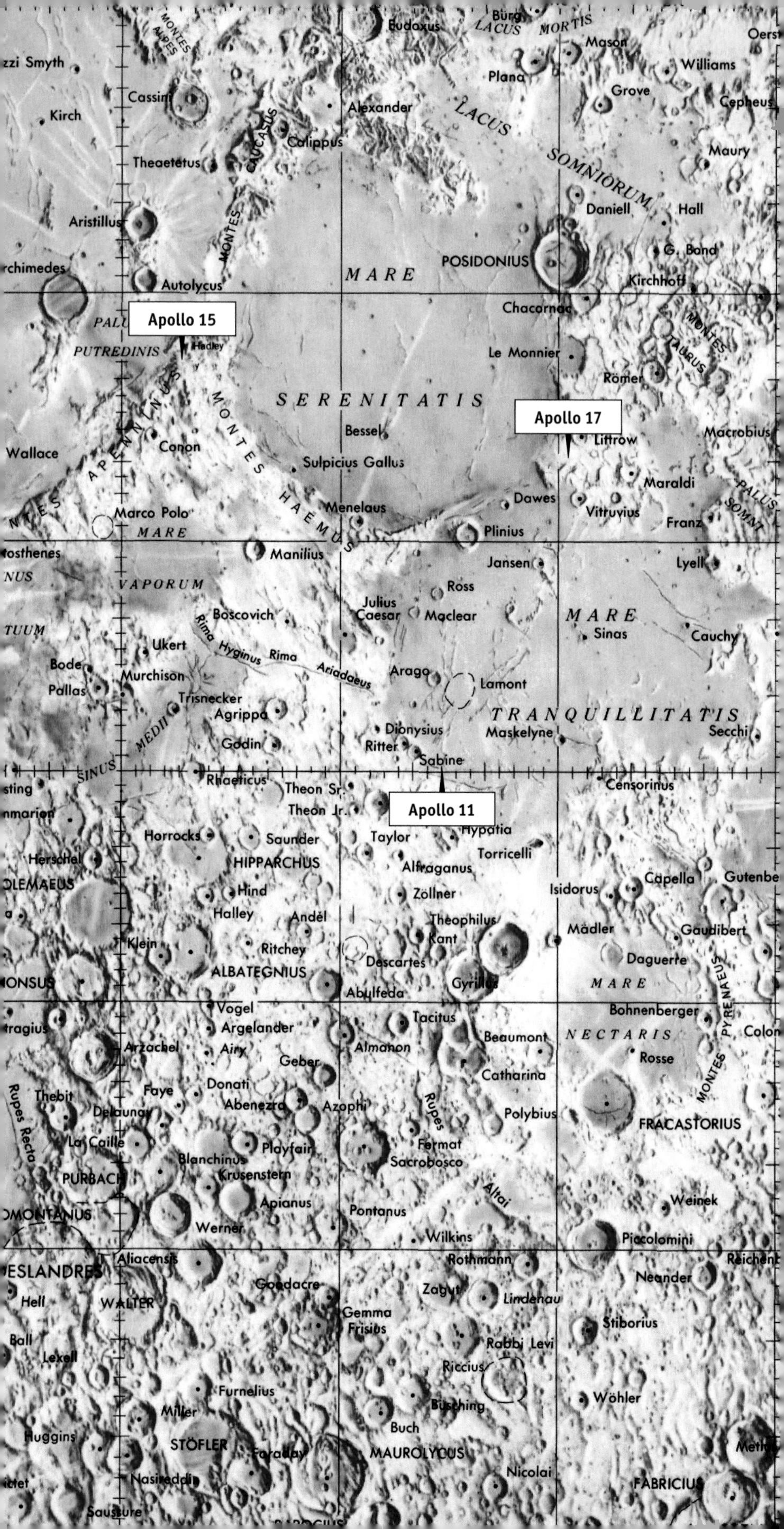

Apollo 15
Apollo 17
Apollo 11
MONTES ALPES
Eudoxus
Bürg
LACUS MORTIS
Mason
Williams
Cassini
Kirch
Alexander
Plana
Grove
Cepheus
LACUS SOMNIORUM
CAUCASUS
Calippus
Theaetetus
Maury
Aristillus
MONTES
Daniell
Hall
G. Bond
POSIDONIUS
MARE
Autolycus
Kirchhoff
Chacornac
MONTES TAURUS
PUTREDINIS
Hadley
Le Monnier
Römer
SERENITATIS
MONTES APENNINUS
MONTES HAEMUS
Bessel
Littrow
Macrobius
Wallace
Conon
Sulpicius Gallus
Maraldi
Dawes
Vitruvius
Franz
Marco Polo
Menelaus
Plinius
MARE VAPORUM
Manilius
Jansen
Lyell
Ross
Julius Caesar
Maclear
MARE TRANQUILLITATIS
Boscovich
Sinas
Cauchy
Ukert
Rima Hyginus
Rima Ariadaeus
Bode
Murchison
Arago
Lamont
Pallas
Trisnecker
Agrippa
SINUS MEDII
Dionysius
Maskelyne
Secchi
Godin
Ritter
Sabine
Rhaeticus
Theon Sr.
Theon Jr.
Censorinus
Horrocks
Saunder
Taylor
Hypatia
Herschel
HIPPARCHUS
Torricelli
Alfraganus
Hind
Zöllner
Isidorus
Capella
Halley
Andĕl
Theophilus
Mädler
Gaudibert
Klein
Kant
Ritchey
Descartes
Daguerre
ALBATEGNIUS
Cyrillus
MARE NECTARIS
Abulfeda
Vogel
Tacitus
Bohnenberger
MONTES PYRENAEUS
Argelander
Beaumont
Rosse
Arzachel
Airy
Almanon
Geber
Catharina
Rupes Recta
Thebit
Donati
Faye
Delaunay
Abenezra
Azophi
Rupes Altai
Polybius
FRACASTORIUS
La Caille
Playfair
Fermat
Sacrobosco
Blanchinus
PURBACH
Krusenstern
Apianus
Weinek
Pontanus
Werner
Wilkins
Piccolomini
Aliacensis
Rothmann
Neander
Goodacre
Zagut
Lindenau
WALTER
Hell
Gemma Frisius
Stiborius
Ball
Rabbi Levi
Lexell
Riccius
Furnelius
Wöhler
Miller
Busching
Huggins
Buch
STÖFLER
Faraday
MAUROLYCUS
Nasireddin
Nicolai
FABRICIUS
Saussure

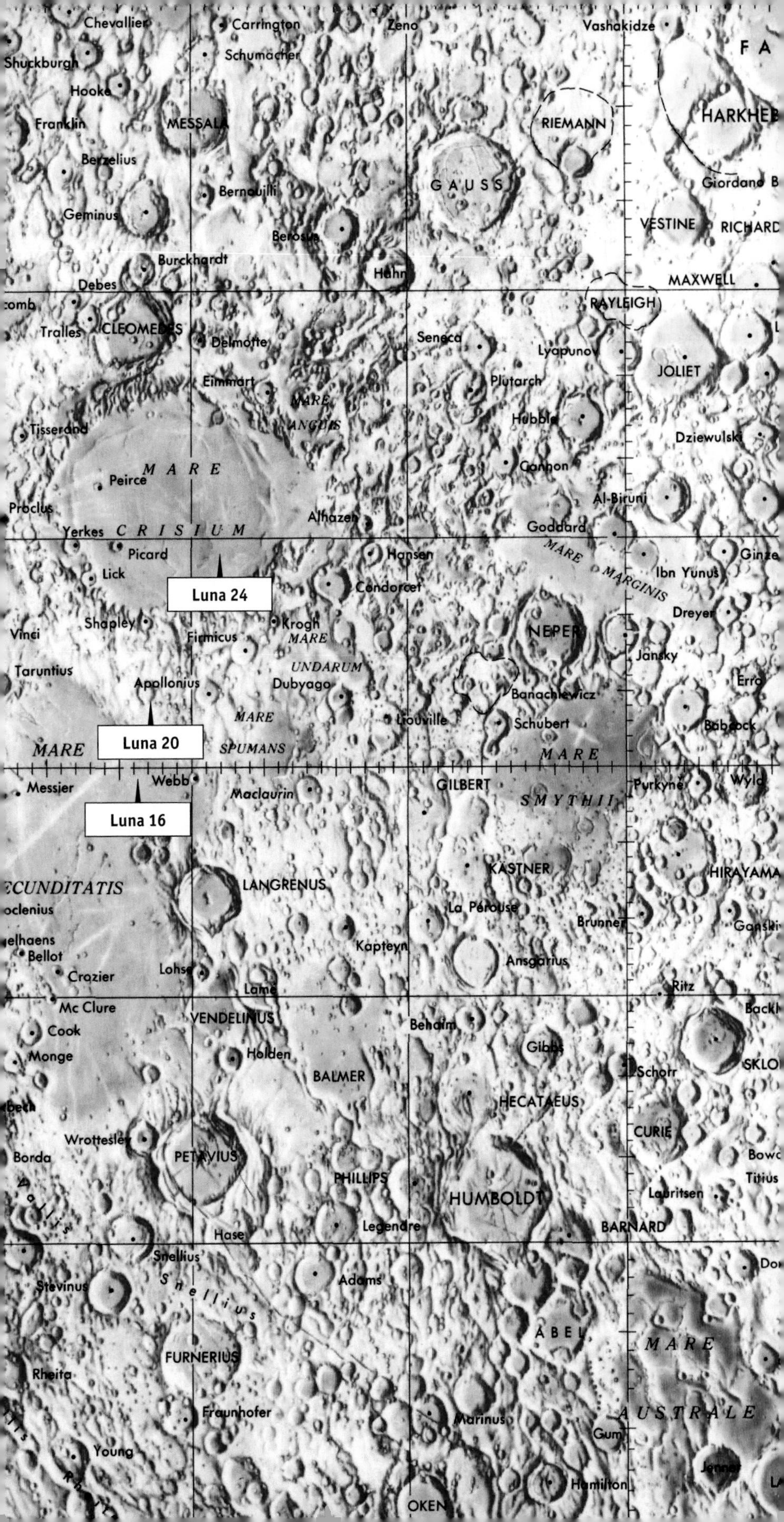
Chevallier
Carrington
Zeno
Vashakidze
Shuckburgh
Schumacher
Hooke
Franklin
MESSALA
RIEMANN
Berzelius
GAUSS
Bernoulli
Giordano
Geminus
VESTINE
Berosus
Burckhardt
Hahn
MAXWELL
Debes
RAYLEIGH
Tralles
CLEOMEDES
Delmotte
Seneca
Lyapunov
JOLIET
Eimmart
Plutarch
MARE
ANGUIS
Hubble
Tisserand
Dziewulski
Peirce
MARE
Cannon
Proclus
Al-Biruni
Alhazen
Yerkes
CRISIUM
Goddard
Picard
Hansen
MARE
MARGINIS
Lick
Ibn Yunus
Luna 24
Condorcet
Dreyer
Shapley
Krogh
NEPER
Vinci
Firmicus
MARE
Jansky
UNDARUM
Taruntius
Apollonius
Dubyago
Erro
Banachiewicz
MARE
Liouville
Schubert
Babcock
Luna 20
MARE
SPUMANS
MARE
Messier
Webb
Maclaurin
GILBERT
SMYTHII
Purkyne
Luna 16
KÄSTNER
HIRAYAMA
LANGRENUS
La Pérouse
Brunner
Bellot
Kapteyn
Ansgarius
Crozier
Lohse
Lamé
Ritz
Mc Clure
VENDELINUS
Behaim
Cook
Holden
Gibbs
Monge
BALMER
Schorr
HECATAEUS
Wrottesley
CURIE
Borda
PETAVIUS
Titius
PHILLIPS
HUMBOLDT
Lauritsen
Hase
Legendre
BARNARD
Snellius
Stevinus
Adams
Snellius
ABEL
MARE
FURNERIUS
Rheita
Fraunhofer
AUSTRALE
Marinus
Gum
Young
Hamilton
OKEN

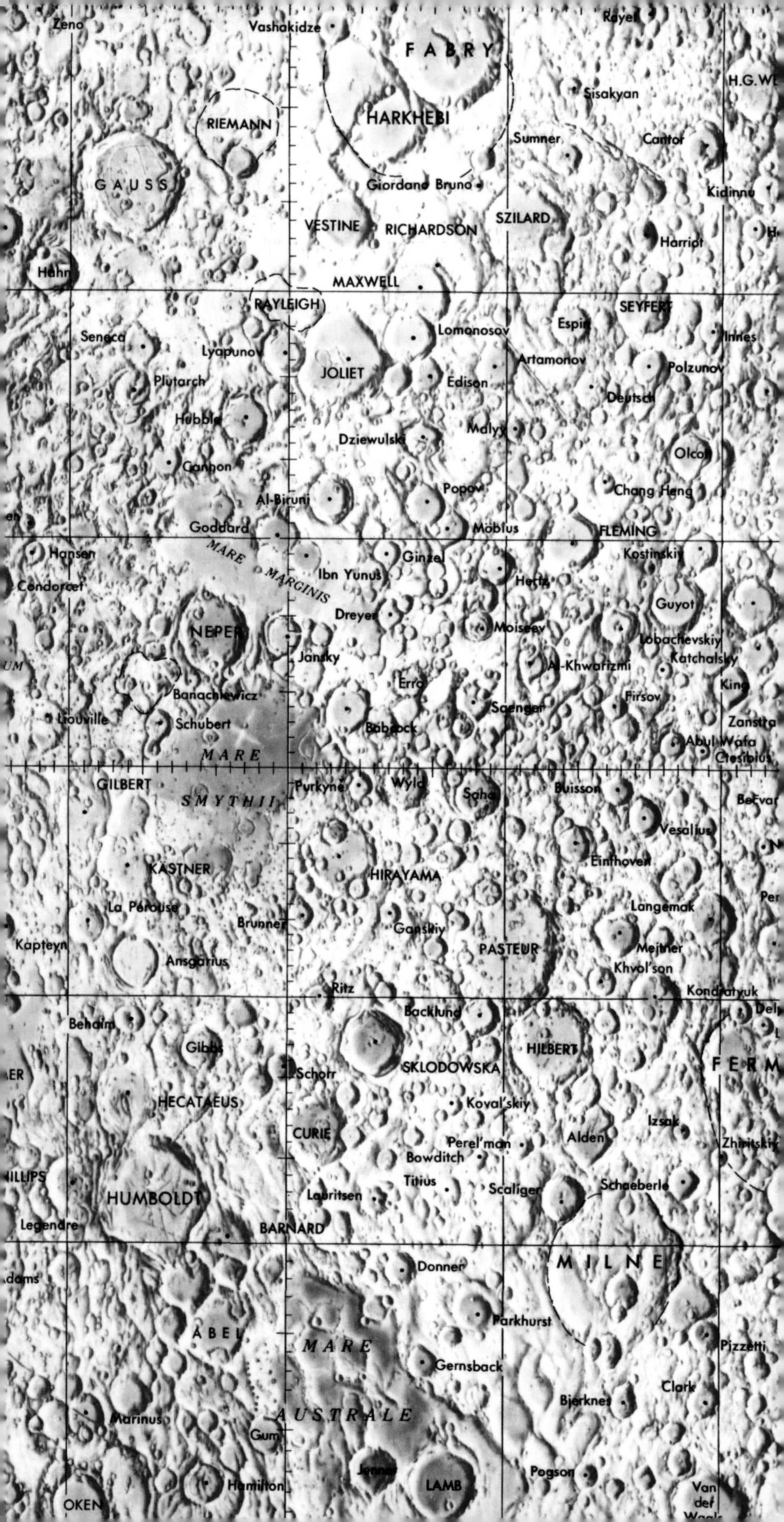

Zeno
Vashakidze
Rayet
FABRY
H.G.We
RIEMANN
HARKHEBI
Sisakyan
Sumner
Cantor
GAUSS
Giordano Bruno
Kidinnu
VESTINE
RICHARDSON
SZILARD
Harriot
Hahn
MAXWELL
RAYLEIGH
SEYFERT
Espin
Innes
Seneca
Lomonosov
Lyapunov
Artamonov
Polzunov
JOLIET
Edison
Plutarch
Deutsch
Hubble
Malyy
Dziewulski
Olcott
Cannon
Popov
Chang Heng
Al-Biruni
Goddard
Möbius
FLEMING
Hansen
MARE
MARGINIS
Ginzel
Kostinskiy
Ibn Yunus
Hertz
Condorcet
Guyot
Dreyer
NEPER
Moiseev
Lobachevskiy
Jansky
Katchalsky
Al-Khwarizmi
Erro
King
Banachiewicz
Firsov
Saenger
Liouville
Schubert
Babcock
Zanstra
Abul Wáfa
MARE
Ctesibius
GILBERT
Purkyne
Wyld
Saha
Buisson
Becvar
SMYTHII
Vesalius
Einthoven
KÄSTNER
HIRAYAMA
La Pérouse
Langemak
Brunner
Ganskiy
Kapteyn
Meitner
PASTEUR
Ansgarius
Khvol'son
Ritz
Kondratyuk
Backlund
Behaim
Gibbs
HILBERT
Schorr
SKLODOWSKA
HECATAEUS
Koval'skiy
Izsak
CURIE
Perel'man
Alden
Zhiritskiy
Bowditch
Titius
Scaliger
Schaeberle
HUMBOLDT
Lauritsen
Legendre
BARNARD
MILNE
Donner
Parkhurst
ABEL
MARE
Pizzetti
Gernsback
Clark
Bjerknes
Marinus
AUSTRALE
Gum
Jenner
Pogson
Hamilton
LAMB
Van der Waals
OKEN

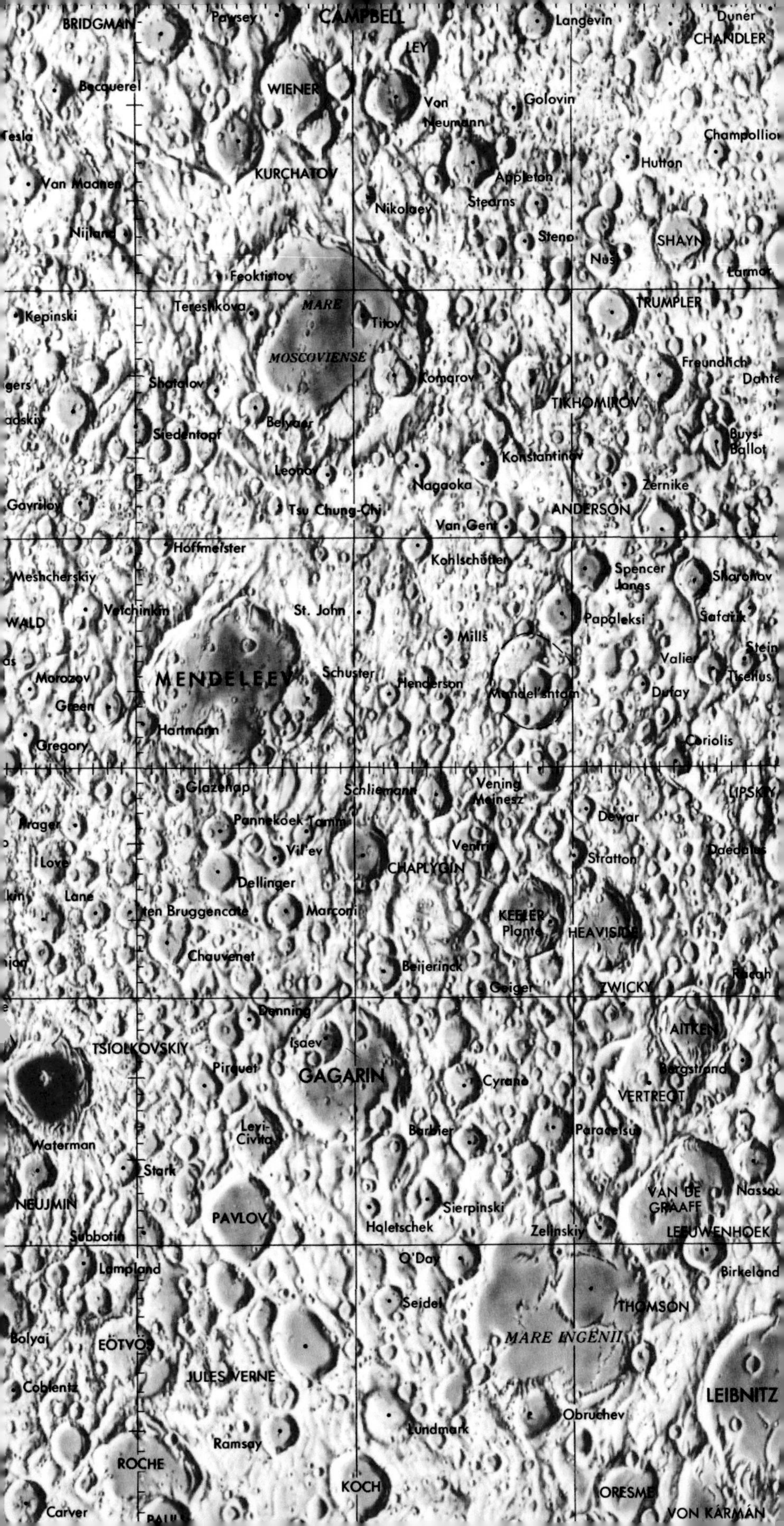
BRIDGMAN
CAMPBELL
Langevin
Dunér
CHANDLER
LEY
Becquerel
WIENER
Von Neumann
Golovin
Tesla
Champollion
Hutton
KURCHATOV
Appleton
Van Maanen
Stearns
Nikolaev
Nijland
Steno
SHAYN
Nušl
Larmor
Feoktistov
TRUMPLER
MARE
Titov
Kepinski
Tereshkova
MOSCOVIENSE
Freundlich
Komarov
Dante
Shatalov
TIKHOMIROV
Belyaev
Siedentopf
Buys-Ballot
Konstantinov
Leonov
Nagaoka
Zernike
Gavrilov
Tsu Chung-Chi
ANDERSON
Van Gent
Hoffmeister
Kohlschütter
Meshcherskiy
Spencer Jones
Sharonov
Vetchinkin
St. John
Papaleksi
Šafařík
WALD
Mills
Morozov
Schuster
MENDELEEV
Henderson
Mandel'shtam
Valier
Tiselius
Green
Dufay
Hartmann
Gregory
Coriolis
Glazenap
Schliemann
Vening Meinesz
LIPSKIY
Dewar
Pannekoek
Tamm
Love
Vil'ev
Ventris
Stratton
Daedalus
CHAPLYGIN
Dellinger
Lane
ten Bruggencate
Marconi
KEELER
Plante
HEAVISIDE
Chauvenet
Beijerinck
Racah
Geiger
ZWICKY
Denning
AITKEN
TSIOLKOVSKIY
Isaev
Pirquet
GAGARIN
Cyrano
Bergstrand
VERTREGT
Levi-Civita
Barbier
Paracelsus
Waterman
Stark
NEUJMIN
Sierpinski
VAN DE GRAAFF
Nassau
PAVLOV
Holetschek
Zelinskiy
LEEUWENHOEK
Subbotin
Lampland
O'Day
Birkeland
Seidel
THOMSON
Bolyai
EÖTVÖS
MARE INGENII
JULES VERNE
Coblentz
LEIBNITZ
Obruchev
Lundmark
Ramsay
ROCHE
KOCH
ORESME
Carver
VON KÁRMÁN

Overview of Lunar Missions

Listed by page number

Overview of Lunar Missions

Listed by page number

Overview of Lunar Missions

Listed by page number

Index of People

Bibliography

The author consulted the works listed below while writing this book. They are presented here as suggestions for further reading.

Ronald Stoyan, Hans-G. Purucker, *Reiseatlas Mond: Krater und andere Mondformationen schnell und sicher finden*, Erlangen 2013.

Lambert Spix, Frank Gaspari, *Der Moonhopper: 20 Mondtouren für Hobby-Astronomen*, Erlangen 2011.

Piers Bizony, *Moonshots: 50 Years of NASA Space Exploration Seen Through Hasselblad Cameras*, Minneapolis 2017.

P. Klushantsev, *Dom na orbite*, Moscow 1975.

Philipp Meuser (ed.), *Architektur für die russische Raumfahrt*, Berlin 2013.

S. Häuplik-Meusburger, O. Bannova, *Space Architecture Education for Engineers and Architects*, New York 2016.

Eugen Reichel, *Projekt Apollo. Die Mondlandungen*, Stuttgart 2016.

C.B. Colby, *Moon Exploration: Space Stations, Moon Maps, Lunar Vehicles*, New York 1970.

Jesco von Puttkammer, *Abenteuer Apollo 11: Von der Mondlandung zur Erkundung des Mars*, Munich 2019.

Gurbir Singh, *The Indian Space Programme*, Manchester 2017.

Brian Harvey, *China in Space: The Great Leap Forward*, New York 2013.

Eugen Reichl, *Moskaus Mondprogramm*, Stuttgart 2017.

S. N. Mintshin, A. T. Ulubekov, *Semlja – Kosmos – Luna*, Moscow 1972.

Neil Leach (ed.), *Space Architecture: The New Frontier for Design Research*, Hoboken 2014.

Cédric Delsaux, *Dark Lens*, Paris 2011.

John Zukowsky, *Space Architecture: The Work of John Frassanito & Associates for NASA*, Stuttgart/London 1999.

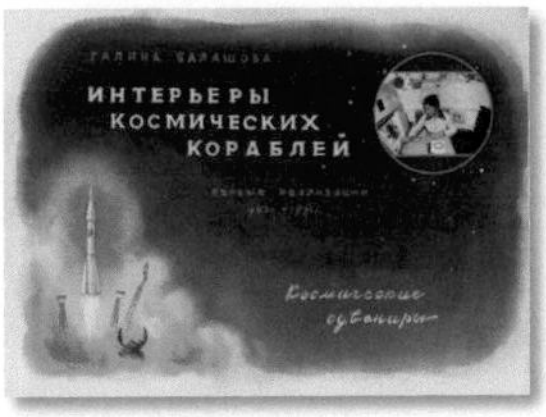

Galina Balashova. *Interiery Kosmitsheskich Korablei. Perviye realisatsii 1963–1991*, Moscow 2011.

Anthony M. Springer, *Aerospace Design. Aircraft, Spacecraft and the Art of Modern Flight*, London/New York 2003.

Wayne Hale, et al. (eds), *Wings In Orbit. Scientific and Engineering Legacies of the Space Shuttle*, Washington 2011.

Philipp Meuser, *Galina Balashova: Architect of the Soviet Space Programme*, Berlin 2015.

S. Häuplik-Meusburger, *Architecture for Astronauts: An Activity-based Approach*, New York 2011.

Rolf Engel, *Russlands Vorstoss ins All: Geschichte der sowjetischen Raumfahrt*, Bonn 1988.

Viorel Badescu (ed.), *Moon: Prospective Energy and Material Resources*, New York 2012.

Alexander Glushko, *Design for Space: Soviet and Russian Mission Patches*, Berlin 2016.

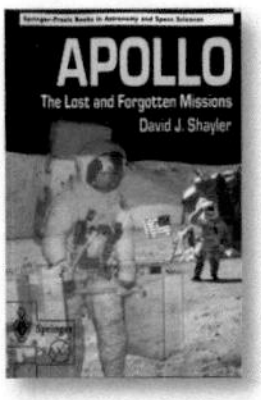

David J. Shayler, *Apollo: The Lost and Forgotten Missions*, Berlin 2002.

A. S. Howe, B. Sherwood, *Out of this World: The New Field of Space Architecture*, Reston 2009.

Yu. Dokutshayev (ed.), *Yuri Gagarin: Vishu Semliu*, Moscow 1968.

Paul Meuser on a research trip in Washington (2018)

Paul Meuser on a research trip in Moscow (2018)

With Gurbir Singh (left) and Brian Harvey (right) during the *International Astronautical Congress* in Bremen (2018)

Interview with Galina Balashova in Korolyov (2017)

RISD Rover Team during the NASA Human Exploration Rover Challenge 2018 in Huntsville/AL.

At work with editor Amelie Ochs

Olga Bannova (2018)

Alexander Glushko (1981)

Alexander Glushko (2018)

Author and Contributors

Paul Meuser
Born in Berlin in 1996. Studied art at the Rhode Island School of Design (RISD) in Providence between 2015 and 2019. Student at the Yale School of Architecture in New Haven since 2019. Participated in the NASA Human Exploration Rover Challenge 2018 at the Marshall Space Flight Center, Huntsville.

Galina Balashova
Born in Kolomna near Moscow in 1931. Architect and space engineer. Studied at the Moscow Architectural Institute (MArchI) between 1949 and 1955. Began her career in Kuybyshev (today: Samara) before joining OKB-1 under the direction of Sergei Korolev. Balashova is considered the first interior architect whose designs were used for Soviet spacecraft. Today she lives in Korolyov near Moscow. DOM publishers published a monograph on her work in 2014 (thereafter translated into English in 2015 and Russian in 2018).

Olga Bannova
Born in Moscow in 1964. Director of the Master of Science degree in Space Architecture at the Sasakawa International Center for Space Architecture (SICSA) of the University of Houston. Vice President of the AIAA Space Architecture Technical Committee. Master in architecture and space architecture from the University of Houston; PhD from Chalmers University of Technology. Also works as an architect in Moscow, focusing on industrial, office, and healthcare buildings.

Alexander Glushko
Born in Moscow in 1972. Head of Russia's Department for the Investigation of Historical Crimes, Commissioner General. Author of many books on the history of Soviet space flight and military uniforms. Son of Valentin Glushko, who served as the head designer of the rocket manufacturer Energia for many years. In 2016, DOM publishers published a collection of mission patches of Soviet and Russian crewed space missions.

Brian Harvey
Born in 1953 in Dublin, where he currently lives. Author and broadcaster focusing on spaceflight. Studied history and political science at Trinity College in Dublin. Has written many books about the Russian, Chinese, European, Indian, and Japanese space programmes, from *Race into Space – the Soviet Space Programme* (1988) to *China's Space Station: Reaching for the Moon and Mars* (2019).

Hans Hollein †
Born in Vienna in 1934, died in Vienna, 2014. Architect, architectural theorist, exhibition curator, designer, and artist. Studied civil engineering (1949–53) in Vienna before studying at the Academy of Fine Arts (until 1956) and Illinois Institute of Technology in Chicago and University of California, Berkeley (until 1960). In the 1960s, was part of the circle of Viennese avant-gardes who questioned the conventional understanding of architecture. Freelance architect in Vienna from 1964 onwards. Published many programmatic texts on architecture.

Gurbir Singh
Born in India in 1958. Living in England since 1966. Consultant in the field of IT security. Was one of 13,000 unsuccessful applicants to become the first British astronaut. Author of *Yuri Gagarin in London and Manchester* (2011) and *The Indian Space Programme* (2017).

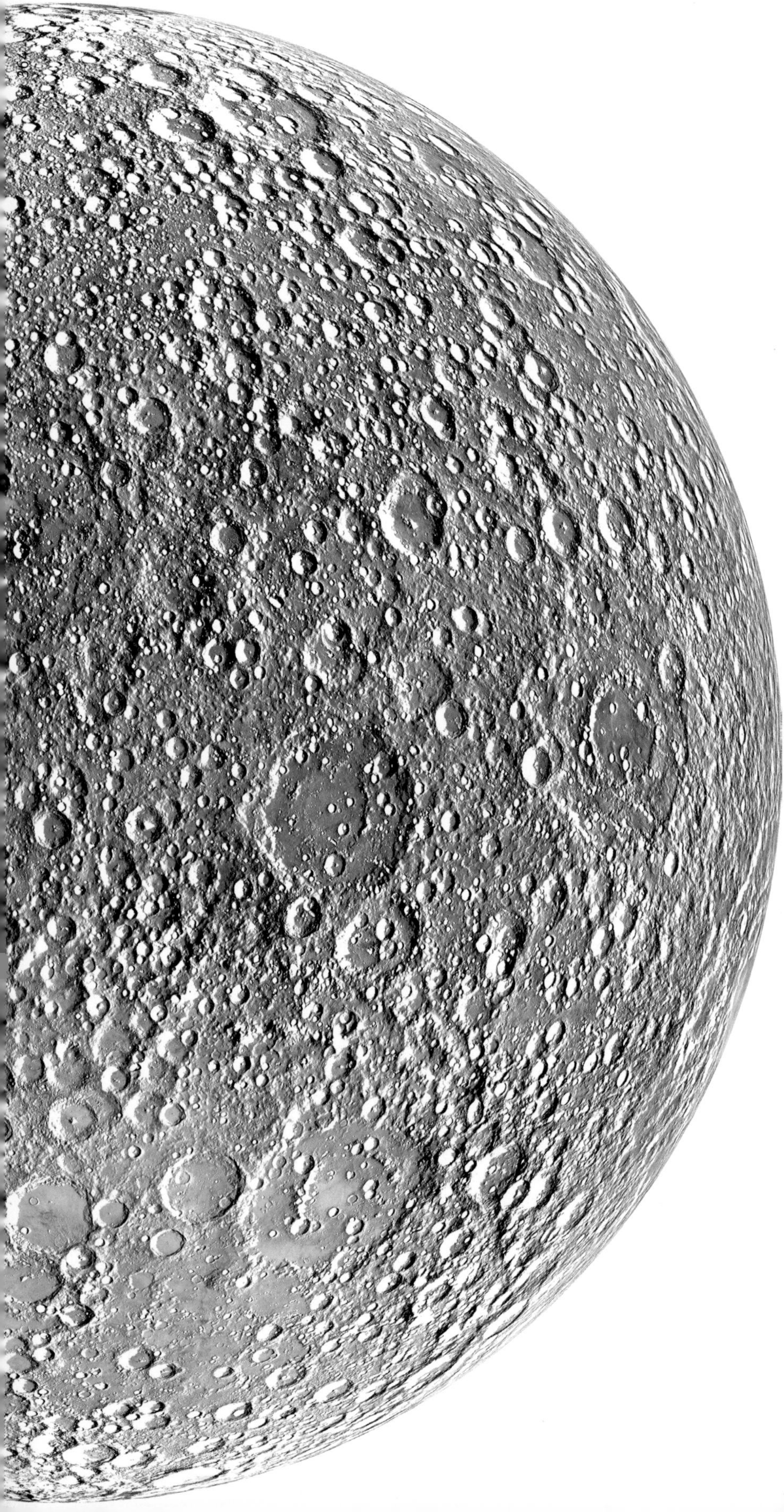

Acknowledgements

The author would like to thank Sophia Xu for her help and contributions while he was researching this project, and Heidi Kraut for her support, which goes far beyond the scope of this book. Furthermore, this architectural guide would not have been the same without Amelie Ochs, who tirelessly edited the original German edition and whose efforts immensely improved the fluidity of the text. He would also like to thank the experts he met during this journey, who contributed their words and knowledge to this project: Olga Bannova, Brian Harvey, Alexander Glushko, Galina Balashova, and Gurbir Singh. Finally, he would like to thank all the staff members of the museums and archives, who kindly provided their help and granted access to valauble materials that made this publication possible.

The publishers would like to thank D.G. Ermakov, the representative of the Russian state archive for scientific and technical documentation, for his help in preparing this book. They would also like to thank the artist A.G. Shlyadinsky for the illustrations he provided, and A.B. Zheleznyakov, the populariser of cosmonautics, for his help in selecting the relevant graphic materials. The publishers also give their special thanks to the authors of the website for the 44th guided missile regiment (military unit 89503), the materials of which became the basis for the book's sections on uncrewed Soviet lunar missions. Finally, the publishers would like to thank Karina Diemer, Tatiana Vinogradova, Olga Vinogradova, Petr Vinogradov, and Yevgeniy Lokhin for all their support.

NASA

This fish-eye camera lens view of the interior of the Apollo Lunar Module Mission Simulator at the Kennedy Space Center is one of several selected by the Apollo 9 crew to appear in *Apollo: Through the Eyes of Astronauts*. The book features images from the Apollo Program that were selected by the crew of each mission. In the foreground is mission commander James McDivitt; in the background is Russell Schweickart, lunar module pilot.